Edited by
Christophe Goupil

Continuum Theory and Modeling of Thermoelectric Elements

Edited by Christophe Goupil

Continuum Theory and Modeling of Thermoelectric Elements

Verlag GmbH & Co. KGaA

Editor

Prof. Christophe Goupil
LIED UMR 8236
Paris Interdisciplinary Energy Research Institute (PIERI)
Paris Diderot University
France

All books published by **Wiley-VCH** are carefully produced. Nevertheless, authors, editors, and publisher do not warrant the information contained in these books, including this book, to be free of errors. Readers are advised to keep in mind that statements, data, illustrations, procedural details or other items may inadvertently be inaccurate.

Library of Congress Card No.: applied for

British Library Cataloguing-in-Publication Data
A catalogue record for this book is available from the British Library.

Bibliographic information published by the Deutsche Nationalbibliothek
The Deutsche Nationalbibliothek lists this publication in the Deutsche Nationalbibliografie; detailed bibliographic data are available on the Internet at <http://dnb.d-nb.de>.

© 2016 Wiley-VCH Verlag GmbH & Co. KGaA, Boschstr. 12, 69469 Weinheim, Germany

All rights reserved (including those of translation into other languages). No part of this book may be reproduced in any form – by photoprinting, microfilm, or any other means – nor transmitted or translated into a machine language without written permission from the publishers. Registered names, trademarks, etc. used in this book, even when not specifically marked as such, are not to be considered unprotected by law.

Print ISBN: 978-3-527-41337-9
ePDF ISBN: 978-3-527-68787-9
ePub ISBN: 978-3-527-68788-6
Mobi ISBN: 978-3-527-68786-2
oBook ISBN: 978-3-527-33840-5

Cover Design Formgeber, Mannheim, Germany
Typesetting SPi Global, Chennai, India
Printing and Binding Markono Print Media Pte Ltd, Singapore

Printed on acid-free paper

To Wolfgang

Contents

List of Contributors XIII
Preface XV
List of Frequently Used Symbols XVII
Glossary XIX

1	**Thermodynamics and Thermoelectricity** *1*	
	Christophe Goupil, Henni Ouerdane, Knud Zabrocki, Wolfgang Seifert,	
	Nicki F. Hinsche, and Eckhard Müller	
1.1	Milestones of Thermoelectricity *1*	
1.1.1	Discovery of the Seebeck Effect *2*	
1.1.2	Discovery of the Peltier Effect *8*	
1.1.3	Discovery of the Thomson Effect *9*	
1.1.4	Magnus' Law *10*	
1.1.5	Early Performance Calculation of Thermoelectric Devices *11*	
1.1.6	First Evaluation of the Performance of a Thermoelectric Device by E. Altenkirch *11*	
1.1.7	Benedicks' Effect *12*	
1.1.8	The Bridgman Effect *13*	
1.1.9	Semiconductors as Thermoelectric Materials *14*	
1.1.10	Thermoelectric Applications – Excitement and Disappointment 1920–1970 *15*	
1.1.10.1	Construction of the First Thermoelectric Generator *15*	
1.1.11	Thermoelectric Industry – Niche Applications 1970–2000 *16*	
1.1.11.1	Thermoelectric Generators for Space *16*	
1.1.12	New Concepts in Thermoelectricity 2000-Present *17*	
1.2	Galvanomagnetic and Thermomagnetic Effects *17*	
1.2.1	The Hall Coefficient *21*	
1.2.2	The Nernst Coefficient *21*	
1.2.3	The Ettingshausen Coefficient *21*	
1.2.4	The Righi–Leduc Coefficient *22*	
1.2.5	Devices Using Galvano- and Thermomagnetic Effects and the Corresponding Figure of Merit *22*	
1.3	Historical Notes on Thermodynamic Aspects *25*	

1.4	Basic Thermodynamic Engine 27
1.5	Thermodynamics of the Ideal Fermi Gas 28
1.5.1	The Ideal Fermi Gas 28
1.5.2	Electron Gas in a Thermoelectric Cell 29
1.5.3	Entropy Per Carrier 30
1.5.4	Equation of State of the Ideal Electron Gas 32
1.5.5	Temperature Dependence of the Chemical Potential $\mu_c(T)$ 34
1.6	Linear Nonequilibrium Thermodynamics 35
1.6.1	Forces and Fluxes 35
1.6.2	Linear Response and Reciprocal Relations 36
1.7	Forces and Fluxes in Thermoelectric Systems 37
1.7.1	Thermoelectric Effects 37
1.7.2	Forces, Fluxes, and Kinetic Coefficients 38
1.7.3	Energy Flux and Heat Flux 39
1.7.4	Thermoelectric Coefficients 40
1.7.4.1	Decoupled Processes 40
1.7.4.2	Coupled Processes 41
1.7.5	The Entropy Per Carrier 41
1.7.6	Kinetic Coefficients and Transport Parameters 42
1.7.7	The Dimensionless Figure of Merit zT 43
1.8	Heat and Entropy 44
1.8.1	Volumetric Heat Production 45
1.8.2	Entropy Production Density 45
1.8.3	Heat Flux and the Peltier–Thomson Coefficient 46
1.8.4	The Peltier–Thomson Term 46
1.8.5	Local Energy Balance 47
1.9	The Thermoelectric Engine and Its Applications 48
1.10	Thermodynamics and Thermoelectric Potential 50
1.10.1	Relative Current, Dissipation Ratio, and Thermoelectric Potential 51
1.10.2	Local Reduced Efficiency and Thermoelectric Potential of TEG, TEC, and TEH 53
1.10.3	Thermoelectric Potential and Nonequilibrium Thermodynamics 56
	References 59
2	**Continuum Theory of TE Elements** 75

Knud Zabrocki, Christophe Goupil, Henni Ouerdane, Yann Apertet, Wolfgang Seifert, and Eckhard Müller

2.1	Domenicali's Heat Balance Equation 75
2.1.1	Tensorial Character of Material Properties 75
2.1.2	Heat Balance and Source Terms 76
2.1.3	Spatial and Temperature Averaging of the Material Properties 79
2.2	Transferred Heat Balance 80
2.3	Ioffe's Description and Performance Parameters of CPM Devices 81
2.3.1	Single-Element Device 82

2.3.2	Performance Parameters of a Thermoelectric Element with Constant Material Properties *84*	
2.3.2.1	Thermoelectric Generator (TEG) *86*	
2.3.2.2	Thermoelectric Cooler (TEC) *89*	
2.3.2.3	Thermoelectric Heater (TEH) *91*	
2.3.3	Inverse Performance Equations and Effective Device Figure of Merit *91*	
2.4	Maximum Power and Efficiency of a Thermogenerator Element *93*	
2.4.1	Load Resistance as Design Parameter for a Thermogenerator *96*	
2.4.2	Efficiency versus Power Approach *97*	
2.4.3	Constant Heat Input (CHI) Model *100*	
2.5	Temperature-Dependent Materials – Analytic Calculations *102*	
2.5.1	Inverse Temperature Dependence of the Thermal Conductivity *103*	
2.5.2	Inverse Temperature Dependence of the Thermal Conductivity and a Variable Electrical Resistivity *105*	
2.5.2.1	Performance Equations *106*	
2.5.2.2	Maximum Power Output *108*	
2.5.2.3	Maximum Efficiency *109*	
2.5.3	Constant Thomson Coefficient – Logarithmic Behavior of the Seebeck Coefficient *111*	
2.5.3.1	Methods of Averaged Coefficients and Geometric Optimization of a TE Couple *114*	
2.5.3.2	Influence of the Thomson Effect on the Performance of a TEG *116*	
2.5.4	Algebraic and General Temperature Dependence *119*	
2.5.5	Constant Thomson Coefficient Combined with Linear Temperature Dependence of Resistivity *121*	
2.5.6	Linear Temperature Dependence of the Resistivity *123*	
2.5.7	Linear Temperature Dependence of the Thermal Conductivity *125*	
2.6	The Influence of Contacts and Contact Resistances on the TE Performance *125*	
2.6.1	Thermoelectric Element with Contacting Bridge *126*	
2.6.2	Numerical Example for the Influence of the Electrical Contact Resistance on the Performance *130*	
2.7	Dissipative Coupling between the TEG and the Heat Baths *133*	
2.7.1	Finite-Time Thermodynamics Optimization *133*	
2.7.2	Thermoelectric Generator Model *133*	
2.7.3	Thermal Flux and Electrical Current *135*	
2.7.4	Calculation of the Temperature Difference across the TEG *136*	
2.7.5	Maximization of Power and Efficiency with Fixed ZT_m *137*	
2.7.5.1	Maximization of Power by Electrical Impedance Matching *137*	
2.7.5.2	Maximization of Power by Thermal Impedance Matching *137*	
2.7.5.3	Simultaneous Thermal and Electrical Impedance Matching *138*	
2.7.5.4	On the Importance of Thermal Impedance Matching *138*	
2.7.5.5	Maximum Efficiency *139*	
2.7.5.6	Analysis of Optimization and Power – Efficiency Trade-Off *139*	

2.8	Shaped Thermoelectric Elements	*140*
2.9	Other Influences on the Performance of TE Devices	*144*
2.9.1	Lateral Heat Losses, Convective and Radiative Heat Transfer	*144*
2.9.1.1	Convection Losses and Benefits	*144*
2.9.1.2	Radiation Losses and Benefits	*146*
2.9.2	Anisotropic Thermoelectric Elements	*147*
	References	*148*
3	**Segmented Devices and Networking of TE Elements**	*157*
	Knud Zabrocki, Christophe Goupil, Henni Ouerdane, Eckhard Müller, and Wolfgang Seifert	
3.1	Segmented Devices	*157*
3.1.1	Double-Segmented Element	*160*
3.1.1.1	Effective Electrical Conductivity	*161*
3.1.1.2	Effective Thermal Conductivity	*163*
3.1.2	Algorithm of Multisegmented Elements	*164*
3.1.2.1	Numerical Parameter Studies on Graded Elements	*166*
3.2	Networks	*169*
3.2.1	Presentation	*169*
3.2.2	Useful Expressions	*171*
3.2.3	Discretization	*171*
3.2.4	Solution: General Millman Theorem	*172*
3.2.5	Implementation	*173*
3.2.6	Numerical Illustration	*174*
	References	*176*
4	**Transient Response and Green's Function Technique**	*177*
	Wolfgang Seifert, Knud Zabrocki, Steven Achilles, and Steffen Trimper	
4.1	Quasi-Stationary Processes	*179*
4.2	Supercooling with a Transient Peltier Cooler	*182*
4.2.1	Steady-State Operation of a Thermoelectric Cooler	*183*
4.2.2	Important Parameters for the Supercooling Case	*187*
4.3	Transient Behavior of a Thermoelectric Generator	*189*
4.4	Dynamic Measurements of the Thermal Conductivity: Laser Flash Analysis	*190*
4.5	Dynamic Measurements of the Thermal Conductivity: Classical Ioffe Method	*193*
4.5.1	Theoretical Basis of Simple Ioffe Method	*194*
4.5.2	Laplace Transformation and Important Properties	*197*
4.5.3	Solution of the Classical Ioffe Method	*198*
4.5.3.1	LT of Original Equation	*198*
4.5.4	Solution of the Temperatures in the s-Domain	*199*
4.5.5	Inverse Laplace Transformation	*202*
4.5.6	Inversion Theorem for the Laplace Transformation	*203*
4.5.7	Inversion of the Temperature Profiles	*205*

4.6	Green's Function Approach in Thermoelectricity	206
4.6.1	Continuity Equations	208
4.6.2	Green's Function Approach in the Steady State	208
4.6.3	One-Dimensional Green's Functions in the Steady State	209
4.6.4	Perturbative Approach to a Full Description (1D)	210
4.7	Linear Transient Approach	212
4.7.1	Relaxation Time	212
4.7.2	Transient Field Equations	213
4.7.3	Transient Linear Response Approximation	214
4.8	Time-Dependent Green's Function Approach	215
	References	217

5 Compatibility 227
Wolfgang Seifert, G. Jeffrey Snyder, Eric S. Toberer, Volker Pluschke, Eckhard Müller, and Christophe Goupil

5.1	Relative Current Density and Compatibility Factors	227
5.2	Compatibility and Segmented Thermogenerators	229
5.3	Reduced Efficiencies and Self-Compatible Performance	232
5.3.1	Performance Integrals for Efficiency and COP	233
5.3.2	Local Efficiency Dependence on Current (TEG)	235
5.4	Power-Related Compatibility	238
5.5	Optimal Material Grading for Maximum Power Output	241
5.6	The Criterion "$u = s$" and Calculus of Variations	243
5.7	Self-Compatibility and Optimum Material Grading	246
5.8	Thermodynamic Aspects of Compatibility	249
5.9	Analytic Results for Self-Compatible TEG and TEC Elements	251
5.9.1	Performance of Self-Compatible TEG and TEC Elements	251
5.9.2	Self-Compatible Elements and Optimal Figure of Merit	253
5.9.3	Optimal Seebeck Coefficients for Self-Compatible Material	255
5.9.4	Temperature Profile for $u = s$ Material	256
5.10	Thermoelectric Thomson Cooler	259
5.10.1	Cooling Performance	262
5.10.2	Thomson Cooler Phase Space	265
5.10.3	Performance Limits	266
5.10.4	Further Characteristics of Self-Compatible Material for Cooling	268
5.11	Compatibility Approach versus Device Optimization	276
	References	277

6 Numerical Simulation 281
Knud Zabrocki and Wolfgang Seifert

6.1	Finite Difference Methods	284
6.2	Finite Volume Method	288
6.3	Finite Element Method	290
6.4	Performance Calculation of a TEG-A Case Study	291

6.4.1	Averages of the Material Properties *294*
6.4.2	Processing Measured Material Properties *295*
6.4.3	Different Averages and the Corresponding Performance Values *297*
6.4.4	Power Factor and Figure of Merit *298*
6.4.5	Optimal Performance Based on Averaged Material Properties *299*
6.4.6	Comparison between CPM, FDM, and FEM Simulations *300*
6.4.6.1	Finite Difference Scheme for a TE Element *300*
6.4.6.2	Performance Parameters Dependent on the Current Density *305*
6.4.7	Calculation for a p-n Thermocouple *309*
6.4.8	Adjustment of Cross-Sectional Areas *310*
6.4.9	FEM Simulation *311*
6.5	Nonlinear Material Parameters *315*
6.5.1	Temperature-Dependent Material Properties *315*
6.5.2	Temperature-Dependent Material Properties: Mathematica Model Using the Compatibility Approach *320*
A	Numerical Data and Illustrative Cases *322*
A.1	Coefficients of the Polynomials of the Material Properties *322*
A.2	Material Properties: p-Type Skutterudite and Averages *323*
A.3	Material Properties: n-Type Skutterudite and Averages *324*
A.4	Power Factor and Figure of Merit of the n-Type Material *325*
A.5	Performance Parameters in dependence on the Current Density *326*
A.5.1	Program code *326*
A.5.2	Results *327*
	References *330*

Index *335*

List of Contributors

Steven Achilles
University Halle-Wittenberg
Institute of Physics – Theoretical Physics
Von-Seckendorff-Platz 1
06120 Halle (Saale)
Germany

Yann Apertet
Université Paris Sud
Institut d'Electronique Fondamentale
CNRS, UMR 8622
Bâtiment 220
Rue André Ampère
91405 Orsay
France

Christophe Goupil
Université Paris Diderot
Laboratoire Interdisciplinaire de Energies de Demain (LIED)
CNRS, UMR 8236
5 rue Elsa Morante
75013 Paris
France

and

Paris Interdisciplinary Energy Research Institute (PIERI)
Paris Diderot University
Paris
France

Nicki F. Hinsche
University Halle-Wittenberg
Institute of Physics – Theoretical Physics
Von-Seckendorff-Platz 1
06120 Halle (Saale)
Germany

Eckhard Müller
Institute of Materials Research
German Aerospace Center (DLR)
Institute of Materials Research – Thermoelectric Materials and Systems
Linder Höhe
51147 Köln Porz-Wahnheide
Germany

and

Justus Liebig University Giessen
Institute of Inorganic and Analytical Chemistry
35392 Giessen
Germany

Henni Ouerdane
Université Paris Diderot
Laboratoire Interdisciplinaire de
Energies de Demain (LIED)
CNRS, UMR 8236
Paris
France

and

Russian Quantum Center
100 Novaya Street
Skolkovo
143025 Moscow
Russia

Volker Pluschke
University Halle-Wittenberg
Institute of Mathematics
06120 Halle
Germany

Wolfgang Seifert[†]
University Halle-Wittenberg
Institute of Physics – Theoretical
Physics
Von-Seckendorff-Platz 1
06120 Halle (Saale)
Germany

G. Jeffrey Snyder
Northwestern University
McCormick School of
Engineering and Applied Science
2220 Campus Drive
Evanston
IL 60208
USA

Eric S. Toberer
Department of Physics
Colorado School of Mines
1500 Illinois St
Golden
CO 80401
USA

Steffen Trimper
University Halle-Wittenberg
Institute of Physics – Theoretical
Physics
Von-Seckendorff-Platz 1
06120 Halle (Saale)
Germany

Knud Zabrocki
Institute of Materials Research
German Aerospace Center (DLR)
Institute of Materials Research –
Thermoelectric Materials and
Systems
Linder Höhe
51147 Köln Porz-Wahnheide
Germany

Preface

With a worldwide growing interest in thermoelectric phenomena and their practical applications in mass products such as thermogenerators in cars, which promise a multimillion-piece production, a multitude of publications on the topic are arising these days. However, the field of thermoelectric continuum theory has not been treated in a comprehensive book for long; the last substantial book on this field dates back to the 1970s or earlier (Anatychuk 1979, Egli 1960). Since then, many progresses and new considerations with respect to future applications have been made; thus, an update on the state of knowledge in this area is more than overdue. The new knowledge in the field is spread over a variety of smaller publications, mainly journal papers, conference presentations, and proceedings but also a few book chapters (CRC Thermoelectric Handbooks) and a review article by the coauthors of this book. The ongoing discussion led to the estimate that the questions of maximizing electrical power production and its efficiency as well as cooling power of Peltier devices and their coefficient of performance seem utmost complicated although being the effect of co-action of very few well-understood basic physical phenomena and have not been solved yet. Several works provide incomplete or sometimes contradicting information, which has to be discussed and reviewed in a consistent approach.

Thus, a comprehensive summary of the reached status is highly desirable and will provide firm orientation among the multitude of available publications in the field to the reader.

In some sections, we have deliberately quoted the original text from published articles in order to document the development in the field of performance optimization by material grading. Copyrights were cleared with EPL (EDP Sciences), MDPI (Switzerland), CRC (Taylor and Francis), as well as APS (USA).

A particular strength of the book lies in the fact that the historic evolution of the understanding of thermoelectric phenomena is presented in a compact manner. An almost complete record of relevant survey publications has also been included. Many quotations to this record facilitate understanding of the origin of important findings in the field. A further unique point is the strongly thermodynamically motivated approach to the thermoelectric phenomena and, in particular, emphasis on the relation to the out-of-equilibrium thermodynamics. In this book, the performance limits of thermoelectric devices are discussed on a continuum theoretical basis in full width and generality including temperature dependence and gradients of the material properties for the first time. The book shows a transparent concept guiding the reader in a compact and understandable way from the fundamentals up to applicable principles and formulae for the optimal design of TE devices. After profound development and explanation of the theoretical fundament based on strict and traceable mathematical and analytical treatments, the engineer will find a collection of helpful tools for practical calculation and numerically based design of optimized TE elements and thermoelectric modules for the application as a thermoelectric generator or Peltier cooler.

Most of the book is written as a summary for the project developer in the field of thermoelectric technology both at academia and in industries, for graduate and advanced undergraduate students (physics, materials science, energy technology, electrical engineering, system and process technology). Some core sections of the book address specialists in the field of thermoelectric energy conversion, providing detailed discussion of key points of the optimization matter. The book will provide a firm base for TE element and module design in technical development process and will thus serve as an indispensable tool for any application development.

The consort of authors is bridging over continents and belonging to leading institutions in the field on an international scale; some of them hold a tradition in TE research over decades, dating back to the early years of development of highly efficient thermoelectric semiconductors. Most of the authors are leading scientists of international reputation, who have created and worked out main lines of research in the field of advanced TE materials and applications over the last decade. A complete list of all contributors is given as follows.

This book addresses a wide and steadily growing scientific and developers community in academia and industrial laboratories. With the growing number of funded research activities in the leading industrial countries and an explosion of activities, a continued spread and rise of interest can be expected.

Over all, the book appears at a time where it will find wide interest because of the actuality and widespread visibility of the topic when many activities are turning to technical implementation in powerful demonstrations and steps for upscaling and production have been taken in many companies all over the world. There are many people who directly or indirectly have contributed to the understanding and development of the subject of this book. The authors of this book particularly wish to thank Cronin B. Vining for his contribution to the thermodynamics of the "thermoelectric process."

Finally, I would like to thank as the chief editor of this book all the coauthors for their contributions to the realization of this project. My special thanks go to Knud Zabrocki for his tireless efforts. The writing of this book has taken a great amount of time and energy, and it is also the loving support, understanding, and patience of the authors' families, which made this project possible. They deserve our most profound gratitude.

Christophe Goupil, Chief Editor

List of Frequently Used Symbols

Symbol	Description	Unit
α	Seebeck coefficient	[V/K]
$\kappa \equiv \kappa_J$	thermal conductivity (zero current)	[W/(Km)]
$\sigma \equiv \sigma_T$	electrical conductivity (isothermal)	$[(\Omega m)^{-1}]$
ϱ	electrical resistivity $\varrho = 1/\sigma$	$[\Omega m]$
ϱ_{el}	charge density	$[As/m^3]$
ϱ_{md}	mass density	$[kg/m^3]$
c	specific heat	[J/(kg K)]
λ	thermal diffusivity $\lambda = \kappa/(\varrho_{md}\, c)$	$[m^2/s]$
τ	Thomson coefficient $\tau = T\, d\alpha/dT$	[V/K]
Π	Peltier coefficient $\Pi = \alpha\, T$	[V]
z	material figure of merit $z = \alpha^2\, \sigma/\kappa$	$[K^{-1}]$
Z	device's figure of merit	$[K^{-1}]$
L	length of the TE element	[m]
A_c	cross-sectional area of the TE element	$[m^2]$
K	thermal conductance	[W/K]
R	electrical resistance	$[\Omega]$
p	pressure	$[N/m^2]$
P	electrical power	[W]
$T = T(x)$	temperature (profile)	[K]
T_1, T_2	boundary temperatures	[K]
T_h, T_c	boundary temperatures (hot/cold side)	[K]
$T_a = T(x=0)$	temperature at the heat absorbing side	[K]
$T_s = T(x=L)$	temperature at the heat sink side	[K]
$\mathbf{E}$	electric field (vectors in bold math)	[V/m]
μ_c, μ_e	chemical potential, electrochemical potential	[Ws]
V	voltage	[V]
I	electric current	[A]
$\mathbf{j}$	electrical current density (vector)	$[A/m^2]$
j_{opt}	scalar value of the optimal current density	$[A/m^2]$
$\dot{Q}$	thermal power	[W]
$\mathbf{j}_Q$ or $\mathbf{q}$	total heat flux density (vector)	$[W/m^2]$
$\mathbf{j}_E$	total energy flux (vector)	$[W/m^2]$

(*continued overleaf*)

List of Frequently Used Symbols

Symbol	Description	Unit
$\mathbf{q}_\pi$	Peltier heat flux density (vector)	[W/m²]
u	relative current density	[V⁻¹]
s	compatibility factor	[V⁻¹]
Φ	thermoelectric potential $\Phi = 1/u + \alpha T$	[V]
$\mathbf{j}_S$	entropy flux density (vector)	[W/(K m²)]
S_N, S_{pc}	entropy per carrier	[Ws/K]
v_S	volumetric entropy production	[W/(K m³)]
v_q	volumetric heat production	[W/m³]
η	efficiency (TEG)	[1]
p	electrical power density (TEG) $p = P/A_c$	[W/m²]
φ, φ_c, COP	coefficient of performance (TEC)	[1]
η_r, φ_r	reduced efficiencies	[1]

Glossary

Shortcut	Declaration
1D	one-dimensional
3D	three-dimensional
BC	boundary condition
COP	coefficient of performance
CPM	constant properties model
FEM	finite element method
FGM	functionally graded material
IC	initial condition
LPM	linear properties model
MBV	mean of the boundary values
ODE	ordinary differential equation
PDE	partial differential equation
PGEC	Phonon glass/electronic crystal
TAGS	Te–Ag–Ge–Sb
TAv	temperature average
TE	thermoelectric
TEC	thermoelectric cooler
TED	thermoelectric device
TEG	thermoelectric generator
TEH	thermoelectric heater
TEM	thermoelectric module
VMT	value at the mean temperature
WAv	weighted average

1
Thermodynamics and Thermoelectricity

Christophe Goupil, Henni Ouerdane, Knud Zabrocki, Wolfgang Seifert†, Nicki F. Hinsche, and Eckhard Müller

1.1
Milestones of Thermoelectricity

Thermoelectric effects result from the interference of electrical current and heat flow in various materials. This interaction, characterized by a coupling parameter called *thermopower*, allows the *direct* conversion of heat to electricity [1–4]; conversely, cooling may be achieved by the application of a voltage across a thermoelectric material.

Almost 200 years after the first discoveries in thermoelectrics by Seebeck in 1823 [5–7], it is now again a very active period of observing thermoelectric (TE) phenomena, materials, and their application in devices. The search for green technologies, for example, converting waste heat generated by industrial facilities and car engines into usable power, pushes scientists to pick up "old" effects with new classes of materials with higher TE efficiency to develop practical applications using the advantages of TE power generation. For an overview of a variety of applications, see, for example, the following books and review articles [8–33].

A similar situation was encountered in the late 1950s when the usage of semiconductors as TE materials was the origin for a revival of thermoelectrics [34]. This is directly connected with the investigations by Goldsmid and Douglas [35] and the Ioffe group [36], who considered both thermodynamics and solid-state approaches. They extended the previous developments to the microscopic area, opening the door for material engineering and practical applications [1, 2, 37–55].

A recommendable overview on the early years of TE research had been given by Finn [52, 56, 57]. Historical facts can be found in several textbooks on thermoelectricity, see, for example, [8, 13, 15, 36, 42, 58]. Clearly, thermoelectricity is influencing and has influenced two branches of physics decisively, see Figure 1.1. On the one hand, it all began with investigations of electromagnetism, but the interest started in thermodynamics as well. Here, we would like to give a retrospective view on some of the milestones in the development of thermoelectricity.

Continuum Theory and Modelling of Thermoelectric Elements, First Edition. Edited by Christophe Goupil.
© 2016 Wiley-VCH Verlag GmbH & Co. KGaA. Published 2016 by Wiley-VCH Verlag GmbH & Co. KGaA.

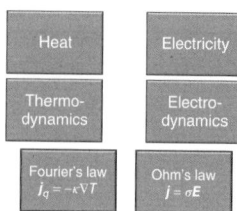

Figure 1.1 Two branches of physics combined.

At the same time, we would like to acknowledge some of those scientists who made an especially valuable contribution to its development in the early years.

1.1.1
Discovery of the Seebeck Effect

Today, the phenomenon in which a temperature gradient in a currentless circuit consisting of different materials produces a primary voltage is known as the Seebeck effect, named after the German-Baltic physicist Thomas Johann Seebeck. However, some hints on TE effects are known from times before Seebeck. Lidorenko and Terekov [59] mentions the studies by the Petersburg academician Aepinus, carried out in 1762 and reviewed in Ref. [60], see also [61]. Sometimes, Volta is also considered as the first person to discover the TE effects. With regard to this point, we refer to several articles by Anatychuk and other authors [62–66]. His bibliographical research indicates that the first experiment on the TE phenomenon had been reported by A. Volta in a letter to A.M. Vassalli written on February 10, 1794:

> "... I immersed for a mere 30 seconds the end of such arc into boiling water, removed it and allowing no time for it to cool down, resumed the experiment with two glasses of cold water. It was then that the frog in the water started contracting, and it happened even two, three, four times on repeating the experiment till one end of the iron previously immersed into hot water did not cool down."[1] More details of the experiments can be found in Refs [67, 68] and references therein.

The role of Alessandro Volta in thermoelectricity is particularly addressed in an article by Pastorino [69]. Volta's observations are based on Galvani's experiments [66, p. 28]: "The novel discovery that eventually led to the recognition of thermoelectricity was first disclosed in 1786 and published in book form in 1791. Luigi Galvani noticed the nerve and muscle of a dissected frog contracted abruptly when placed between dissimilar metal probes. Alessandro Volta, in 1793, concluded that the electricity which caused Galvani's

Alessandro Giuseppe Antonio Anastasio Volta (1745–1827)

---

1) Translation of a letter to professor Antonio Maria Vassalli (accademia delle scienze di torino).

frog to twitch was due to the interaction of the tissue with metals that were dissimilar. This observation, though not of the Seebeck effect eventually did lead indirectly to the principle of the thermocouple that also uses dissimilar conductors (but in a quite different way) to create an emf[2] as a measure of temperature. Pioneers of thermoelectricity built on Volta's observation." Volta's contributions are highlighted and concluded in Ref. [64] as follows:

1) *"A. Volta organized these experiments deliberately, bearing in mind the discovery of thermoelectromotive forces arising due to temperature difference. This distinguished A. Volta's experiments from those of Seebeck who rejected the electrical nature of thermomagnetism effect that he had discovered."*
2) *"A. Volta discovered immediately and directly the origination of thermoelectromotive forces, unlike Seebeck, who observed the magnetic effect of TE current excited by thermoelectromotive forces in a closed circuit."*
3) *"A. Volta as talented and experienced experimenter first excluded the distortions that might take place in his experiments due to galvanic EMF caused by inhomogeneity of metal wires. For this purpose he first selected inhomogeneous arcs which in the absence of temperature difference in the wire did not result in the origination of electromotive forces even on such primitive EMF indicator as prepared frog."*
4) *"Direct observation of thermoEMF by A. Volta took place 27 years earlier than Seebeck observed the thermomagnetism effect."*

Korzhuyev and Katin have a different opinion on the contribution of Galvani and Volta to thermoelectricity, see [65, p. 16]:

> *Due to the above ambiguity of terms, present-day specialists in history of physics in their attempts to determine the researcher pioneer of thermoelectricity encounter difficulties. As the trailblazers of thermoelectricity "prior to T. Seebeck," apart from the above-mentioned Volta, Ritter, Shweiger, Aepinus, the names of Ch. Oersted (1822), J. Fourier (1822) et al. [8][3] are referred to. Having analyzed the effects observed by the above authors, we have concluded that the process of discovery of thermoelectricity as phenomenon in its modern "physical" understanding can be represented by the schematic*
>
> **Aepinus, G&V ⇒ Seebeck ⇒ Peltier, Thomson,**
>
> *where Aepinus, Galvani and Volta are the "forerunners" of discovery, and Peltier and Thomson – scientific "successors" of Seebeck who obtained decisive results for the formation of the respective division of physical science.*

---

2) "Electromotive force."
3) Citation of [70].

They conclude with two points:

1) *"Thermoelectric effect observed by Volta in his experiments with the iron wire is of a complicated nature and is classified in the paper as predominantly* galvanothermal effect (GTE) *related to temperature dependence of electrode potentials of the cell Fe/H_2O/Fe."*
2) *"Heuristic analysis of formulae of G&V effects that included substitution of media and actions, as well as compiling of effects combinations according to procedure [2],*[4]) *indicates the possibility of existence of new phenomena related to G&V effects that are partially represented in this paper."*

Johann Wilhelm Ritter (1776–1810)

The German Scientist Franz Peters stated in his review book [58] that Ritter and Schweigger observed electrical currents created by temperature gradients before Seebeck. Johann Wilhelm Ritter (1776–1810), a German physicist, is nowadays known for the discovery of the ultraviolet light (by chemical means) and the invention of the first dry battery. From his discoveries, it can be seen that his primary interest had focused on electricity, in particular on electrochemistry and electrophysiology. In 1799, he investigated and carried out the electrolysis of water. One year later, he did experiments on electroplating. In 1801, he observed TE currents and investigated artificial electrical excitation of muscles. With the first observation, he paved way for the scientific foundation of thermoelectricity by Seebeck.

Johann Salomo Christoph Schweigger (1779–1857)

Another scientist who observed TE currents before Seebeck is supposed by Peters to be Johann Salomo Christoph Schweigger (1779–1857), who was a professor of chemistry at the university of Halle[5]) and editor of the "Journal für Chemie und Physik." Schweigger's physical work was also concentrated on the investigation of electricity. In 1808, he developed and constructed an electrometer to measure the electrical force by the magnetic one. He is famous for having developed the electromagnetic multiplicator named after him. Peters dated his observation of TE currents to the year 1810 [58].

Despite all these facts, Thomas Johann Seebeck is regarded today as the discoverer or scientific founder of thermoelectricity. He called the phenomena he investigated "thermomagnetism" as he reported for the first time on August 16, 1821, at a session at the Berlin Academy of Sciences. The following facts are collected

---

4) Author's note: Citation of [71].
5) Author's note: Today: Martin-Luther-University Halle-Wittenberg, Saxony-Anhalt, Germany.

mainly from the two publications of Velmre [6, 7] containing more detailed biographical information about Seebeck.

Thomas Johann Seebeck (1770–1831)

Thomas Johann Seebeck was born on April 9, 1770, in Tallinn[6], the capital and largest city of Estonia. His father, Johann Christoph, was a wealthy merchant. His father's wealth gave him a kind of independence in his later studies. As discoverer of the Seebeck effect, he is a key person in the development of thermoelectricity as a scientific branch, but there are other things he is less known for, such as the discovery of the piezooptic effect and photoelasticity in 1813. By investigating tempered glass blocks/plates, he described how mechanical stress in an amorphous transparent material (plastics/glass) can cause the material to be birefringent. Photoelasticity, which had been later rediscovered by the English physicist David Brewster, is concerned with the determination of residual stress in glasses. Seebeck summarized all of his early research on optics in a contribution to Goethe's work in natural sciences "Zur Farbenlehre." Seebeck also contributed to the development of photography, as he described the solar spectrum in the natural colors of silver chloride in 1810. With that, he is a forerunner of interference color photography, which led to the invention of holography. But his now most famous impact in science is the widespread investigation of TE materials although he misinterpreted the results of his experiments as "thermomagnetism." He investigated materials we call today semiconductors and stated that the following materials, such as *Bleiglanz* (Galena, PbS), *Schwefelkies* (pyrite, FeS_2), *Kupferkies* (chalcopyrite, $CuFeS_2$), *Arsenikkies* (arsenopyrite, AsFeS), *Kupfernickel* (nickeline NiAs), *weisser Speiskobalt* (white skutterudite, $(Co,Ni)As_{3-x}$), all of which display Bismuth-like behavior, and *Kupferglas* (chalcocite, Cu_2S), *Buntkupfererz* (bornite, Cu_5FeS_4), *blättriger Magnetkies* (pyrrhotite, $Fe_{1-x}S$), which display an Antimony-like behavior, exhibit a stronger "thermomagnetism" when they are in contact with copper than other materials he had investigated.

After his graduation in 1788 at Reval Imperial Grammar School, Seebeck went to Berlin and later on to Göttingen to begin studies in medicine, which he finished in 1792 with excellent marks. Later, he shifted to Bayreuth to study physics. This was possible because of his father's inheritance, which allowed him independent studies without practicing medicine. In March 1802, Seebeck received the doctor of medicine degree in Göttingen. Afterward, he shifted to Jena, which had a great influence on him, because he found a very stimulating intellectual atmosphere with the famous "Jenaer Romatikerkreis" in which natural philosophers were active, like Oken, Schelling, the Schlegel brothers, and Ritter, the "romantic physicist" who founded the discipline of electrochemistry. This was also the time when the long-lasting friendship with Goethe and the philosopher Hegel began.

6) Author's note: Formerly known as Reval.

In 1818, Seebeck went to Berlin and became a member of the Prussian Academy of Sciences. On August 16, 1821, he gave a first report on "thermomagnetism" of a galvanic circuit in a session at the Berlin academy. Three further talks on this topic followed on October 18 and 25, 1821, as well as on February 22, 1822, the results of which were published in Refs [5, 72]. One key topic in physics of the nineteenth century was the discovery of electromagnetism by Oersted in 1820 and further investigation of the phenomenon afterward. Oersted sent out a circular letter written in Latin to his colleagues and institutes "*Experimenta circa effectum confliktus electrici in acum magneticam*" [73]. Seebeck, recognizing this, gave up his studies in optics to go over to investigate electricity and magnetism and re-examined the Volta theory [74].

Even though Alessandro Volta may have been the discoverer of the TE effect, Seebeck was the first to carry out a series of detailed investigations on TE materials, see Table 1.1 taken from Ioffe's textbook [36, 42]. On December 10, 1831, Seebeck died in Berlin.

Parallel to Seebeck, there were several scientists doing similar experiments. Independently of Seebeck, Julius Conrad von Yelin discovered on March 1, 1823, thermomagnetic currents in metals [77, 78]. He was very confident about his results, as he expressed in his letter [79]:[7]

> *From a letter of Yelin's (Munich May, 6th 1823)*
> *You will receive in a few days an essay in which I explain the whole thermo-magnetism from the known laws of electro-magnetism in a very simple way. I am curious how I come along with these statements with Mr. Dr. Seebeck, who as Mr. Prof. Oersted informed in the recent February issue of the "Annalen der Chemie" that he also and indeed as is evident earlier than me found that an unequal heating makes all metals to electro-magnets (I observed these phenomena first at a simple copper arc on March, 14th). Since Mr. Seebeck had concealed his discovery, then this is not depriving my discovery's merit, and as Galilei with Jupiter's moons, Kleist with the gain-bottle [Leyden bottle] and Kunkel the phosphorus have been discovered for the second time, I will give myself credit as the discoverer of the new, very important phenomena of the thermo-magnetism and their effects on the geogeny and crystal formation.*

In Figure 1.2, Seebeck's classical experiment is sketched. In a closed circuit of two dissimilar metallic conductors, one of the soldered junctions between the conductors is heated. A magnetic needle is positioned near this arrangement and then deviated. The declination angle is proportional to the temperature difference between the two (hot and cold) junctions. As a result of his experiments, Seebeck proposed his "thermomagnetische Reihe"[8], which is actually a TE series, see [36, 42] and Table 1.1. Finally, it was Oersted who provided a physically correct explanation of the phenomenon that the *electric current* in the circuit is due to the

---

7) Translation from the German original.
8) *engl.* Thermomagnetic series.

Table 1.1 Seebeck series, Justi's, and Meissner's thermoelectric series taken from Refs [36, 42, 75, 76]

Seebeck 1822	Justi 1948		Meißner 1955			
			Metals		Semiconductors	
PbS	Bi	−80	Bi	−70	MoS	−770
Bi	Co	−21	Co	−17.5	ZnO	−714
Bi-amalgam	Ni	−20	Ni	−18	CuO	−696
Ni	K	−14	K	−12	Fe_2O_3 (400 °C)	−613
Co	Pd	−8	Pd	−6	FeO	−500
Pd	Na	−7	Na	−4.4	Fe_3O_4	−430
Pt Nr 1	Pt	−5	Pt	−3.3	FeS_2	−200
U	Hg	−5	Hg	−3.4	MgO_3H_2	−200
Au Nr 1	C	−3.5			SnO	−139
Cu Nr 1	Al	−1.5	Al	−0.6	Fe_2O_3 (50 °C)	−60
Rh	Mg	−1.5	Mg	−0.4	CdO	−41
Au Nr 2	Pb	−1.0	Pb	−0.1	CuS	−7
Ag	Sn	−1.0	Sn	+0.1	FeS	+26
Zn	Cs	−0.5	Cs	+0.2	CdO	+30
C	Y	−1.0	Y	+2.2	$GeTiO_3$	+140
Cu Nr 3	Rh	+1.0	Rh	+2.5	NiO	+240
Pt Nr 4	Zn	+1.5	Zn	+2.9	Mn_2O_3	+385
Cd	Ag	+1.5	Ag	+2.4	Cu_2O	+474
Steel	Au	+1.5	Au	+2.7	Cu_2O	+1000
Fe	Cu	+2.0	Cu	+2.6	Cu_2O	+1120
As	W	+2.5	W	+1.5	Cu_2O	+1150
Sb	Cd	+3.5	Cd	+2.8		
SbZn	Mo	+6.5	Mo	+5.9		
Fe	Fe	+12.5	Fe	+16		
	Sb	+42	Sb	+35		
	Si	+44				
	Fe	+49	Fe	+400		
	Se		Se	+1000		

heat and coined the term "thermo-électricité"[9] instead of Seebeck's chosen term [80, 81]. Oersted was the one [80] who brought thermoelectricity to the attention of several French scientists, such as father and son Becquerel, Fourier, Melloni, Pouillet, and Nobili.

Fourier and Oersted made the first thermopile for TE power generation in collaboration, see [82].

From a mathematical point of view, a voltage, the Seebeck voltage, can be measured due to a temperature difference

$$V_\alpha = -\int_{T_c}^{T_h} \alpha(T)\,dT, \tag{1.1}$$

---

9) engl. Thermoelectricity.

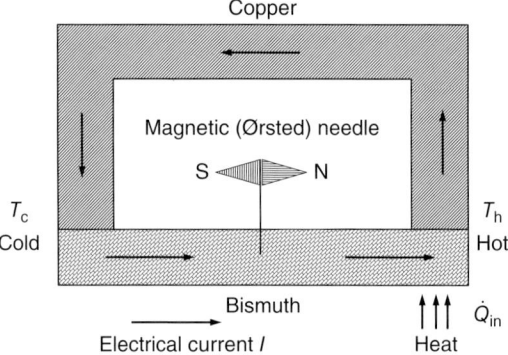

Figure 1.2 Schematics of Seebeck's classical experiment.

where $\alpha(T)$ is the Seebeck coefficient of a material. If this coefficient is supposed to be independent of temperature or if small temperature differences are assumed, the relation (1.1) simplifies to

$$V_\alpha = -\alpha\,(T_h - T_c). \tag{1.2}$$

1.1.2
Discovery of the Peltier Effect

Jean C. A. Peltier
(1785–1845)

Jean Charles Athanase Peltier was born in 1785 in a modest family in northern France. As apprentice watchmaker from an early age, he showed great technical skills. He then developed his self-taught scientific knowledge in the fields of human physiology, electrical phenomena, and later meteorology, of which he is now considered a precursor. More about his life and his scientific work can be found in a book written by his son Ferdinand Athanase Peltier [83].

His great experimental skills allowed him to develop galvanometers and electrometers of very high precision, which were essential to the discovery of heat absorption measurement by circulating electric current. This effect, thoroughly studied by Peltier, soon became a subject of communication to the French Academy of Sciences (April 21, 1834).

Additional experimental proof of the effect had been given by Emil Lenz in 1838, as he observed that water could be frozen at an electrical junction by passage of an electric current. Lenz also found that if the electric current is reversed, the ice can be melted again [84].

A schematic of the Peltier effect is shown in Figure 1.3, where two different materials (different electronic heat capacities) are in contact. After this, Peltier continued his work on thermoelectricity and coupled it with his other works in

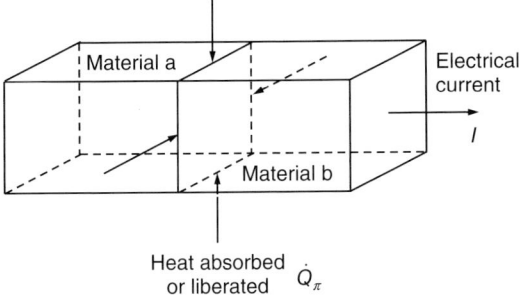

Figure 1.3 Schematics for the Peltier effect.

electromyology and meteorology. As an ending of his large contribution to science, his latest work on "electrical meteorology" was published in 1844, see [85].

Besides the irreversible Joule heat, which is found in all conductors, the Peltier heat flow $\dot{Q}_\pi \propto I$ occurs if a current passes through the material, which is reversed if the orientation of the current is reversed. The proportionality factor between the current and the absorbed heat released is the Peltier coefficient, which is in relation to the Seebeck coefficient α given by the first Thomson (Kelvin) relation $\Pi = \alpha T$, where T denotes the absolute temperature. So, the Peltier heat power is found as

$$\dot{Q}_\pi = \alpha T I. \tag{1.3}$$

1.1.3
Discovery of the Thomson Effect

Lord Kelvin (1824–1907)

A first theoretical description of TE effects had been given by William Thomson (later known as Lord Kelvin) in 1851, as he brought the observed effects in harmony with the two laws of thermodynamics [86]. He combined the descriptions of the Seebeck and Peltier effects into a single expression by using thermodynamic arguments and providing decisive arguments in favor of a compact and complete description of all phenomena [86–90]. Furthermore, he found in this theoretical analysis of the relationship between both effects that an additional effect has to occur, which is named after him. The Thomson effect describes the generation or absorption of heat along a homogeneous conductor that is under a thermal gradient **and** carries an electric current. It is a distributed Peltier effect due to the temperature dependence of the Seebeck coefficient, leading to the Thomson coefficient, that is, $\tau = T\, d\alpha/dT$, which manifests the second Thomson (Kelvin) relation. If $\alpha(T) = $const., then $\tau = 0$.

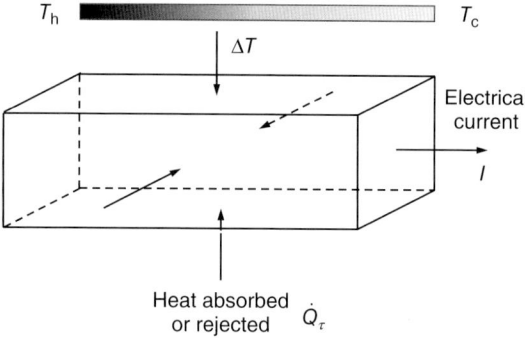

Figure 1.4 Schematics of the Thomson effect.

In Figure 1.4, a schematic of the Thomson effect is shown. It is found that

$$d\dot{Q}_\tau = I \cdot \tau \cdot \frac{\partial T}{\partial x} \cdot dx.$$

Thomson spoke of convection of heat in the nominal direction of the current in a conductor and introduced the term "specific heat of electricity" [89]. His theory of thermoelectricity can be considered as the first reasonable theory in nonequilibrium thermodynamics [90, 91]. A comprehensive overview of Thomson's work in his historical background is given by Finn [52, 56, 57].

1.1.4
Magnus' Law

Heinrich Gustav Magnus
(1802–1870)

In 1851, the German chemist and physicist (Heinrich) Gustav Magnus discovered that the Seebeck voltage does not depend on the distribution of the temperature along the metals between the junctions, see, for example, [92]. There he wrote: "*There is no way that in a homogeneous conductor a power-supplying potential difference is caused by a temperature difference alone.*"[10] This is an indication of the fact that thermopower is a thermodynamic state function. Obviously, the direct correlation is the physical basis for a thermocouple, which is often used for temperature measurements, see [93].

Magnus was born on May 2, 1802, in Berlin, and he died there on April 4, 1870. Magnus also made his contributions to the field of chemistry, but today he is better known for his contributions to physics and technology [94]. He became especially famous for his physical explanation of a phenomenon in the field of

---

10) Translation from German: "Im homogenen Leiter können auf keine Weise allein durch Temperaturdifferenzen stromliefernde Potentialdifferenzen hervorgerufen werden," see [75, pp. 74–75].

fluid mechanics. This effect, the Magnus effect, named after him, describes the occurrence of a shear force effect (force) of a rotating body (cylinder or sphere) in a fluid flow.

In 1822, Magnus studied chemistry, physics, and technology at the "Berliner Universität" (today Humboldt University). In 1827, he received his PhD on a topic on the chemical element tellurium. Afterward, he visited Berzelius' laboratory at the Stockholm Academy of Sciences and then the "Sorbonne" in Paris, where he stayed with Gay-Lussac and Thénard. He returned to Berlin, and after completing his "Habilitationsschrift" on mineral analysis in 1831, he started lecturing physics and technology at the university.

In Kant's article [95], Magnus' work is acknowledged as follows[11]: *"Particularly Magnus worked in the physical area on some questions of the expansion of the gases caused by heat and fluid mechanics (Magnus effect). His works were often inspired by practical problems: for example, he designed a special mercury thermometer – the geothermometer – to investigate the change in temperature with depth of mine shafts. He was above all a brilliant experimentalist and in the end he was not satisfied until all feasible attempts have been 'tried out' to get a result."*

1.1.5
Early Performance Calculation of Thermoelectric Devices

In 1885, it was Lord John William Rayleigh (fka J.W. Strutt) who suggested power generation using the Seebeck effect [96]. Although there are erroneous results in this work, it can be considered as an origin of the concept of direct energy conversion. Thermoelectricity is just a principle of direct energy conversion [1–4, 36, 45, 47, 97], and the application of this principle to build solid-state devices recently regained new interest among scientists and technologists. Currently, the focus is the development of novel and advanced materials reaching higher performance in devices. Until now, the lack of sufficiently high performance is one of the main reasons why there is no broad commercial application of TE devices yet [10–12]. There was a long time period following the exploration of the TE effects in which there was a great interaction of experiment and theory in the field.

1.1.6
First Evaluation of the Performance of a Thermoelectric Device by E. Altenkirch

The German scientist Edmund Altenkirch (1880–1953), one of the pioneers of technological thermoelectricity, studied mathematics and physics in Berlin and was interested in theoretical investigations at an early stage. He was stimulated by the lectures of Max Planck. Later, he turned his focus to problems in technical physics.[12]

---

11) As a translation from German.
12) An overview is given in the German book by Unger/Schwarz "Edmund Altenkirch-Pionier der Kältetechnik" [98].

Two theoretical papers by E. Altenkirch, who was certainly inspired by former investigations, especially those by Lord Rayleigh [96], are particularly worth mentioning. In Reference [99], he formulated the efficiency of a thermopile from the material properties that are relevant for practical devices [98, 100].

By comparing his results with the efficiency of the Carnot cycle, Altenkirch called the thermopile[13] in his article a "rather imperfect thermodynamic engine." In a subsequent paper, he described the effectiveness of TE cooling [100]. Altenkirch gave the first evidence that a good TE material should have a large Seebeck coefficient α, a high electrical conductivity σ (low electrical resistivity ϱ) to minimize the Joule heat, and a low thermal conductivity κ to retain the heat at the junctions and maintain a large temperature gradient. A patent in collaboration with Gehlhoff demonstrates that Altenkirch not only conducted theoretical investigations but was also interested in practical applications [102].

Early thermal conductivity measurements by Eucken [103–106] on solids quickly revealed that point defects found in alloys significantly reduce lattice thermal conduction – a fact that became important for the improvement of TE materials.

1.1.7
Benedicks' Effect

Carl Axel Fredrik Benedicks
(1875–1958)

The TE effects described by Carl Benedicks directly contradict the law of Magnus, which is valid for chemically and physically homogeneous thermoelements or thermocouples. A discussion about this can be found in Section I of [107]. His contributions to the field of thermoelectricity have been valued in Ref. [108] as follows: *"In the light of modern physics, some of Benedicks theories and interpretations of TE phenomena and some of his solutions to problems concerning physical properties of steel seem to be outdated and are subject to criticism. Nevertheless, during his lifetime they arouse great interest in international circles."*

The Swedish physicist Carl Axel Fredrik Benedicks was born on May 27, 1875 in Stockholm and died there on July 16, 1958. Already in his early years, his thoughts turned to the theoretical studies of minerals and metals. He studied natural sciences at the University of Uppsala, where he received his PhD in 1904 on the work *"Recherches physiques et physiochimiques sur l'acier au carbone,"* see [109].[14] In 1910, he became a professor of physics at the University of Stockholm. He is known as the father of Swedish metallography, as he established a special research laboratory. He pioneered in the field of metal microscopy.

---

13) Thermopiles have been used since their invention by Oersted and Fourier around 1823 to generate electrical energy for various purposes. A great deal of historical information about thermopiles can be found in Refs [58, 101].

14) *engl.* Physical and physicochemical research on carbon steel.

Benedicks and his coworkers claimed [110–114] that Magnus' law was not valid even if the element was perfectly homogeneous physically as well as chemically. Fuschillo opposed to Benedicks' conclusions and findings [115]. Domenicali [107] also mentioned that, in order to explain Benedicks' effect, one has to go beyond Onsager theory. The transport entropy per particle/carrier should then depend not only on the temperature gradient but also on the temperature itself, and a higher order approximation has to be taken into account. In summary, it is stated in Ref. [107]: *"In any case the effects reported by Benedicks and his co-workers seem to be very small and not particular reproducible, so that until affirmative evidence is produced we shall not deal with these reported effects any further."*

There is a variety of works dedicated to the topic as follows. In a review on Benedicks' book [113] in Nature [116], the Benedicks effect is explained as *"the homogeneous electrothermic effect."* The Czech scientist Jan Tauc reported on this in several publications, see, for example, [51, p. 147] and [38] as well as references therein. A theoretical description of the Benedicks effect based on a nonlocal theory was given by Mahan [117]. Mahan points in his work to Tauc [38, 118–120] and Grigorenko *et al.* [121]. Piotrowski and coworkers revisited the problem by the application of a nonlinear theory [122]. In a recent review article by Martin *et al.* [26] it was stated that *"the Magnus law has been verified for extremely homogeneous metals."*[15] On the other hand, it was pointed out that an effect can be observed in case of steep thermal gradients in nondegenerate semiconductors.

1.1.8
The Bridgman Effect

Percy Williams Bridgman (1882–1961)

The U.S. physicist Percy Williams Bridgman, who was born on April 21, 1882, in Cambridge, Massachusetts, and died on August 21, 1961, in Randolph, New Hampshire, is famous for his work on properties of matter under high pressure: 'The Nobel Prize in Physics 1946 was awarded to Percy W. Bridgman *"for the invention of an apparatus to produce extremely high pressures, and for the discoveries he made therewith in the field of high pressure physics".*', see [123].

He is further known for his studies on the electrical conductivity of metals and crystal properties. Two methods for single crystal growth, named after him, are the *horizontal* and *vertical* Bridgman methods. More about his life and a list of publications can be found in Refs [124, 125]. He contributed numerous works on thermoelectricity and its connection to thermodynamic concepts, see, for example, [126–136]. The effect that he proposed in Ref. [127] is defined:

'*The Bridgman effect (named after P. W. Bridgman), also called the internal Peltier effect, is a phenomenon that occurs when an electric current passes*

---

15) Citation of [115].

through an anisotropic crystal - there is an absorption or liberation of heat because of the nonuniformity in current distribution.'

Two good reviews and calculations of TE effects in anisotropic materials and systems can be found in Refs [107, 137]. In Reference [137], it is stated that *"the thermoelectric anisotropy has important consequences in the form of thermoelectric eddy currents and the Bridgman effect"*.

1.1.9
Semiconductors as Thermoelectric Materials

The contribution of Russian scientists to the advancement of thermoelectricity is unmitigated. One of the most widely known names is Abram Fedorovich Ioffe.

After graduating from Saint Petersburg State Institute of Technology in 1902, A.F. Ioffe spent 2 years as Wilhelm Conrad Roentgen's assistant in his Munich laboratory. Ioffe completed his PhD at Munich University in 1905. From 1906, Ioffe worked in the Saint Petersburg (from 1924 Leningrad) Polytechnical Institute. In 1952–1954 A.F. Ioffe headed the Laboratory of Semiconductors of the Academy of Sciences of the USSR, which in 1954 was reorganized into the Institute of Semiconductors.

In the late 1950s and in the 1960s, Ioffe made extensive investigations on semiconductors as TE materials and initiated with that a revival of thermoelectrics [34, 36, 39, 138–144]. His achievements can be summarized as follows:

1) Altenkirch already showed in his works [99, 100] what Ioffe denoted as the material's or intrinsic figure of merit [36], which is defined as

$$z = \frac{\alpha^2 \sigma}{\kappa}. \tag{1.4}$$

Ioffe was the first to use z extensively for characterizing the quality/efficiency of thermoelectrics.

2) The concentration of charge carriers in a TE material, which is optimal for large z, was determined. It turned out to be rather high: $\approx 10^{19} \text{cm}^{-3}$ at room temperature. This means that the best thermoelectrics should be degenerated semiconductors or semimetals.

3) The first estimation of the efficiency of semiconductors for TE refrigeration and heating was made. The efficiency appeared to be of interest for practical applications.

4) Ioffe and his coworkers built thermogenerators based on PbS in 1942–1945 [139].[16]

The "conventional" figure of merit as defined by Ioffe is exactly valid merely for Seebeck and Peltier devices made of materials with constant parameters depending neither on temperature nor on position [146]. It explicitly appears in formulae

---

16) A historical overview was given by Glen Slack at ICT 2011 in Traverse City, Michigan, USA [145].

for the efficiency η, which contain only the material properties but no geometrical parameters of the device, see Section 2.3. However, if the material properties are no longer constant (temperature/position dependence), other factors will appear, such as the working conditions, which can be taken into account in "effective" performance parameters as the effective power factor and effective TE figure of merit of a device [147]. As a scientific leader, Ioffe directed a whole group including Stil'bans and further coworkers dealing with problems of TE energy conversion [138, 139].

1.1.10
Thermoelectric Applications – Excitement and Disappointment 1920–1970

It is a known fact that during and after the world wars, thermoelectricity was actively studied for use in valuable technologies, primarily for cooling and power generation for military as well as for civilian uses. The political and economic importance of such devices made advances more difficult and slow to publicize particularly between the Eastern European and Western countries.

1.1.10.1 Construction of the First Thermoelectric Generator

One of the early pioneers of thermoelectricity in the United States was Maria Theresa Telkes (later nicknamed the "Sun Queen"). She was born in Budapest, Hungary, on December 12, 1900. She worked as a biophysicist in the United States, after completing her PhD in physical chemistry in Hungary.

From 1939 to 1953, she was engaged in solar energy research at Massachusetts Institute of Technology. In the 1930s, Maria Telkes made a thorough study on the materials Pb S and Zn Sb, which had already been observed by Seebeck more than a century ago, and in her report [148], these materials *"were stated to produce the best couple for thermoelectric energy conversion"* [149]. Telkes is known for creating the first (solar) TE power generator in 1947 [41, 150, 151] and the first TE refrigerator in 1953 using the principles of semiconductor thermoelectricity.[17]

Maria Theresa Telkes
(1900–1995)

Telkes's generator efficiencies had reached about 5%, and by the 1950s, cooling from ambient below 0°C was demonstrated, which has ultimately led to some viable industries. Many thought thermoelectrics would soon replace conventional heat engines and refrigeration, and interest and research in thermoelectricity grew rapidly at major appliance corporations such as Westinghouse, universities, and national research laboratories [36, 37, 44, 45, 47, 49, 54, 152]. However, by the end

---

17) A "Preliminary Inventory of the Maria Telkes Papers" can be found at the library of the Arizona State University. A list of the contents is provided in the "Arizona Archives Online" under http://www.azarchivesonline.org/xtf/view?docId=ead/asu/telkes_acc.xml.

of the 1960s, the pace of progress had slowed with some discussion that the upper limit of zT might be near 1 and many research programs were dismantled (despite several reports of $zT > 1$).

Ioffe's contributions were already mentioned as he promoted the use of semiconductors in thermoelectrics and semiconductor physics to analyze the results and optimize the performance. Materials with high TE figures of merit are typically heavily doped semiconductors, the best known are the tellurides of antimony, bismuth, and lead. Ioffe and his institute in Saint Petersburg actively pursued TE research and development in the USSR, leading to some of the first commercial TE power generation and cooling devices. Ioffe was one of the first to promote the use of alloying to reduce the lattice thermal conductivity by point defects.

Without doubt, H. Julian Goldsmid is an outstanding personality in thermoelectricity in the twentieth century. One of the first verifications of 0 °C TE cooling was demonstrated by Goldsmid in 1954 using thermoelements based on Bi_2Te_3 [35]. Furthermore, he was one of the first to utilize the TE quality factor, identifying the importance of high mobility and effective mass combination and low lattice thermal conductivity in semiconductors that when properly doped produce good TE materials. Now Goldsmid is active a lifespan since more than 60 years in the field and authored many introductory books [33, 37, 152].

In the search for high zT materials, a general strategy guided by the quality factor has been to look for small band-gap semiconductors made of heavy elements. Slack summarized the material requirements succinctly in the "phonon-glass electron-crystal" (PGEC) concept that the phonons should be disrupted as in a glass, but electrons should have high mobility as they do in crystalline semiconductors [153].

1.1.11
Thermoelectric Industry –Niche Applications 1970–2000

The reliability and simplicity of thermoelectricity enable niche applications for this solid-state technology even while conventional processes are more efficient. Besides thermocouples, a small but stable industry to produce Peltier coolers based on $Bi_2Te_3-Sb_2Te_3$ was formed, which now produces coolers for a variety of products ranging from optoelectronics, small refrigerators, and seat/cooling systems. The need for reliable, remote sources provides some niche applications for TE power generation. The advancement in the science, technology, and commercial use of thermoelectricity has led to a number of focused scientific meetings and organizations, the largest of which is the "International Thermoelectric Society" with meetings since 1970 [154].

1.1.11.1 Thermoelectric Generators for Space
For space exploration missions, particularly beyond the planet Mars, the light from the sun is too weak to power a spacecraft with solar panels. Instead, the electrical power is provided by converting the heat from a ^{238}Pu heat source into electricity using TE couples. Such radioisotope thermoelectric generators (RTG)

have been used by NASA in a variety of missions such as Apollo, Pioneer, Viking, Voyager, Galileo, and Cassini. With no moving parts, the power sources for Voyager are still operating, allowing the spacecraft to continue to make scientific discoveries after over 35 years of operation. The Curiosity rover on Mars is the first rover powered by thermoelectrics using a multi-mission RTG (MMRTG).

1.1.12
New Concepts in Thermoelectricity 2000-Present

Interest in thermoelectricity was renewed in the 1990s with the influx of new ideas. The hope that engineered structures will improve zT, particularly at the nanometer scale, has reinvigorated research in TE materials [31, 155, 156]. While some of these ideas have shown to be ineffective, others have led to entirely new classes of complex TE materials [11].

The global need for alternative sources of energy has revived interest in commercial applications [157] and stimulated interest in developing inexpensive and environmentally friendly TE materials.

For spreading thermoelectrics throughout Europe, another outstanding researcher and communicator of thermoelectricity, David Michael Rowe, OBE (1939–2012), has to be mentioned. Rowe greatly contributed to the dissemination of thermoelectricity, particularly by editing the "*CRC Handbooks of Thermoelectrics*" 1995, 2006, and 2012 [15–18].

This section concludes with a timeline of TE and related research (see Table 1.2).

1.2
Galvanomagnetic and Thermomagnetic Effects

If the material is anisotropic or an additional magnetic field is applied, the theoretical description of the additionally appearing phenomena becomes more complicated in comparison to the pure "thermoelectric" case in an isotropic medium. By taking into account both electric fields **E** and magnetic fields **B**, a generalization of Maxwell's equations of electromagnetic fields under the influence of temperature gradients is obvious, see, for example, [20, 196–203]. Sometimes, the phenomenon described in these works is called "*thermoelectric induction.*" Here we will not go into detail of these phenomena because it is beyond the scope of this book, but at least we want to provide some references where such details can be found. The thermodynamic theory of galvano- and thermomagnetic effects was reviewed in great detail by Fieschi in several articles [204–207]. A bunch of textbooks, chapters, and review articles are also dedicated to the topic [54, 75, 76, 208–217].

The most influential scientists who investigated the effects in the late nineteenth century are shown in Figure 1.5. The works they did have been focused on metals as in thermoelectricity as well. The classical works [218–228] are discussed and reviewed in Ref. [208].

1 Thermodynamics and Thermoelectricity

Table 1.2 Milestones in thermoelectricity

Date	Event	Source(s)
1762	*"Memoirs comprising precise description of tests with tourmaline"* by Aepinus, first observations in the field of thermoelectricity	[59, 60, 158]
1800	Voltaic pile	[159]
1801	Observation of TE currents by J. W. Ritter	[58]
1810	Observation of TE currents by J. S. C. Schweigger	[58]
1820	Oersted's letter *"Experimenta circa effectum confliktus electrici in acum magneticam"* – Discovery of electromagnetism	[73]
1821	Seebeck's first talk and report on the phenomenon *"thermomagnetism"*	[6, 7, 58]
1822	Heat conduction theory by J. B. J. Fourier	[160–162]
1823	First correct explanation of the phenomenon by Oersted coining the term *"thermo-électricité"*	[80, 81]
1824	Carnot cycle – heat engine efficiency	[163, 164]
1827	Ohm's law investigated with help of a thermopile	[165, 166]
1834	Discovery of the Peltier effect by J. C. A. Peltier – DC current through a Bi–Sb junction	[167]
1838	Experiment – freezing a water droplet via the Peltier effect by E. Lenz	[84]
1843	Studies on DC current producing heat investigated by J. P. Joule	[168]
1851	Discovery of the Thomson effect, derived by W. Thomson (Lord Kelvin)	[86]
1851	Magnus' law	[92]
1866	E. Becquerel studied Zn–Sb and Cd–Sb thermoelectrics	[169]
1885	First proposal of a TE generator by Lord Rayleigh	[96]
1909	Comprehensive theory of TE generation by E. Altenkirch	[99]
1910	PbTe, Bi_2Te_3, and Te as thermoelectrics by W. Haken	[170]
1911	Theory of TE refrigeration by E. Altenkirch	[100]
1911	Crystal thermal conductivity measured by A. Eucken	[103–105]
1914	Discovery of the relation of lattice thermal conductivity to heat capacity by P. Debye	[171]
1918	Reports on the Benedicks effect	[110–114]
1926	Discovery of the Bridgman effect	[127]
1928	Electron band theory of crystals by M. J. O. Strutt	[172]
1928	Mass fluctuation decreases lattice thermal conductivity	[106]
1945	Phonon drag effect by Gurevich	[173–175]
1947	Construction of a TE generator by M. Telkes	[150]
1949	Theory of semiconductor thermoelements (published) by A. F. Ioffe	[138–140]
1954	Cooling from ambient temperature down below 0°C with thermoelectricity by Goldsmid and Douglas	[35]

Table 1.2 (Continued)

Date	Event	Source(s)
1956	Solid solutions, Si–Ge alloys by Ioffe	[36]
1957	Thermoelectric book by Ioffe and coworkers	[36]
1957	Discovery of $CoSb_3$ as thermoelectrics by Dudkin and Abrikosov	[176, 177]
1972	TAGS by Skrabek/Trimmer	[178, 179]
1979	Concept of minimal lattice thermal conductivity by G. A. Slack	[180]
1982	Observation of the decrease of lattice thermal conductivity in ice by Ross and Anderson	[181]
1993	Application of nanotechnology leading to advances in the efficiency of TE materials	[182, 183]
1995	Phonon glass/electron crystal (PGEC) concept by G. A. Slack	[153]
1995	Filled skutterudites by JPL, GM, UM, ORNL	[184–186]
1997	Oxide thermoelectrics by Teresaki	[187–189]
2001	Bi_2Te_3/Sb_2Te_3 superlattice by Venkatasubramanian	[190]
2002	Pb Te–Pb Se quantum dots by Harman	[191]
2004	LAST material by Kanatzidis	[192]
2007	Bulk nanocomposites by Tang	[193]
2008	Nanocomposites $(Bi_xSb_{1-x})_2Te_3$ by Poudel	[194]
2008	Tuning resonances in Pb Te by Heremans	[195]

Edwin Herbert Hall (1855–1938)

Walther Nernst (1864–1941)

Augusto Righi (1850–1920)

Figure 1.5 The name givers of galvanomagnetic and thermomagnetic effects. In addition, Albert Freiherr von Ettingshausen (1850–1932) and Sylvestre Anatole Leduc (1856–1937) have to be mentioned.

As for TE materials, the galvano- and thermomagnetic effects can be found to be more pronounced in semiconductor materials. In the two articles [229, 230], the effects found in semiconductor materials are illustratively described in comparison to TE effects. For a more scientific way of description of these effects in semiconductor materials, the reviews [212, 231] are recommended for study.

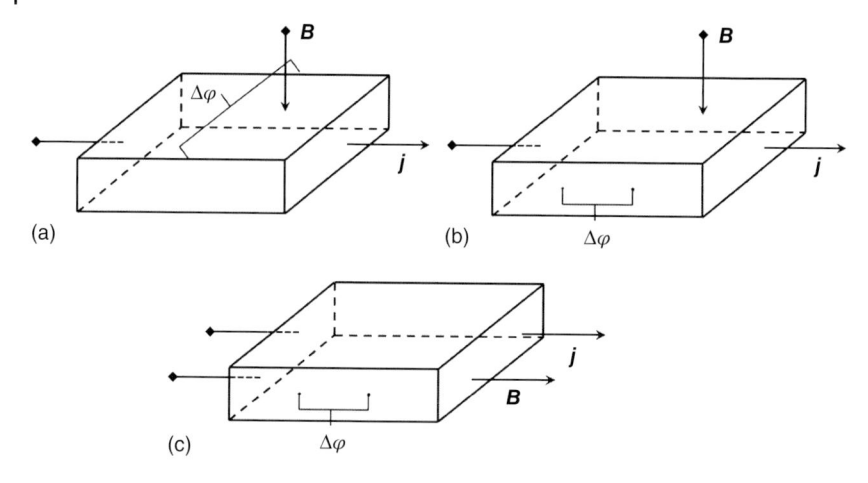

Figure 1.6 Schematics of the different effects: (a) transversal effects, (b) longitudinal effects in a transverse field, and (c) longitudinal effects in a longitudinal field (adapted from Figures 154–156 in Ref. [76, pp. 311–312]).

There is a clear definition of the effects to be studied given by Meissner, as *galvanomagnetic effects* are caused by an electric current on the one hand, and *thermomagnetic effects* are caused by a heat flow on the other hand, see [76]. A distinction is made between *transversal* and *longitudinal* effects. In the first case, the primary current (either heat or electric current) is perpendicular to the produced effect, whereas in the longitudinal case, both variables are supposed to have the same direction.

A more detailed distinction is made between *"transversal," "longitudinal effects in the transverse field"* and *"longitudinal effects in the longitudinal field."* In the transversal effect, see Figure 1.6(a), the magnetic field is perpendicular to the original electric current (density) or heat flux. This effect is found to be perpendicular to both the current/flux and the magnetic field. For the longitudinal effects in the transverse field, see Figure 1.6(b), the magnetic field is perpendicular to the original electric current (density) or heat flux, but the effect is parallel to the current/flux. For longitudinal effects in the longitudinal field, the magnetic field and the effect are parallel to the original electrical current (density) or heat flux, see Figure 1.6(c).

Another possibility to distinguish between different processes is led the thermal condition. The effect is called *isothermal* if there is no temperature gradient present in the direction perpendicular to the primary current. An *adiabatic* effect is characterized by the fact that there is no heat flow in the direction perpendicular to the primary current. Various local effects are theoretically possible. In Reference [206], a number of 560 is proposed. This is found in a situation where the transport coefficients are tensors, that is, they are anisotropic, see Section 21.7 in Ref. [215, p. 240ff] as well as Table 21.5 on [215, p. 242]. Fourteen effects with a transverse magnetic field and four with a longitudinal magnetic field out of these 560 theoretically possible effects and their corresponding phenomenological coefficients between these coefficients are discussed in detail in Ref. [207] (summarized in Table 1 of this reference).

Here, we want to discuss some of them to give a brief overview. First, we assume that the material is isotropic. The primary flow or gradient is applied in x-direction and the magnetic field is directed in the transverse z-direction. The already discussed "thermoelectric" coefficients show in general a dependence on the magnetic field. They can be seen as longitudinal thermogalvanomagnetic coefficients [213]. Now, definition of a number of transversal thermogalvanomagnetic coefficients shall be given without any claim to completeness.

1.2.1
The Hall Coefficient

If there is no temperature gradient in any direction, then a transverse electric field E_y can be observed if an electrical current density j_x is applied while a magnetic field B_z is present. The corresponding coefficient is the so-called Hall coefficient R_H, which is defined as follows:

$$|R_H| = \frac{E_y}{j_x B_z}. \tag{1.5}$$

The signs of the Hall coefficient and the other transverse coefficients can be determined from Figure 1.7.

1.2.2
The Nernst Coefficient

The Nernst effect can be seen as transversal equivalent of the Seebeck effect. If a longitudinal temperature gradient $\frac{dT}{dx}$ is present with a magnetic field B_z, a transverse electric field E_y results. The corresponding coefficient, the Nernst coefficient $\mathcal{N}$, is defined as

$$|\mathcal{N}| = \frac{E_y}{B_z} / \frac{dT}{dx}. \tag{1.6}$$

Note that there is no electrical current present in any direction.

1.2.3
The Ettingshausen Coefficient

The Ettingshausen effect is identified to be equivalent to the Peltier effect. Although this is the case, the Ettingshausen coefficient P_E is defined in terms of the transversal temperature gradient and not the heat flow, that is,

$$|P_E| = \frac{1}{j_x B_z} \frac{dT}{dy}. \tag{1.7}$$

1.2.4
The Righi–Leduc Coefficient

If a longitudinal temperature gradient, dT/dx, and a transverse magnetic field B_z are present, not only a transverse electric field E_y but also a transversal temperature gradient dT/dy result. This observed effect is called Righi–Leduc effect, and the corresponding coefficient S_{RL} is defined as

$$S_{RL} = \frac{1}{B_z} \frac{dT}{dy} \Big/ \frac{dT}{dx}. \tag{1.8}$$

The introduced effects and coefficients are highlighted and illustrated in the following Table 1.3 and Figure 1.9.

The correlation between the introduced effects can be seen in Figure 1.8

1.2.5
Devices Using Galvano- and Thermomagnetic Effects and the Corresponding Figure of Merit

The galvano- and thermomagnetic effects can be used in solid-state devices as in TE elements for either cooling [230] or power generation [232] (Figure 1.9).

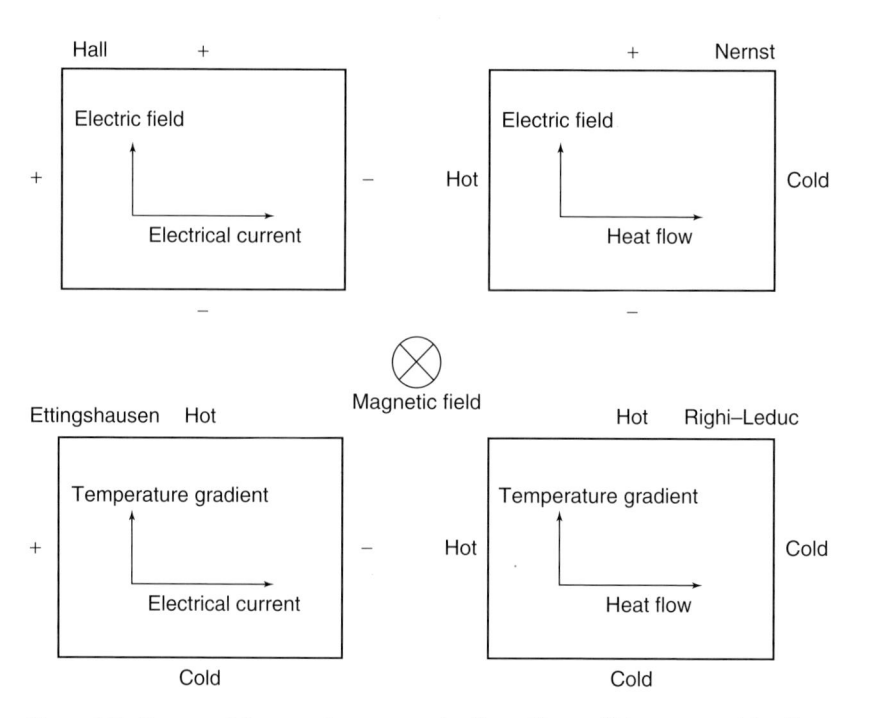

Figure 1.7 Transversal thermogalvanomagnetic effects. The coefficients are positive if the effects have the same directions as shown in the schematics (adapted from Ref. [152, Figure 4-1, p. 83]).

1.2 Galvanomagnetic and Thermomagnetic Effects

Table 1.3 Transport coefficients in isotropic conductors.

Name of the coefficient	Symbol	Definition	Conditions
Electrical conductivity	ϱ	$\dfrac{E_x}{j_x}$	$j_y = j_z = 0,\ \nabla T = 0$
Thermal conductivity	κ	$-\dot{q}_x / \dfrac{dT}{dx}$	$\mathbf{j} = 0,\ \dfrac{dT}{dy} = \dfrac{dT}{dz} = 0$
Seebeck coefficient	α	$E_x / \dfrac{dT}{dx}$	$\mathbf{j} = 0,\ \dfrac{dT}{dy} = \dfrac{dT}{dz} = 0$
Peltier coefficient	Π	$\dfrac{\dot{q}_x}{j_x}$	$j_y = j_z = 0,\ \nabla T = 0$
Hall coefficient	R_H	$\dfrac{E_y}{j_x B_z}$	$j_y = j_z = 0,\ \nabla T = 0$
Nernst coefficient	$\mathcal{N}$	$\dfrac{E_y}{B_z} / \dfrac{dT}{dx}$	$\mathbf{j} = 0,\ \dfrac{dT}{dy} = \dfrac{dT}{dz} = 0$
Ettingshausen coefficient	P_E	$\dfrac{1}{j_x B_z}\dfrac{dT}{dy}$	$j_y = j_z = 0,\ \nabla T = 0$
Righi–Leduc coefficient	S_{RL}	$\dfrac{1}{B_z}\dfrac{dT}{dy} / \dfrac{dT}{dx}$	$\mathbf{j} = 0,\ \dfrac{dT}{dz} = 0$

$\mathbf{j}$ is the electrical current density; $\dot{\mathbf{q}}$ is the heat flux; T is the (absolute) temperature; $\mathbf{E}$ is the electric field; $\mathbf{B}$ is the magnetic field (if $\mathbf{B}$ is nonzero, then it is supposed to lie in the z-direction).

	Galvanomagnetic effects		Thermomagnetic effects	
Cause	Electrical current I	Magnetic field $\mathbf{B}$	Heat flux $\dot{q}$	Magnetic field $\mathbf{B}$
Effect	Potential difference $\Delta\varphi$	Temperature difference ΔT	Potential difference $\Delta\varphi$	Temperature difference ΔT
Transverse effect	Hall effect	Ettingshausen effect	Ettingshausen Nernst effect	Righi–Leduc effect
Longitudinal effect	Thomson effect	Nernst effect	2. Ettingshausen Nernst effect	2. Righi–Leduc effect

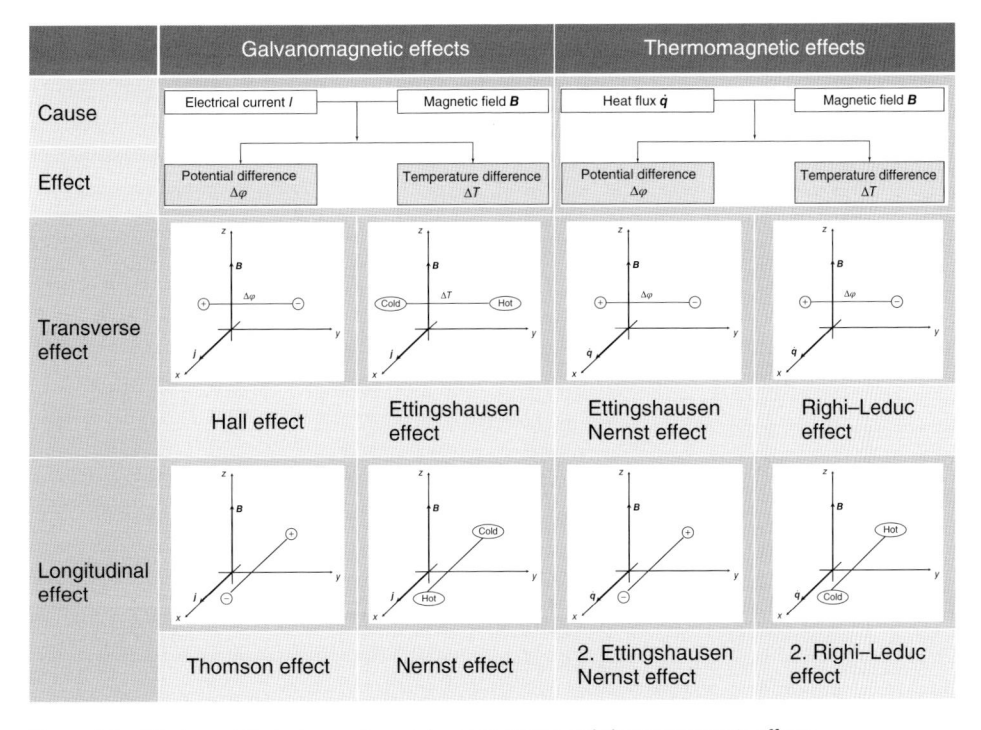

Figure 1.8 Effects and their causes: Four galvanomagnetic and thermomagnetic effects (require a magnetic field) and two TE effects and their correlations (adapted from Ref. [229, p. 126]).

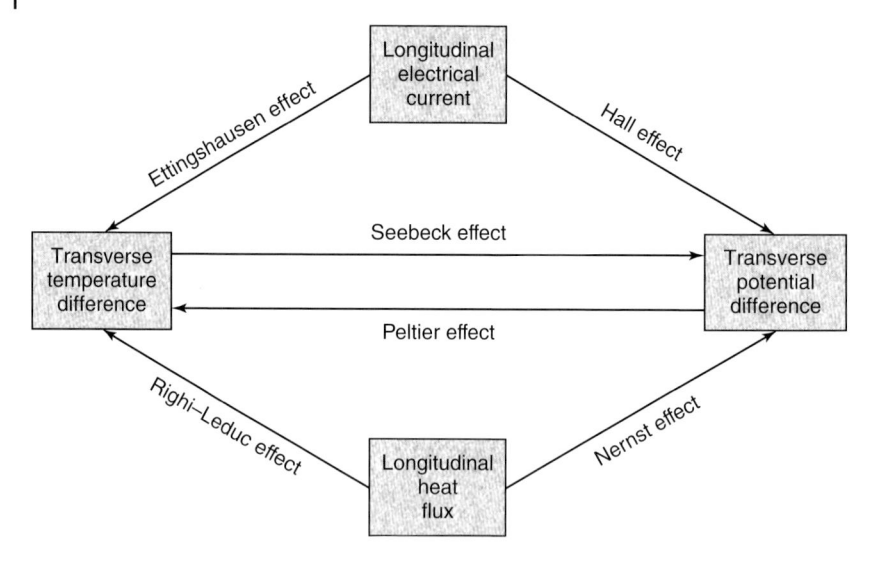

Figure 1.9 Galvanomagnetic and thermomagnetic effects (adapted from Ref. [217, Figure 9.66, p. 846]).

Several works are dedicated to investigate such devices and their performance [232–254]. Analogous to the TE case, the so-called *"thermomagnetic figure of merit"* can be defined (Table 1.4). In general, the calculations for thermomagnetic devices have to be performed in 3D as in an anisotropic crystal, the orientation of the fields and the current densities have to be specified. In Cartesian coordinates, that is, $(x, y, z) \equiv (x_1, x_2, x_3)$, the *isothermal thermomagnetic figure of merit* is given by

$$Z_{ji}^{iso} = \frac{(\mathcal{N}_{ji}^{iso} B_k)^2}{\kappa_{ii}^{iso} \varrho_{jj}^{iso}}, \tag{1.9}$$

see [247]. Note that there are different definitions of the thermomagnetic figure of merit in the literature [253], for example, taking the adiabatic resistivity

$$\varrho_{jj}^{ad} = \varrho_{jj}^{iso} + (E_{ij}^{iso} B_k)(\mathcal{N}_{ji}^{iso} B_k), \tag{1.10}$$

where

$$\varrho_{jj}^{ad} = E_i/j_j; \quad j_i = j_k = \partial T/\partial x_j = \partial T/\partial x_j = \dot{q}_i = 0 \tag{1.11}$$

and

$$E_{ij}^{iso} B_k = (\partial T/\partial x_i)/j_j; \quad j_i = j_k = \partial T/\partial x_j = \partial T/\partial x_j = \dot{q}_i = 0 \tag{1.12}$$

a relation between isothermal and adiabatic thermomagnetic figure of merit can be found

$$Z_{ji}^{ad} = \frac{Z_{ji}^{iso} T}{1 - Z_{ji}^{iso} T} \quad \text{or} \quad Z_{ji}^{iso} = \frac{Z_{ji}^{ad} T}{1 + Z_{ji}^{ad} T}, \tag{1.13}$$

see [247, 253].

Table 1.4 Comparison between thermoelectric and thermomagnetic figure of merit.

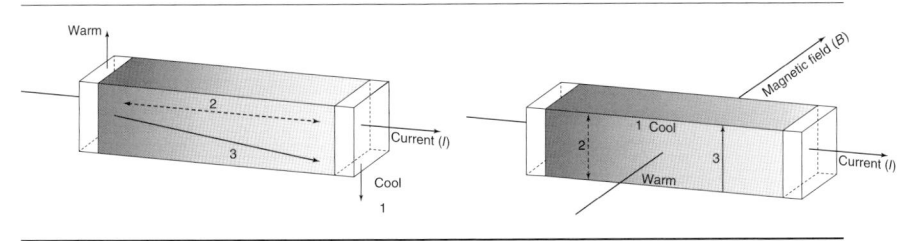

Effect	Thermoelectric figure of merit $Z_{TE} = \frac{\alpha^2}{\varrho \kappa}$ Requirements	# Effect	Thermomagnetic figure of merit $Z_{TM} = \frac{\mathcal{N}^2 B^2}{\varrho_L \kappa_T}$ Requirements
Thermoelectric cooling (Peltier effect)	Large Seebeck coefficient (α)	1 Thermomagnetic cooling (Ettingshausen effect)	Large thermomagnetic coefficients ($\mathcal{N}$) Large magnetic field (B)
Joule heating	Low resistivity (ϱ)	2 Joule heating	Low longitudinal resistivity (ϱ_L)
Heat conduction	Low thermal conductivity (κ)	3 Heat conduction	Low transverse thermal conductivity (κ_T)

Both of which contain three effects: (1) Peltier (TE) or Ettingshausen (TM) cooling, (2) Joule heating, and (3) Fourier heat conduction. The electrical resistivity of the material produces heat (Joule heating). It is observed that half of this Joule heat flows to the cold end. Heat transported via conduction flows from the warm to the cold end of the material. In a cooling device, both effects, that is, Joule heating and Fourier heat conduction, reduce the cooling effectiveness. (Table and Figures adapted from Ref. [230, p. 76])

1.3 Historical Notes on Thermodynamic Aspects

Although Max Planck was never directly involved in the history of thermoelectricity, his contribution to the development of thermodynamic concepts and, especially, the second principle of thermodynamics is notable. Strongly influenced by the work of Rudolf Clausius, he studied in detail the concept of entropy in his doctoral thesis defended in 1879 [255]. His thesis, similarly to his later works, remained unknown for a long time while his contemporaries, Gibbs and Boltzmann, found more echo. By his large and fierce correspondence with Wilhelm Ostwald, "Energétiste convaincu," Max Planck also contributed to the end of the violent conflict that Boltzmann had with the positivists, headed by Ernst Mach.

In the 1930s, Lars Onsager proposed a theoretical description of linear nonequilibrium thermodynamic processes where the coupled thermodynamic forces and fluxes are described in a very general form. In two major articles, the fundamentals of thermodynamics of dissipative transport were developed in a consistent manner [256, 257]. A summary can be found in a later work of Onsager [258]. Onsager expressed his initial thoughts on the dissipation function and the principle of least dissipation of energy, see [256, S 5, p. 420] and also [257, S 5, p. 2276].

The thermodynamic theory of TE phenomena in isotropic media was first worked out by Callen [210, 259] and is presented in more detail in de Groot's monograph [260]. Usually denoted as Onsager–de Groot–Callen theory, it might be called a "first approximation" theory of TE transport, giving a coherent thermodynamic description of TE processes on a phenomenological level. Domenicali's fundamental article [107] summarizes the principles of steady-state thermodynamics applied to a TE system out of thermostatic equilibrium. He pointed out that a complete description of the state includes the determination of the "electrochemical potential" from the overall electronic and crystalline structure of all phases constituting the TE system (see Table 1.4).

The introduction of the irreversible entropy production in the form of an equality is a very old problem mentioned by Lord Kelvin himself. Tolman and Fine [261] were probably the first to point out that the entropy production of a TE process can be considered as a measure of the total irreversibilities. Before this, Bridgman discussed the relation between thermodynamics and thermoelectrics in several articles [126, 129, 131, 133]. At the beginning of the 1930s, Sommerfeld and Frank provided a review about the statistical theory of TE phenomena in metals, but without considering entropy production [262]. Their calculation was based on Darrow's report [263]. In 1952, Haase [264] presented a review about the thermodynamic phenomenological theory of irreversible processes containing considerations about thermoelectrics as well. During the 1950s and 1960s, a very active period of thermoelectrics, there were many works dedicated to the topic of this review. For a small selection, we draw the reader's attention to [107, 204–206, 265–279]. Another work should be particularly emphasized: Sherman *et al.* stated in Ref. [271] that the conditions that maximize the efficiency of a TE generator are precisely the conditions that minimize the irreversibility process, allowing a closer approach to the Carnot cycle where entropy production is zero. This concept has been extended by Clingman [273, 274] for TE generator and TE cooler.

After a very active period, the interest in thermoelectricity collapsed under the weight of inflated hopes, because there had been no significant advances in practically achieved material efficiency after the mid-1960s. As basic research in thermoelectrics lay stagnant for 30 years after that, meanwhile some materials and commercial applications, in particular of customized Peltier coolers, were still developed. In this period, there were a few works produced on this topic [280–288].

New ideas and materials in the mid-1990s brought thermoelectrics back into the scope of research. The search for green technologies, for example, for converting

$$P = \dot{W} = \dot{Q}_1 - \dot{Q}_2$$

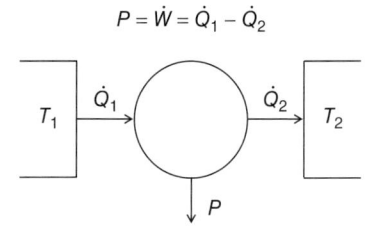

Figure 1.10 Carnot engine.

waste heat generated by car engines into usable power, pushes scientists to pick up "old" effects with new classes of materials with higher TE efficiency to enable practical applications using the advantages of TE power generation [15, 16, 289]. An overview of different applications is given by Riffat and Ma [9]. The reader will find the thermodynamic theory of TE materials and devices in the period from the 1990s to the present day in, for example, [290–315] and references therein. In the next sections, we provide a summary from our point of view.

1.4
Basic Thermodynamic Engine

Let us consider a basic thermodynamic system composed of two thermostats at temperatures T_1 and T_2, respectively (see Figure 1.10). The engine located in between these thermostats receives the entering heat flow $\dot{Q}_1$ from the source (at temperature T_1), and the sink absorbs the rejected heat flow $\dot{Q}_2$. Since the power produced, or received, by the engine is P, then the energy budget of the complete system is given by:

$$P = \dot{W} = \dot{Q}_1 - \dot{Q}_2.$$

The efficiency goal for any thermodynamic engine is the reduction of the entropy production. Then, for a perfect engine called endoreversible engine, the entering entropy flow, $\dot{Q}_1/T_1$, and outgoing entropy flow, $\dot{Q}_2/T_2$, should be equal, so,

$$\frac{\dot{Q}_1}{T_1} = \frac{\dot{Q}_2}{T_2}.$$

By combining these two expressions, we get the relation between the power and the entering heat flow,

$$\dot{W} = \dot{Q}_1 \left(1 - \frac{T_2}{T_1}\right).$$

For an endoreversible engine, *the Carnot expression of the efficiency* is then

$$\eta_C = \frac{\dot{W}}{\dot{Q}_1} = \left(1 - \frac{T_2}{T_1}\right).$$

In order to achieve an efficient system, we need to reach these two goals:

1) Find a good working fluid that produces as little dissipation as possible under standard working conditions.
2) Find a good system that permits quasi-perfect transport of the entropy between both thermostats.

1.5
Thermodynamics of the Ideal Fermi Gas

1.5.1
The Ideal Fermi Gas

The ideal Fermi gas, which shall be considered here, is a physical model assuming a vast number of noninteracting *identical* particles with half-integer spin, in equilibrium at temperature T. The essential difference between the ideal Fermi gas and its classical counterpart is rooted in the half-integer electron spin angular momentum, which governs the statistical distribution of the single-particle energies of the many-body system. The relationship between fermion spin and statistics derives from the canonical anticommutation rules for the second-quantized creation and destruction operators that act in the occupation-number space. Satisfying these anticommutation rules amounts to satisfying the Pauli exclusion principle.

The ideal Fermi gas can be characterized by intensive parameters such as pressure, temperature, and chemical potential. The zero temperature condition is interesting since it puts forth the consequence of the quantum nature of the constituents of the Fermi gas. At zero temperature, the single-particle energies are distributed up to a maximum called the Fermi level, which is the value then taken by the chemical potential (zeroth-order term of the Sommerfeld expansion). The Fermi level also defines a temperature below which the system is considered as degenerate, that is, when quantum effects become dominant. A consequence of the quantum nature of the Fermi particles is that even at zero temperature, there is a nonzero pressure whose origin is defined by the Pauli exclusion principle: the fermions cannot condense neither in momentum space nor in configuration space, which implies that maintaining a spatial separation between the fermions involves the necessary existence of a pressure.

Now let us briefly comment on the interacting Fermi system. An *adiabatic* switching on of the interparticle pair interaction that conserves spin, particle number, and momentum yields some changes to the ideal Fermi system. With the assumption that one particle with well-defined spin and momentum is added to the system, which just before the interaction is switched on is in its ground state, the particle becomes dressed[18] by the interaction with the other particles. The new state is that of an excited particle with the *same* spin and momentum, but characteristic of a different Hamiltonian since inclusion of a new particle modified

---

18) One of the most important changes in the particle properties is mass renormalization.

the particle number and energy. The elementary excitations of the interacting Fermi system are the so-called quasiparticles, which are characterized by their spin, effective mass, momentum, and magnetic moment. The new state induced by switching on the interaction may decay into other states; in other words, the quasiparticles have a finite lifetime. Because of the presence of a nonzero interaction between its constituents, the interacting Fermi system is often called a Fermi liquid in analogy with an ordinary classical liquid characterized by finite interactions and nonnegligible correlations between its constituents. Therefore, there cannot be a phase transition between a Fermi gas and a Fermi liquid since the words "gas" and "liquid" are not related here to actual phases of the Fermi system, but rather to the absence or presence of an interaction; in other words, there is no change in any intensive parameter of a Fermi system, which may switch on or switch off an interparticle interaction. Considering the free electron gas as a working fluid does not imply that operation of the heat engine constituted by a TE cell could be based on a liquid–gas phase transition.

1.5.2
Electron Gas in a Thermoelectric Cell

It is possible to obtain a schematic but useful description of how a TE cell operates by using an analogy with a closed volume that contains a working fluid, which here is an electron gas assumed to be ideal. The electron gas thus considered is a noninteracting system albeit elastic collisions ensure that the microscopic velocity distribution allows the definition of a temperature. The electrochemical potential of the electron system is the analogue of the partial pressure, p_{part}, and reads:

$$\mu_e = \mu_c + e\,\mathcal{V}, \tag{1.14}$$

where μ_c is the chemical potential, e is the electric charge, and $\mathcal{V}$ the electric potential. In the classical description, the gas is suitably characterized by a Maxwell–Boltzmann distribution, but this is not possible for the electron gas: electrons satisfy the Pauli exclusion principle, and hence, as an assembly of indistinguishable particles, they obey the Fermi–Dirac quantum statistics at equilibrium. In terms of intensive variables, the correspondence between the classical gas and the Fermi gas is:

Classical gas: $p_{part}, T \longleftrightarrow$ Fermi gas: $\mu(r, T), T$,

where r is the local position. Thanks to the aforementioned correspondence, one may define schematically a TE cell by considering a charged gas enclosed in a volume, see Figure 1.11. Each end of the volume is maintained at another temperature: T_h on the hot side and T_c on the cold side.

From the kinetic theory of gases, we infer that on the hot side, the system is characterized by a large average particle velocity and a low gas density; conversely, on the cold side, the density is high and the average velocity is low. There is a clear dissymmetry between the carrier populations at both ends of the cell; this dissymmetry is maintained owing to the presence of diffusion currents. As a result,

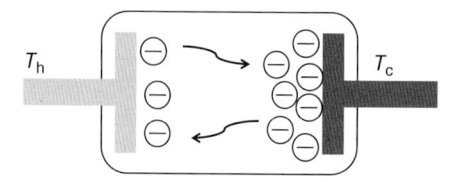

Figure 1.11 Schematic of a thermoelectric cell with a charged gas.

a difference in the electrochemical potentials $\Delta\mu_e$ is *directly* triggered by the temperature difference ΔT. One may then define the thermopower or Seebeck coefficient as the ratio $\alpha = -\frac{1}{e}\Delta\mu_e/\Delta T$, and it appears that the two intensive variables, temperature and electrochemical potential, are coupled.

The system thus described is not in thermodynamic equilibrium since a heat current goes through the gas; however, with the system being closed, the average matter current is zero on average, which means that the two convection currents, cold-to-hot side and hot-to-cold side, compensate each other. Note that heat transport is performed by conduction in the absence of an average particle transport, but also by convection. To simplify the present analysis, we neglect conduction through the walls of the volume. This approximation, rather common in the description of thermodynamic engines, neglects heat leaks through the walls. In TE systems, this leakage exists since the crystal lattice acts as a finite thermal resistance in parallel to the Fermi gas and, hence, deteriorates rather drastically the device ability to convert heat into work. This degradation is due to the phonons, which act as damped oscillators. However, note that there is a phonon-driven mechanism called phonon drag that contributes, though modestly, to the dissymmetry between the hot and cold populations, thus increasing the thermopower.

Here, we focus on the properties of the working fluid. From a thermodynamic point of view, the process includes two isothermal steps and two adiabatic steps. This idealized description puts forth the fact that the performance of the working fluid will be optimized if the particles interact neither with each other nor with the crystal lattice in order to ensure adiabaticity of two of the steps. Kinetic equations should not be introduced here. This would go beyond of the scope of this book.

1.5.3
Entropy Per Carrier

To characterize the behavior of the working fluid, we consider the entropy transported between the two ends of the cell. We consider that the closed volume behaves as two connected compartments with volumes V_1 and V_2, which, respectively, contain N_1 and N_2 particles. The total particle number is fixed: $N = N_1 + N_2$.

Now we assume that p number of particles (with $p \ll N$) move from one compartment to the other.

The total volume remains unchanged. The statistical entropy of the system is given by the accessible configurations defined as follows:

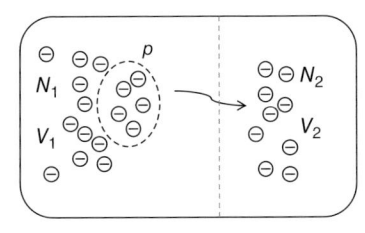

Figure 1.12 Schematic view of the basic entropy per carrier calculation.

- Ω denotes the number of configurations with N_1 particles in compartment 1 and N_2 particles in compartment 2, with probability $P(N_1) \propto \Omega$. Ω' denotes the configurations with $N'_1 = N_1 - p$ particles in compartment 1 and $N'_2 = N_2 + p$ particles in compartment 2, with probability $P(N'_1) \propto \Omega'$.

The resulting variation of entropy reads:

$$dS = k_B \ln \frac{\Omega'}{\Omega} = k_B \ln \frac{P(N'_1)}{P(N_1)}, \qquad (1.15)$$

where k_B is the Boltzmann constant. The binomial law gives:

$$P(N_1) = \frac{N!}{N_1! N_2!} \left(\frac{V_1}{V}\right)^{N_1} \left(\frac{V_2}{V}\right)^{N_2} \qquad (1.16)$$

so

$$dS = -k_B \ln \frac{N'_1!}{N_1!} + k_B \ln \frac{N'_2!}{N_2!} + k_B \ln \left(\frac{V_2}{V_1}\right)^p, \qquad (1.17)$$

which, after using Stirling's approximation,[19] becomes

$$dS = -p\, k_B \ln \frac{n_2}{n_1}, \qquad (1.18)$$

where n_1 and n_2 are the carrier densities in the respective volumes, that is, $n_1 = N_1/V_1$ and $n_2 = N_2/V_2$. Within the framework of statistical mechanics, the entropy per carrier, S_N, is defined by dividing the given expression by p:

$$S_N = -k_B \ln \frac{n_1}{n_2} \qquad (1.19)$$

and, as a result, the free enthalpy becomes:

$$dG = -T\, dS = p k_B T [\ln(n_2) - \ln(n_1)] = p(\mu_2 - \mu_1), \qquad (1.20)$$

an expression from which we obtain the chemical potential:

$$\mu_c(T) = k_B T \ln \frac{n(T)}{n_0}, \qquad (1.21)$$

where n_0 is a constant that depends on the choice of the zero on the energy scale.

Several conclusions from the elementary model, see also Figure 1.12, may be drawn at this stage:

---

19) An approximation for factorials: $\ln(n!) \approx n \ln n - n + \mathcal{O}(n)$. It's named after James Stirling though it was first stated by Abraham de Moivre [316–318].

- The population dissymmetry becomes larger as the gas density decreases. As a consequence, an insulating material, which has very few free carriers, boasts a thermopower much higher than that of a metal, host of a dense gas.
- The adiabatic behavior corresponds to the free, that is, ballistic, transport of the carriers.
- The thermal conductivity of the gas has two contributions: one is the conduction that results from the microscopic energy transfer as particles collide in the gas; the other is the convection.
- The conduction does not contribute to the entropy transport and thus only acts as a thermal leak process, because it does not contribute to any motion of the carriers, in contrast with the convective contribution.
- The convection contributes to the transport of entropy, which must be maximal.
- The carrier mobility thus must be maximal.
- As a consequence, the electrical conductivity is optimal if the carrier mobility is maximal.
- The electrical conductivity may become maximal as the electron gas density increases, but in this case, the ability to transport entropy falls drastically.

1.5.4
Equation of State of the Ideal Electron Gas

Though they are similar at a first glance, the Fermi gas is different from the classical ideal gas because of the quantization of the energy levels that are accessible to the carriers: the electrons are distributed on the energy levels with the additional constraint that they have to fulfill the Pauli exclusion principle. The energy distribution at equilibrium is that of Fermi–Dirac, which ensures the allocation of electrons in the accessible energy states.

Let W_i be the number of ways to distribute N_i indistinguishable particles over S_i states of energy E_i. This number defined at equilibrium is constrained by the laws of conservation of energy and matter:

$$\sum_i N_i = N, \tag{1.22}$$

$$\sum_i E_i N_i = E_{\text{total}}. \tag{1.23}$$

As for all isolated systems in thermodynamic equilibrium, the carriers' distribution is that which yields a maximal entropy:

$$dS = 0 = k_B \, d \ln W = k_B \, d \left(\sum_i \ln \frac{S_i!}{(S_i - N_i)! N_i!} \right) \tag{1.24}$$

with the constraints of conservation of matter and energy that read:

$$\sum_i dN_i = 0, \tag{1.25}$$

$$\sum_i E_i \, dN_i = 0. \tag{1.26}$$

The problem is then defined by the three equations (1.24)–(1.26). An elegant way to solve it is to build a unique expression from these. This can be done by the introduction of Lagrange multipliers. Then, both constraints are introduced in the expression of the entropy using the two multipliers β_1 and β_2:

$$\sum_i \left[\ln\left(\frac{S_i}{N_i} - 1\right) - \beta_1 - \beta_2 E_i \right] dN_i = 0. \tag{1.27}$$

Note that dS is not modified since $\beta_1 \sum_i dN_i = 0$ and $\beta_2 \sum_i E_i dN_i = 0$. As a result, the following equality

$$\ln\left(\frac{S_i}{N_i} - 1\right) - \beta_1 - \beta_2 E_i = 0 \tag{1.28}$$

must be satisfied for all i.

From a thermodynamic point of view, the Lagrange multipliers act as two thermodynamic potentials of the system under consideration. As for all potentials, their equilibrium values are reached when entropy is maximal. In the present case, one of these multipliers is associated to the particle number, and the other to the energy:

$$\beta_1 = -\frac{\mu_c}{k_B T} \quad \text{and} \quad \beta_2 = \frac{1}{k_B T}.$$

Specifically, this means that β_2 is linked to the temperature and $\beta_1 = -\mu_c \beta_2$ is linked to the chemical potential. With these multipliers, the energy distribution function takes the following form:

$$f(E) = \left.\frac{N_i}{S_i}\right|_{E=E_i} = \frac{1}{1 + \exp[(E-\mu_c)/(k_B T)]}, \tag{1.29}$$

which is naturally the Fermi–Dirac distribution applicable for fermions (electrons or holes). At low temperatures, the function $f(E)$ takes values close to 1 if $E < \mu_c$ and close to 0 if $E > \mu_c$. At zero temperature, the chemical potential is identically equal to the Fermi energy ϵ_F. At equilibrium, the chemical potential μ_c and the temperature T are constant, as the pressure and temperature are in a classical fluid.

If one deals with a two-component Fermi gas, the electrochemical potentials have to be distinguished as one would do for the partial pressures of a multicomponent classical fluid.[20] The number of free carriers is obtained by the integral of the product of the occupation probability $f(E)$ and the density of states $g(E)$. By choosing the bottom of the conduction band as the reference for the energy, it is found that for a nondegenerate gas,

$$n(T) = n_0 \exp[\mu_c/(k_B T)] \tag{1.30}$$

for the electrons, where $n_0 = 2(2\pi m_c^\star k_B T/h^2)^{3/2}$, with $m_c^\star$ being the effective mass of the electrons in the material. In accordance with Gibbs' formula, the chemical potential of the free electron gas reads as defined in Eq. (1.21).

---

20) In the case of an electron/hole system, two chemical potentials, $\mu_{c,n}$ and $\mu_{c,p}$, may be defined.

In the case of silicon, the equation of state for the free electron gas can be written as

$$N = AT^{3/2} \exp[\mu_c/(k_B T)] \qquad (1.31)$$

with

$$A = \frac{2.5 \times 10^{19}}{300^{3/2}} \left(\frac{m_c^\star}{m}\right)^{3/2} V, \qquad (1.32)$$

where V is the volume of the gas.

1.5.5
Temperature Dependence of the Chemical Potential $\mu_c(T)$

As already implied within Eq. (1.21), the chemical potential is a function of the temperature. This dependence can already be rather complex in basic model systems. In an ideal Fermi gas, with Fermi energy defined as $\epsilon_F = \mu_c(T = 0)$, the temperature dependence of the chemical potential is approximately given by a Sommerfeld expansion of the energy to

$$\mu_c(T) = \epsilon_F \left[1 - \frac{\pi^2}{12}\left(\frac{k_B T}{\epsilon_F}\right)^2 - \mathcal{O}(T^4)\right]. \qquad (1.33)$$

However, as shown in Figure 1.13, this approximation is already not valid for degenerate semiconductors at elevated temperatures and its failure is even more pronounced in the nondegenerate limit of charge carrier concentration. As a solution, the temperature-dependent chemical potential has to be obtained by an iterative solution of $n[\mu_c(T)]$ under the constraint of conserved number of additional carriers N [319, 320]. Precisely, it is

$$n = \frac{N}{\Omega} = \frac{1}{\Omega} \int_{\mu_c-\Delta E}^{VB^{\max}} dE\, g(E)[f(\mu_c, T) - 1] + \int_{CB^{\min}}^{\mu_c+\Delta E} dE\, g(E) f(\mu_c, T), \qquad (1.34)$$

where n is the extrinsic or inherent charge carrier concentration and Ω is the unit cell volume. The density of states is given by $g(E)$. The Fermi–Dirac function $f(\mu_c, T)$ is defined as $f(\mu_c, T) = [\exp((\epsilon-\mu_c)/(k_B T)) + 1]^{-1}$. The limits of integration are given by valence and conduction band edge energies ($VB^{\max}$ and $CB^{\min}$, respectively) and the chemical potential μ_c itself. The parameter $\Delta E \geq 10\, k_B T$ has to be set sufficiently accurate to ensure convergence of the integrals even in the limit of low carrier concentrations. By changing the temperature T, the chemical potential $\mu_c(T)$ has to be adapted to fulfill the constraint $N(T) = N(T = 0\text{ K})$, that is, the total charge has to be conserved. By iterative solution, the exact temperature dependence of the chemical potential $\mu_c(T)$ is introduced, as depicted in Figure 1.13.

For cooler applications feasible, at low temperatures ($T \leq 100$ K) and low temperature differences ($\Delta T \leq 30$ K), the Sommerfeld approximation gives an accurate description of the temperature-dependent chemical potential and $\Delta \mu_c(\Delta T) \leq 5$ meV. However, this crude approximations fails for the TE generator

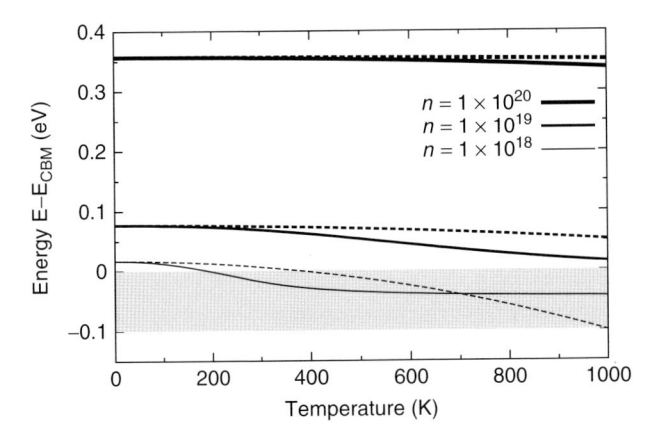

Figure 1.13 Dependence of the chemical potential on the applied temperature as calculated within a parabolic two-band semiconductor model without approximation [solid lines, using Eq. (1.34)] and within the Sommerfeld approximation [dotted lines, using Eq. (1.33)]. The temperature-dependent chemical potential is calculated for three different charge carrier concentrations. An incomplete description within the Sommerfeld approximation can be clearly seen for charge carrier concentrations below 1×10^{20} cm^{-3}. For the sake of clarity, the effective masses of the conduction band edge and valence band edge are assumed to be identical ($m_{VB}^{max} = m_{CB}^{min} = m_{el}$). The band gap (light gray shaded area) is chosen to be 100 meV.

case at higher temperatures ($T \geq 300$ K) and large temperature differences ($\Delta T \geq 100$ K). Here, the chemical potential varies in the order of 50 meV with a monotonic dependence on the temperature. Especially, the saturation of $\Delta \mu_c(\Delta T)$ at higher temperatures due to bipolar carrier conduction cannot be described within a Sommerfeld approximation and demands for full analysis by using Eq. (1.34).

1.6 Linear Nonequilibrium Thermodynamics

Classical thermodynamics, which is useful for describing equilibrium states, provides very incomplete information on the actual physical phenomena, which are characterized by irreversibility and nonequilibrium states. Accounting for the rates of the physical processes, irreversible thermodynamics thus extends the equilibrium analyses and establishes links between the measurable quantities, while nonequilibrium statistical physics provides the tools to compute these.

1.6.1 Forces and Fluxes

Let us assume that at the macroscopic scale, the states of a thermodynamic system may be characterized by a set of extensive variables X_i. At equilibrium, these variables assume values that yield the maximum of the entropy $S \equiv S(\{X_i\})$. Now, we are interested in situations where the system has been put under nonequilibrium conditions and is allowed to relax. The response of a system upon which constraints are applied is the generation of fluxes, which correspond to transport phenomena. When the constraints are lifted, relaxation processes drive the system to an equilibrium state. Energy dissipation and entropy production are associated to transport and relaxation processes, which have characteristic times. As the dynamics of the variables X_i is typically much slower than that of the microscopic variables, one may define an *instantaneous entropy*, $S(\{X_i\})$, at each step of their slow relaxation. The differential of the function S is:

$$dS = \sum_i \frac{\partial S}{\partial X_i} dX_i = \sum_i F_i dX_i, \tag{1.35}$$

where each quantity F_i is the intensive variable conjugate of the extensive variable X_i.

The notions of forces and fluxes are best introduced in the case of a discrete system: one may imagine, for instance, two separate homogeneous systems prepared at two different temperatures and placed in thermal contact through a thin diathermal wall, which implies that they are weakly coupled. The thermalization process triggers a flow of energy from one subsystem to the other, to which an extensive variable taking the values X_i and X'_i is associated, so that $X_i + X'_i = X_i^{(0)} = $ constant and $S(X_i) + S(X'_i) = S(X_i^{(0)})$. The equilibrium condition maximizing the total entropy is given by:

$$\left.\frac{\partial S^{(0)}}{\partial X_i}\right|_{X_i^{(0)}} = \left.\frac{\partial(S+S')}{\partial X_i}dX_i\right|_{X_i^{(0)}} = \frac{\partial S}{\partial X_i} - \frac{\partial S'}{\partial X'_i} = F_i - F'_i = 0, \tag{1.36}$$

which implies that if the difference $\mathcal{F}_i = F_i - F'_i$ is zero, the system is in equilibrium; otherwise, an irreversible process takes place and drives the system to equilibrium. The quantity $\mathcal{F}_i$ thus acts as a *generalized force* (or affinity) permitting the evolution of the system toward equilibrium.

The rate of variation of the extensive variable X_i characterizes the response of the system to the applied force:

$$J_i = \frac{dX_i}{dt} \tag{1.37}$$

so that a given flux cancels if its conjugate affinity cancels and, conversely, a finite affinity yields a finite conjugated flux. It thus appears that the relationship between affinities and fluxes characterizes the changes due to irreversible processes.

1.6.2
Linear Response and Reciprocal Relations

Consider a continuous medium in local equilibrium, where at a given point in space and time (r, t), the flux J_i is mathematically defined as dependent on the force $\mathcal{F}_i$, but also on the other forces $\mathcal{F}_{j \neq i}$:

$$J_i(r, t) \equiv J_i(\mathcal{F}_1, \mathcal{F}_2, \cdots). \tag{1.38}$$

The given definition implies that the nonequilibrium dynamics is governed by direct effects: each flux depends on its conjugate affinity, but *also* by indirect effects: each flux depends on the other affinities as well. Not far from equilibrium, $J_i(r,t)$ can be obtained by a Taylor expansion:

$$J_k(r,t) = \sum_j \frac{\partial J_k}{\partial \mathcal{F}_j} \mathcal{F}_j + \frac{1}{2!} \sum_{i,j} \frac{\partial^2 J_k}{\partial \mathcal{F}_i \partial \mathcal{F}_j} \mathcal{F}_i \mathcal{F}_j + \cdots$$

$$= \sum_j L_{jk} \mathcal{F}_j + \frac{1}{2} \sum_{i,j} L_{ijk} \mathcal{F}_i \mathcal{F}_j + \cdots, \quad (1.39)$$

where the quantities $L_{jk} \equiv \partial J_k / \partial \mathcal{F}_j$ are the elements of the matrix $[\mathcal{L}]$ of the *first-order* kinetic coefficients; they are obtained by the equilibrium values of the intensive variables F_i. The matrix $[\mathcal{L}]$ thus characterizes the *linear response* of the system.

In the linear regime, the source of entropy reads:

$$\sigma_S = \sum_{i,k} L_{ik} \mathcal{F}_i \mathcal{F}_k. \quad (1.40)$$

Since $\sigma_S \geq 0$, the kinetic coefficients satisfy

$$L_{ii} \geq 0 \quad \text{and} \quad L_{ii} L_{kk} \geq \frac{1}{4}(L_{ik} + L_{ki}). \quad (1.41)$$

In 1931, Onsager put forward the idea that there exist symmetry and antisymmetry relations between kinetic coefficients [256, 257]: the so-called *reciprocal relations*, $L_{ik} = L_{ki}$, must exist in all thermodynamic systems for which transport and relaxation phenomena are well described by linear laws. Note that Onsager's results have been generalized to account for situations where a magnetic field and/or a Coriolis field may play a rôle; in this case, one must check whether the studied quantity is invariant under time reversal transformation or not [321].

1.7
Forces and Fluxes in Thermoelectric Systems

1.7.1
Thermoelectric Effects

A naive definition would state that thermoelectricity results from the coupling of Ohm's law and Fourier's law. The TE effect in a system may rather be viewed as the result of the mutual interference of two irreversible processes occurring simultaneously in this system, namely heat transport and charge carrier transport. In thermoelectricity, three effects are usually described:

1) The Seebeck effect, which is the rise of an electromotive force in a thermocouple, that is, across a dipole composed of two conductors forming two junctions maintained at different temperatures, under zero electric current.

2) The Peltier effect, which is a thermal effect (absorption or production of heat) at the junction of two conductors maintained at the same temperature with a current flowing.
3) The Thomson effect, which is a thermal effect due to the existence of a temperature gradient along the material. The effect only exists if the Seebeck coefficient is a function of the temperature.

It is important to realize here that these three "effects" all boil down to the same process: At the microscopic level, an applied temperature gradient causes the charges to diffuse,[21] so the Seebeck, Peltier, and Thomson effects are essentially the same phenomenon, that is, thermoelectricity, which manifests itself differently as the conditions for its observation vary. Broadly speaking, when a temperature difference is imposed across a TE device, it generates a voltage, and when a voltage is imposed across a TE device, it generates a temperature difference. The TE devices can be used to generate electricity, measure temperature, cool or heat objects. For a thermocouple composed of two different materials A and B, the voltage is given by:

$$V_{AB} = \int_{T_c}^{T_h} (\alpha_B - \alpha_A)\, dT, \tag{1.42}$$

where the parameters $\alpha_{A/B}$ are the Seebeck coefficients or thermopowers.

1.7.2
Forces, Fluxes, and Kinetic Coefficients

The main assumption of Onsager's work is based on the hypothesis that the system evolution is driven by a minimal production of entropy where each fluctuation of any intensive variable undergoes a restoring force to equilibrium [322]. Though the system itself produces dissipation, one may use well-defined thermodynamic potentials at each time step for the analysis of the quasistatic processes, bringing the system back to equilibrium, so that the classical quasistatic relation between heat and entropy variation $dS = \delta Q_{qs}/T$ may be extended to finite time response thermodynamics in the following flux form:

$$\mathbf{j}_S = \frac{\mathbf{j}_Q}{T}. \tag{1.43}$$

The Onsager force–flux derivation is obtained from the laws of conservation of energy and matter. If we consider the complete energy flux, then the first principle of thermodynamics gives the expression of the total energy flux $\mathbf{j}_E$, heat flux $\mathbf{j}_Q$, and particles flux $\mathbf{j}_N$,

$$\mathbf{j}_E = \mathbf{j}_Q + \mu_e \mathbf{j}_N. \tag{1.44}$$

These fluxes are conjugated to their thermodynamic potential gradients, which, as general forces, derive from the thermodynamic potentials. The question of

---
21) One may see an analogy with a classical gas expansion.

the correct expression of these potentials is out of the scope of this chapter, but it can be shown that the correct potentials for energy and particles are $1/T$ and μ_e/T, respectively, see [321]. The corresponding forces can be expressed as their gradients

$$\mathbf{F}_N = \nabla\left(-\frac{\mu_e}{T}\right), \quad \mathbf{F}_E = \nabla\left(\frac{1}{T}\right), \tag{1.45}$$

and the linear coupling between forces and fluxes can simply be derived by a linear set of coupled equations with kinetic coefficient matrix $[L]$,

$$\begin{bmatrix} \mathbf{j}_N \\ \mathbf{j}_E \end{bmatrix} = \begin{bmatrix} L_{NN} & L_{NE} \\ L_{EN} & L_{EE} \end{bmatrix} \begin{bmatrix} \nabla\left(-\frac{\mu_e}{T}\right) \\ \nabla\left(\frac{1}{T}\right) \end{bmatrix},$$

where $L_{NE} = L_{EN}$. The symmetry of the off-diagonal term is fundamental in the Onsager description since it is equivalent to a minimal entropy production of the system under nonequilibrium conditions [322]. A first experimental verification of the Onsager reciprocal relations had been given by Miller for different materials [270]. As we already pointed out, the minimal entropy production is not a general property of nonequilibrium processes. However, under steady-state conditions, a fluctuating thermodynamic potential will undergo a restoring force due to the presence of another potential. This mechanism has to be symmetric, and so do the off-diagonal terms of the kinetic matrix.[22] From a microscopic point of view, this equality also implies the time reversal symmetry of the processes.[23] By extension, processes at microscopic scale should be "microreversible." Since the irreversibility is a statistical consequence of the number of particles inside the considered system, then, at a microscopic scale, "irreversible thermodynamics" simply becomes a "reversible dynamics."

1.7.3
Energy Flux and Heat Flux

In order to treat properly heat and carrier fluxes, it is more convenient to rewrite the second equation of the given matrix formulation for $\mathbf{j}_Q = \mathbf{j}_E - \mu_e \mathbf{j}_N$. By doing this, it is advantageous to change slightly the first force in order to let μ_e appear explicitly and not only $\nabla(-\mu_e/T)$. By using the development

$$\nabla\left(-\frac{\mu_e}{T}\right) = -\frac{1}{T}\nabla\mu_e - \mu_e \nabla\left(\frac{1}{T}\right),$$

a straightforward calculation gives

$$\begin{bmatrix} \mathbf{j}_N \\ \mathbf{j}_Q \end{bmatrix} = \begin{bmatrix} L_{11} & L_{12} \\ L_{21} & L_{22} \end{bmatrix} \begin{bmatrix} -\frac{1}{T}\nabla\mu_e \\ \nabla\left(\frac{1}{T}\right) \end{bmatrix} \tag{1.46}$$

22) The off-diagonal terms of the kinetic matrix are symmetric, only if the correct thermodynamic potentials of the system have been chosen. In the case of a Fermi gas, the correct potentials are μ_e/T and $1/T$.

23) This time reversal symmetry is broken under the application of Coriolis or magnetic force.

with $L_{12} = L_{21}$ and the kinetic coefficients become

$$L_{11} = L_{NN}, \quad L_{12} = L_{NE} - \mu_e L_{NN}, \quad L_{22} = L_{EE} - \mu_e(L_{EN} + L_{NE}) + \mu_e^2 L_{NN}.$$

By derogation from electronic systems, which are described based on a charge density distribution, the electric field derives from the electrochemical potential μ_e

$$\mathbf{E} = -\frac{\nabla \mu_e}{e} = -\frac{\nabla \mu_c}{e} - \nabla V, \tag{1.47}$$

where V is the electrical potential.

1.7.4
Thermoelectric Coefficients

Depending on the thermodynamic working conditions, the TE coefficients can be derived from the two expressions of particle and heat flux density.

1.7.4.1 Decoupled Processes
By using expression (1.46) under isothermal conditions, we get the electrical current density in the form

$$\mathbf{j} = \frac{-eL_{11}}{T} \nabla \mu_e, \tag{1.48}$$

where $\mathbf{j} = e\mathbf{j}_N$. This is an expression of Ohm's law, then the isothermal electrical conductivity is

$$\sigma_T = \frac{e^2}{T} L_{11}. \tag{1.49}$$

Alternatively, if we consider the heat flux density in the absence of any particle transport, or under zero electric current, then we now get

$$\mathbf{j} = 0 = -L_{11}\left(\frac{\nabla \mu_e}{T}\right) + L_{12} \nabla\left(\frac{1}{T}\right), \tag{1.50}$$

and the heat flux density becomes

$$\mathbf{j}_{Q_J} = \frac{1}{T^2}\left(\frac{L_{21}L_{12} - L_{11}L_{22}}{L_{11}}\right) \nabla T, \tag{1.51}$$

which is the Fourier law, where the thermal conductivity under zero electric current (open circuit) is

$$\kappa_J = \frac{1}{T^2}\left(\frac{L_{11}L_{22} - L_{21}L_{12}}{L_{11}}\right). \tag{1.52}$$

Finally, we can also consider the thermal conductivity under zero electrochemical gradient (closed circuit), then we get

$$\mathbf{j}_{Q_E} = \frac{L_{22}}{T^2} \nabla T \quad \text{with} \quad \kappa_E = \frac{L_{22}}{T^2}. \tag{1.53}$$

1.7.4.2 Coupled Processes

Let us now consider the TE coupling in more detail. In the absence of any particle transport, the basic expression is already known since it is given by Eq. (1.50). We now define the Seebeck coefficient as the ratio between the two forces, electrochemical gradient and temperature gradient, then the Seebeck coefficient expression is given by

$$-\frac{1}{e}\nabla\mu_e \equiv \alpha\nabla T = \mathbf{E}|_{j=0} \qquad (1.54)$$

for the electric field relation. By using Eq. (1.50), we finally find for Seebeck

$$\alpha = \frac{1}{eT}\frac{L_{12}}{L_{11}}. \qquad (1.55)$$

If we consider now an isothermal configuration, we can derive the expression of the coupling term between current density and heat flux, which is nothing more than the Peltier coefficient

$$\mathbf{j} = eL_{11}\left(-\frac{1}{T}\nabla\mu_e\right), \quad \mathbf{j}_Q = L_{21}\left(-\frac{1}{T}\nabla\mu_e\right), \qquad (1.56)$$

we get

$$\mathbf{j}_Q = \frac{1}{e}\frac{L_{12}}{L_{11}}\mathbf{j}, \qquad (1.57)$$

and the Peltier coefficient Π is given by

$$\mathbf{j}_Q = \Pi\,\mathbf{j}, \quad \Pi = \frac{1}{e}\frac{L_{12}}{L_{11}}. \qquad (1.58)$$

As one can see, we have the equality

$$\Pi = \alpha T.$$

The close connection between Peltier and Seebeck effects is illustrated by this compact expression. In a later section, we show that a similar connection can be derived for the Thomson effect. From a fundamental point of view, this shows that all of these effects are in fact different expressions of the same quantity, called the "entropy per carrier," defined by Callen [323, 324]. It will be considered first, followed by the definitions of the transport parameters.

1.7.5
The Entropy Per Carrier

By using a classical approach of thermodynamic cycle, we can consider a carrier traveling through the different step of the Carnot cycle. With focus on the two adiabatic branches of the thermodynamic cycle, it appears that a certain amount of entropy is driven from the hot reservoir to the cold one, but also from the cold reservoir to the hot side. In this convective process, the carrier acts as if it was carrying some entropy. Let us derive this by considering the entropy flux density. From the heat flux density expression, we can write

$$\mathbf{j}_S = \frac{1}{T}\left[L_{21}\left(-\frac{1}{T}\nabla\mu_e\right) + L_{22}\nabla\left(\frac{1}{T}\right)\right]. \qquad (1.59)$$

According to Ohm's law, see Eq. (1.48), it can be simplified into

$$\mathbf{j}_S = \frac{L_{21}}{TeL_{11}} \mathbf{j} + \frac{1}{T} L_{22} \nabla\left(\frac{1}{T}\right). \tag{1.60}$$

We see here that the entropy flux contains two terms, one with an electrochemical origin and the other with a thermal origin. The first term shows that a fraction of the entropy is transported by the flux of carriers. Then the entropy transported per carrier (or per particle) is given by

$$S_N = \frac{L_{21}}{TL_{11}}. \tag{1.61}$$

We remark that the Seebeck coefficient is directly proportional to S_N since we have

$$S_N = e\alpha. \tag{1.62}$$

It is important to note that the entropy per particle is a fundamental parameter from which all the TE effects derive. Nevertheless, the reader should take care not to attribute a specific entropy to each carrier. Since thermodynamics never considers isolated particle but only a large number of particles, the definition of the entropy per particle refers to an averaged property of the fermion gas, as a statistical definition. This is also valid for the $S_N = e\alpha$ expression where the Seebeck effect cannot be reduced to the direct summation of the individual contribution of the carriers. As an illustration, one can see that S_N is a function of the electrical conductivity through the term L_{11} and the conductive models, similarly to the Drude model [325–327], cannot be derived at the scale of a carrier, with the attribution of a specific electrical conductivity to each carrier.

1.7.6
Kinetic Coefficients and Transport Parameters

By using the entropy per carrier S_N defined in Eq. (1.62), we can obtain now a complete correspondence between the kinetic coefficient and the transport parameters. We have

L_{11}	$L_{12} = L_{21}$	L_{22}
$\frac{T}{e^2}\sigma_T$	$\frac{T^2}{e^2}\sigma_T S_N$	$\frac{T^3}{e^2}\sigma_T S_N^2 + T^2 \kappa_J$

and the Onsager expressions become

$$\mathbf{j} = -\sigma_T \left(\frac{\nabla \mu_e}{e}\right) + \frac{\sigma_T S_N T^2}{e^2} \nabla\left(\frac{1}{T}\right), \tag{1.63}$$

$$\mathbf{j}_Q = -T\sigma_T S_N \left(\frac{\nabla \mu_e}{e}\right) + \left(\frac{T^3}{e^2}\sigma_T S_N^2 + T^2 \kappa_J\right) \nabla\left(\frac{1}{T}\right). \tag{1.64}$$

Finally, we distinguish between the thermal conductivity under zero electrochemical gradient and under zero particle transport,

$$\kappa_E = \frac{L_{22}}{T^2}, \quad \kappa_J = \frac{1}{T^2}\left(\frac{L_{11}L_{22} - L_{21}L_{12}}{L_{11}}\right), \quad (1.65)$$

leading to the equality

$$\kappa_E = T\alpha^2\sigma_T + \kappa_J. \quad (1.66)$$

It should be mentioned here that the present description only considers the electronic gas contribution to the thermal conductivity. An additional contribution κ_{lat} arises from the lattice.

By inserting Eq. (1.63) into Eq. (1.64) and using $\mathbf{E} = -\nabla\mu_e/e$ and the local expansion $\nabla(1/T) = -1/T^2 \nabla T$, the "classical" constitutive relations

$$\mathbf{j} = \sigma_T \mathbf{E} - \sigma_T \alpha \nabla T \quad \text{and} \quad \mathbf{j}_Q = \alpha T \mathbf{j} - \kappa_J \nabla T \quad (1.67)$$

are reproduced, see also [328]. Then, it follows that $\mathbf{E} = \alpha \nabla T + \varrho \mathbf{j}$ with electrical resistivity $\varrho = 1/\sigma_T$.

1.7.7
The Dimensionless Figure of Merit zT

We have seen from the kinetic matrix $[L]$ that the off-diagonal terms represent the coupling between the heat and the electrical fluxes. The question is now to consider the way to optimize a given material in order to get an efficient heat pump driven by an input electric current or an efficient TE generator driven by the heat flow supplied. The procedure can be derived for both applications, but we propose here to consider a thermogenerator application.

Let us first look at the optimization of the fluxes. Since a TE material is an energy conversion device, the more heat flows into the material, the more electrical power may be produced. In order to achieve this, we expect a large thermal conductivity for the material. Unfortunately, this will also lead to a very small temperature difference and, consequently, small electrical output voltage and power. This configuration can be called the short-circuit configuration since the fluxes are maximized and the potential differences are minimized.

Now we consider the coupled processes from a potential point of view. In order to obtain a larger voltage, the material should exhibit a large temperature difference. Then the thermal conductivity of the material should be as small as possible, leading to a very small heat flux and, consequently, again, a small electrical power output. This configuration can be called the open-circuit configuration since the potential differences are maximized and the fluxes are minimized.

It is worth noticing that both short-circuit and open-circuit configurations lead to a nonsatisfactory solution. Moreover, they are in contradiction since the thermal conductivity is expected to be maximal in the short-circuit configuration and minimal in the open-circuit one! This contradiction can be resolved if we consider the expression of the thermal conductivities previously given by Eq. (1.66), that is,

$\kappa_E = T\alpha^2 \sigma_T + \kappa_J$. Since it is established under zero current, the κ_J corresponds to the open-circuit configuration while κ_E corresponds to the short-circuit configuration. From the previous developments, see Eq. (1.66), we expect $\frac{\kappa_E}{\kappa_J}$ to be maximal in order to obtain the maximal output electrical power. Then we can write

$$\frac{\kappa_E}{\kappa_J} = 1 + zT \tag{1.68}$$

with the figure of merit zT defined by

$$zT = \frac{\alpha^2 \sigma_T}{\kappa_J} T. \tag{1.69}$$

This relation was also found by Zener, see [50].

As one can notice from Eq. (1.68), the zT term should be maximal in order to obtain an optimal material. The TE properties of the material are summarized in the zT expression, firstly proposed by Ioffe [36]. zT enables a direct measurement of the quality of the material for practical applications, and the figure of merit is clearly the central term for material engineering research.

At first glance, the presence of the temperature in the expression of the figure of merit may be strange since T is not a material property, but an intensive parameter, which partly defines the working conditions. Nevertheless, one should notice that, in terms of thermodynamic optimization, the material properties are negligible without considering the available exergy of the working system. This is achieved by introducing the temperature in the expression of the figure of merit, which provides a reference to the exergy evaluation.

1.8
Heat and Entropy

Let us consider the coupled Onsager expressions:

$$\mathbf{j} = -\sigma_T \left(\frac{\nabla \mu_e}{e}\right) + \frac{\sigma_T S_N T^2}{e^2} \nabla\left(\frac{1}{T}\right), \tag{1.70}$$

$$\mathbf{j}_Q = -T\sigma_T S_N \left(\frac{\nabla \mu_e}{e}\right) + \left[\frac{T^3}{e^2}\sigma_T S_N^2 + T^2 \kappa_J\right] \nabla\left(\frac{1}{T}\right). \tag{1.71}$$

We can combine both equations to get

$$\mathbf{j}_Q = T\, S_N \mathbf{j} + T^2 \kappa_J \nabla\left(\frac{1}{T}\right), \tag{1.72}$$

where we identify a conductive term, proportional to $\nabla\left(\frac{1}{T}\right)$, and a "Peltier" term, proportional to $\mathbf{j}$:

$$q_\kappa = T^2 \kappa_J \nabla\left(\frac{1}{T}\right) = -\kappa_J \nabla T, \tag{1.73}$$

$$q_\pi = T\frac{S_N}{e} \mathbf{j}. \tag{1.74}$$

1.8 Heat and Entropy

This Peltier heat transported because of the TE effects results in the effect commonly attributed to Peltier: the heat observed at an inhomogeneous junction due to the TE effects.

1.8.1 Volumetric Heat Production

The volumetric heat production can be estimated from the total energy flux

$$\mathbf{j}_E = \mathbf{j}_Q + \frac{\mu_e}{e} \mathbf{j}.$$

According to energy and particle conservation, we have, under steady state,

$$\nabla \cdot \mathbf{j}_E = 0 \quad \text{and} \quad \nabla \cdot \mathbf{j} = 0. \tag{1.75}$$

Then,

$$\nabla \cdot \mathbf{j}_Q = -\frac{\nabla \mu_e}{e} \cdot \mathbf{j}.$$

Since the electrical field is $\mathbf{E} = -\frac{\nabla \mu_e}{e}$, we find

$$\nabla \cdot \mathbf{j}_Q = \mathbf{E} \cdot \mathbf{j}. \tag{1.76}$$

This summarizes the possible transformation of the energy since it shows that heat can be produced by the degradation of the electrochemical potential μ_e and that electrical power can also be extracted from heat.

1.8.2 Entropy Production Density

If we consider the entropy flux density, we can calculate the entropy production v_S from

$$v_S = \nabla \cdot \mathbf{j}_S = \nabla \left(\frac{\mathbf{j}_Q}{T} \right) = \nabla \left(\frac{1}{T} \right) \cdot \mathbf{j}_Q + \frac{1}{T} \nabla \cdot \mathbf{j}_Q$$

to get

$$v_S = \nabla \left(\frac{1}{T} \right) \cdot \mathbf{j}_Q - \frac{\nabla \mu}{eT} \cdot \mathbf{j}. \tag{1.77}$$

As shown earlier, the entropy production is due to nonisothermal heat transfer and electrical Joule production. The previous expression can be rewritten in the form

$$v_S = \nabla \left(\frac{1}{T} \right) \cdot \mathbf{j}_E + \nabla \left(-\frac{\mu_e}{T} \right) \cdot \mathbf{j}_N. \tag{1.78}$$

In this form, we obtain the illustration of one major result of the Onsager description: The total entropy production is given by the summation of the force–flux products,

$$v_S = \nabla \cdot \mathbf{j}_S = \Sigma \, \overrightarrow{\text{force}} \cdot \overrightarrow{\text{flux}}. \tag{1.79}$$

This is a very general result of the Onsager theory. When deriving the entropy production according to Onsager kinetic expressions, the constraint of minimal entropy production leads to a final expression where the overall entropy production is directly given by the sum of the products of forces and fluxes.

1.8.3
Heat Flux and the Peltier–Thomson Coefficient

In the previous sections, we considered the volumetric heat transformation from the calculation of the divergence of the heat flux $\nabla \cdot \mathbf{j}_Q$. We now propose to analyze its different terms. First, by elimination of the electric field $\mathbf{E}$ from the previous set of equations, we get

$$\mathbf{j}_Q = \alpha T \mathbf{j} - \kappa_J \nabla T, \tag{1.80}$$

and the divergence of the heat flux becomes

$$\nabla \cdot \mathbf{j}_Q = \nabla \cdot (\alpha T \mathbf{j} - \kappa_J \nabla T)$$
$$= T \mathbf{j} \cdot \nabla \alpha + \alpha \, \nabla T \cdot \mathbf{j} + \alpha T \, \nabla \cdot \mathbf{j} + \nabla \cdot (-\kappa_J \nabla T), \tag{1.81}$$

where we find four terms, which can be identified:

- $\alpha T \nabla \cdot \mathbf{j}$: equals zero due to particle conservation,
- $T \mathbf{j} \cdot \nabla \alpha$: "Peltier–Thomson" term,
- $\mathbf{j} \cdot \alpha \nabla T = \mathbf{j} \cdot (\mathbf{E} - \mathbf{j}/\sigma_T) = \mathbf{j} \cdot \mathbf{E} - j^2/\sigma_T$: electrical work production and dissipation,
- $\nabla \cdot (-\kappa_J \nabla T)$: change in thermal conduction due to heat produced or absorbed.

To sum up, the sources of the heat flux are

$$\nabla \cdot \mathbf{j}_Q = T \mathbf{j} \cdot \nabla \alpha + \mathbf{j} \cdot \mathbf{E} - \frac{j^2}{\sigma_T} - \nabla \cdot (\kappa_J \nabla T). \tag{1.82}$$

Most of these terms are common, but less intuitive is the Peltier–Thomson term, which is now considered.

1.8.4
The Peltier–Thomson Term

Now we show that the $T \mathbf{j} \cdot \nabla \alpha$ term contains both the Thomson contribution (local temperature gradient effect) and the Peltier contribution (isothermal spatial gradient effect). By using the equivalence $\Pi = \alpha T$, we obtain

$$T \mathbf{j} \cdot \nabla \alpha = T \mathbf{j} \cdot \nabla \left(\frac{\Pi}{T}\right) = T \mathbf{j} \cdot \left(\frac{1}{T} \nabla \Pi - \frac{1}{T^2} \Pi \nabla T\right) = \mathbf{j} \cdot (\nabla \Pi - \alpha \nabla T). \tag{1.83}$$

Then, the traditional separation of the Peltier and Thomson contribution is artificial since they both refer to the same physics of the gradient of the entropy per particle, temperature-driven gradient or spatially driven gradient. The isothermal configuration leads to the Peltier expression, meanwhile a spatial gradient leads to the Thomson result.

- Pure Peltier, isothermal junction between two materials:

$$\mathbf{j} \cdot (\nabla \Pi - \alpha \nabla T) = \mathbf{j} \cdot (\nabla \Pi),$$

- Thomson, homogeneous material under temperature gradient:

$$\mathbf{j} \cdot (\nabla \Pi - \alpha \nabla T) = \mathbf{j} \cdot \left(\frac{d\Pi}{dT} - \alpha\right) \nabla T = \tau \mathbf{j} \cdot \nabla T, \quad (1.84)$$

with

$$\nabla \Pi = \frac{d\Pi}{dT} \nabla T \quad \text{and} \quad \tau = \frac{d\Pi}{dT} - \alpha, \quad (1.85)$$

and the heat flux divergence takes the form

$$\nabla \cdot \mathbf{j}_Q = \tau \mathbf{j} \cdot \nabla T + \mathbf{j} \cdot \mathbf{E} - \frac{j^2}{\sigma_T} - \nabla \cdot (\kappa_J \nabla T). \quad (1.86)$$

If we consider a configuration where $\kappa_J \neq f(T)$, then Eq. (1.86) reduces to

$$\nabla \cdot \mathbf{j}_Q = \tau \mathbf{j} \cdot \nabla T + \mathbf{j} \cdot \mathbf{E} - \frac{j^2}{\sigma_T} - \kappa_J \nabla^2 T. \quad (1.87)$$

As one can notice, the Peltier and Thomson terms both refer to the gradient $\nabla \alpha$. It is worth noticing that the isothermal configuration for the Peltier expression and the temperature gradient configuration for the Thomson effect correspond to specific chosen conditions. With another set of conditions, one can obtain other definitions. For example, Peltier heat can be considered to be absorbed or released inside the active material due to the position-dependent Seebeck coefficient. It is then referred to as *"distributed Peltier effect"* or *"extrinsic Thomson effect"* [329–331].

1.8.5
Local Energy Balance

By using the expression $\nabla \cdot \mathbf{j}_Q = \mathbf{E} \cdot \mathbf{j}$, see Eq. (1.76), the local energy balance can be expressed from Eq. (1.87) [332]:

$$\nabla \cdot \mathbf{j}_Q - \mathbf{E} \cdot \mathbf{j} = \kappa_J \nabla^2 T + \frac{j^2}{\sigma_T} - \tau \mathbf{j} \cdot \nabla T = 0. \quad (1.88)$$

It should be noticed that this derivation does not need any assumption concerning the behavior of the system, whether in equilibrium or not. In the case of transient configuration, the energy balance equation should be corrected using $\varrho_{md} C_\mu$, where C_μ is the heat capacity (thermal mass, thermal capacity) and ϱ_{md} is the mass density:

$$\varrho_{md} C_\mu \frac{\partial T}{\partial t} + \nabla \cdot \mathbf{j}_Q = \mathbf{E} \cdot \mathbf{j} \longrightarrow$$

$$\kappa_J \nabla^2 T + \frac{j^2}{\sigma_T} - \tau \mathbf{j} \cdot \nabla T - T \mathbf{j} \cdot (\nabla \alpha)_T = \varrho_{md} C_\mu \frac{\partial T}{\partial t}. \quad (1.89)$$

In this form, the local energy balance has the general form of a continuity equation [54]. Besides the classical terms of heat equation due to Fourier heat transfer, see [333], there are contributions due to Joule's heat, Thomson heat, and distributed Peltier heat, which lead to a general form of heat equation in the framework of thermoelectricity.

One-dimensional models are often used, see, for example, [15, 334, 335]. Even in one dimension, the addition of time dependence can induce additional effects. For example, the spatial separation of Peltier cooling from Joule heating enables additional transient cooling when a cooler is pulsed [336, 337], see also Section 4.2. The reader may find some more information and insights about transient effects in thermoelectricity in Refs [46, 308, 309, 338–349] and in Chapter 4 of this book.

1.9
The Thermoelectric Engine and Its Applications

In a first approach, as usually carried out for traditional steam engines, only the fluid is considered and the walls of the enclosure containing this fluid are not taken into account. These contributions of the walls to the global efficiency are not considered, neither the boiling walls of the steam engine nor the lattice vibrations (phonons) of the TE material. Then we have a similar picture of the two systems, not only for the fluid (steam or electronic gas) but also for the thermal leak (boiling walls or lattice vibrations) as symbolically shown in Figure 1.16.

For every thermodynamic engine, the TE can work as a generator or a consumer. As a thermoelectric generator (TEG), the engine is driven by the entering heat flux and converts a fraction of it into electrical power through an electric current. As a receptor, the engine is driven by an electric current and acts as a pump for the heat flux. Two modes can be distinguished when operated as a consumer, a thermoelectric pump for heating or cooling. Heating and cooling modes are defined with a unique description. The difference comes from the useful part of the system that should be cooled or heated. In the cooling mode, we refer to a thermoelectric cooler (TEC), and in the heating mode, we refer to thermoelectric heater (TEH) (Figure 1.14).

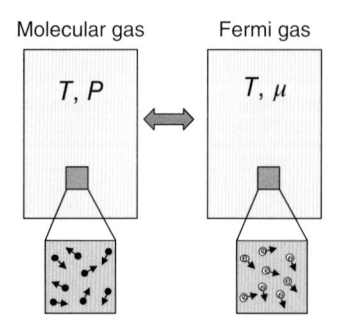

Figure 1.14 Comparison between steam and thermoelectric engine.

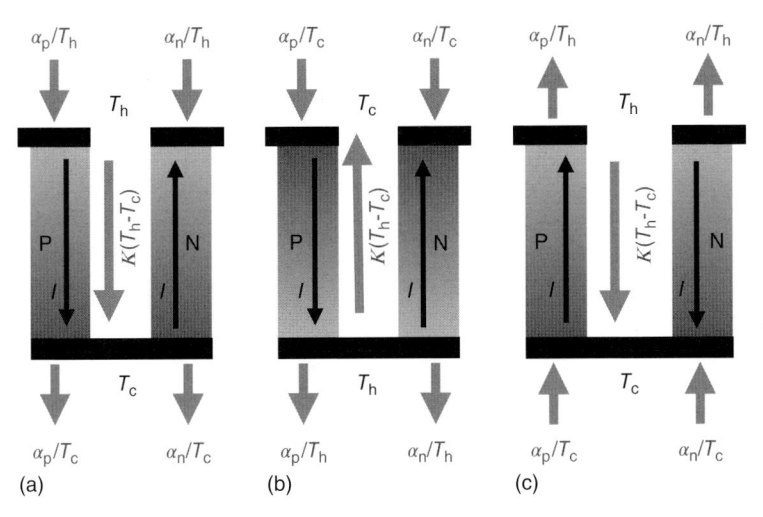

Figure 1.15 Thermoelectric applications: (a) thermogenerator (TEG), (b) TE cooler (TEC), (c) TE heater (TEH), see Figure 1.18 for details.

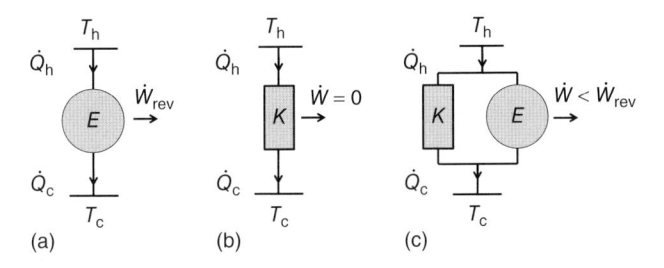

Figure 1.16 Thermodynamic system: (a), reversible; (b) fully dissipative, (c) reversible in parallel with a pure leakage. The latter case obtains a correct model of the thermodynamic fluid and its convective contribution (E), and the leakages, lattice and conductive contribution of the fluid (K).

Since heat flux and electric current are coupled, it is not possible to separate these three modes strictly, and, in fact, a thermogenerator also acts as a heat pump. On the other hand, a TEC can be understood as a TEG working with an electric current larger than its short-circuit current. All three configurations are summarized in Figure 1.15, where the thermal processes are shown in gray and the electrical processes in black.

As we observe, and regardless of the working mode, the efficiency of the engine is reduced by the presence of a heat leak. As a consequence, materials with very low lattice thermal conductivity are highly required for TE applications.

Let us consider now a sample of TE material where one end is maintained at temperature T_h and the other at temperature T_c with $T_h > T_c$. If we consider the Fermi gas inside the sample, then we achieve from elementary statistical arguments a large velocity combined with a small gas density at the hot end and a small velocity combined with a large gas density at the cold end. It should be noted that since heat flows from the hot to the cold end, the system cannot be considered under equilibrium conditions although the averaged carrier flux is zero: particles move from the cold to the hot side and from the hot to the cold side, but the two counter-directed fluxes are equal since the cell is closed. We also see that the gradient of carrier density is directly driven by the temperature gradient. Since the carriers are charged particles, we obtain an electrochemical potential difference, commonly called voltage difference, which is induced by the application of a temperature difference. This illustrates the coupling of the electrochemical potential gradient and the temperature. Next, since the averaged carrier flux is zero, but heat is transported at the same time, we obtain the same values of particle fluxes from the hot to the cold side and from the cold to the hot side. From this observation, we can conclude that heat and carrier fluxes are coupled. While being very simple, this description contains the main contributions to TE processes. In the ideal, reversible case without entropy production, this would be a Carnot cycle containing two "isothermal" processes of heat exchange at the hot and cold sides, and two "adiabatic" processes of noninteracting motion from the hot to the cold side and from the cold to the hot side.

Actually, since the TE process is not ideal, we can then estimate the principle sources of entropy of the working system that are the nonisothermal heat transfer and the nonadiabatic travel of the carriers from cold to hot terminals and from hot to cold terminals. This entropy production in the nonadiabatic branches is due to the collisions between carriers and the interactions with the crystal lattice of the material. Nonideal situations are shown in Figure 1.16. In the case of a reversible system (case a), the entropy flux is conserved so we have $\dot{Q}_h/T_h = \dot{Q}_c/T_c$. The output power is then given by $\dot{W}_{rev} = \dot{Q}_h - \dot{Q}_c = \eta_C \dot{Q}_h$ where $\eta_C = (1 - T_c/T_h)$ is the Carnot efficiency. In the case of a fully dissipative system (case b), we have $\dot{Q}_h = \dot{Q}_c$, then $\dot{W} = 0$. For most of the systems, we have the configuration (c) and the output power is strictly smaller than $\dot{W}_{rev}$. Finally, let us stress again that the current description does not explicitly mention the atoms of the crystal lattice that provide parasitic thermal conduction due to phonons or other thermal conduction mechanisms. This is due to the Onsager description, which follows the so-called linear response theory where the linear-response Fourier's law is used. Fourier's law is valid for thermal conduction due to phonons as well as electrons and is therefore included in the phenomenological description. If these two processes are independent, then it is common to describe the thermal conductivity in the absence of carrier transport as $\kappa_J = \kappa_{el} + \kappa_{lat}$, where κ_{el} is the electronic contribution and κ_{lat} is the lattice contribution.

1.10
Thermodynamics and Thermoelectric Potential

Until now, we have not really taken into account the working conditions of the TE system. Similarly to any working engine, a TE device should be correctly driven in order to provide work in the best conditions. Then a precise control of the applied

1.10 Thermodynamics and Thermoelectric Potential

thermodynamic potentials (or fluxes) is needed in order to get correct use of the potentialities of the TE material. Since the TE process implies the coupling of the heat flux and electric current, these two fluxes should be driven optimally. This important question has been addressed by Snyder et al. in 2002/2003. They derived two key parameters of the "compatibility approach": the relative current and the TE potential [350, 351]. Both are reduced variables that can be used as a mathematical basis to analyze the local performance of TE material related to the working conditions.

1.10.1 Relative Current, Dissipation Ratio, and Thermoelectric Potential

The relative current density is defined by the ratio of electrical current density j to purely conductive fraction of the heat flux q_κ

$$u = -\frac{j^2}{\kappa \nabla T \cdot \mathbf{j}} \quad \text{or} \quad \frac{1}{u}\mathbf{j} = -\kappa \nabla T. \tag{1.90}$$

Note the writing of the ratio on the basis of two vectors $\mathbf{j}$ and $\mathbf{q}_\kappa = -\kappa \nabla T$.

Instead of using relative current density u, one could have introduced the dimensionless function r defined by

$$r = \frac{1}{u\alpha T} = \frac{-\kappa \nabla T \cdot \mathbf{j}}{j^2 \alpha T} = \frac{-\kappa \nabla T \cdot \mathbf{n}}{j\alpha T}, \tag{1.91}$$

which represents the ratio of dissipative to reversible heat fluxes. In the last term, we assume parallel (or antiparallel) fluxes with flow direction $\mathbf{n} = \mathbf{j}/j$.

In this and in section 1.10.2, we write some formulae both in terms of u and as function of r; in the latter case, they may appear more transparent.

Another possible variant was published by Kedem and Caplan [352], who described two coupled flows by their "degree of coupling functions." This approach reduces both the TE flows and forces to dimensionless numbers, an approach that is very similar to that applied in numerical fluid dynamics where dimensionless numbers (e.g., the Prandtl number) are used for nondimensionalizing the field equations [353] Such an approach had been firstly proposed by Clingman [273] in 1961, using a dimensionless heat flux c. For a detailed introduction of the ratios r and c, we refer to Chapter 5, Section 5.8.

From the point of view of local equilibrium thermodynamics, all local flows are considered as dependent on quasistatic local intensive variables. By relating this to the relative current, we must state that u is a function of the "potentials" T and μ_e in general. However, since the coupling between T and μ_e is weak (but nonzero) in the available TE materials (see Section 1.5.5), it is possible under certain conditions to focus on $u(T)$ as achieved in the compatibility approach, see Chapter 5 of this book.

From Eq. (1.80), the total heat flow $\mathbf{q}$ becomes

$$\mathbf{q} = \alpha T \mathbf{j} + \frac{\mathbf{j}}{u} = (1+r)\,\alpha T \mathbf{j} =: \Phi \mathbf{j} \quad \text{with}$$

$$\Phi = \alpha T + \frac{1}{u} = (1+r)\,\alpha T, \tag{1.92}$$

where Φ is the "thermoelectric potential" [351, 354]. Equation (1.92) points to the coupling of the two potentials Φ and T because we find

$$y(\Phi, T) = \frac{1}{u(\Phi, T)} = \Phi - \alpha T = r\alpha T. \tag{1.93}$$

Note that $y = 1/u$ can be defined as relative heat flux (as, e.g., achieved by Sherman et al. [271]).

The previous definitions make it clear that the TE coupling in isotropic media is described by

$$\mathbf{j} = \frac{1}{\Phi} \mathbf{q} = u\,\mathbf{q}_\kappa \quad \text{with} \quad \mathbf{q}_\kappa = r\alpha T \mathbf{j}, \tag{1.94}$$

whereby we generally assume parallelity of electric current and heat flow.[24] The fact that the total heat and carrier fluxes are directly connected by the TE potential is fundamental since it allows us to derive the principle results of the thermodynamics of thermoelectricity directly from it. According to the previous definitions, the volumetric heat production v_q becomes (with $\nabla \cdot \mathbf{j} = 0$)

$$v_q = \nabla \cdot \mathbf{q} = \nabla \cdot (\Phi\,\mathbf{j}) = \nabla \Phi \cdot \mathbf{j} = \mathbf{j} \cdot \nabla\left(\alpha T + \frac{1}{u}\right) = \mathbf{j} \cdot \nabla[(1+r)\,\alpha T]. \tag{1.95}$$

Note that $\nabla \cdot \mathbf{q} = \mathbf{j} \cdot \mathbf{E}$, see Eq. (1.76), and $\mathbf{E} = -\nabla \mu_e/e$, see Eq. (1.47). Then, we find

$$\mathbf{E} = \nabla \Phi = -\nabla \mu_e / e \implies \mu = -e\,(\Phi - \Phi_0), \tag{1.96}$$

which means that the electric field $\mathbf{E}$ can be calculated on a phenomenological level from the gradient of the TE potential Φ, or alternatively, by the negative gradient $-\nabla \mu_e / e$ when referring to a TE system by taking into account its chemical nature or solid-state physics. For details, we recommend the reading of Domenicali's review [107].

Since the heat production term $\mathbf{j} \cdot \nabla\left(\frac{1}{u}\right)$ directly reduces the efficiency, it now becomes evident that the maximum efficiency coincides with the minimization of $\nabla(1/u)$. This is obtained for a specific value of $u_{\text{opt}} = s$, where s is called the "compatibility factor" (see the next section). For optimization of r and thermodynamic aspects of compatibility, see Section 5.8.

By considering the entropy flux, we obtain $\mathbf{j}_S = \frac{1}{T}(\alpha T + 1/u)\mathbf{j} = \Phi/T \mathbf{j}$, and the expression of the volumetric entropy production becomes

$$v_S = \nabla \cdot \mathbf{j}_S = \mathbf{j} \cdot \nabla\left(\frac{\Phi}{T}\right) = \frac{v_q}{T} + \mathbf{q} \cdot \nabla\left(\frac{1}{T}\right) = \mathbf{j} \cdot \nabla[(1+r)\,\alpha\,]. \tag{1.97}$$

---

24) A more general definition of u seems possible when writing the relative current density in terms of fluctuating currents, which are indeed 3D. In this context, particle and heat flow should be considered in an anisotropic medium where the material parameters are tensors, see, for example, [355].

1.10 Thermodynamics and Thermoelectric Potential

This expression is in agreement with the Onsager formulation of the entropy production as the summation of the flux–force products, see Eq. (1.78), which here reduces to a single product since Φ is a compact expression of the thermodynamic potentials. For a given material, the TE potential enables a direct measurement of the total volumetric heat and entropy production by the respective degradation of Φ and Φ/T at a preset current flux $\mathbf{j}$.

1.10.2
Local Reduced Efficiency and Thermoelectric Potential of TEG, TEC, and TEH

Following [289, Section 9.2.2, 335], we can conclude that the local performance of an infinitesimal segment of a TE element can be defined as

$$\eta_{\text{loc}} = \frac{dT}{T}\eta_r \quad \text{and} \quad \varphi_{\text{loc}} = \frac{T}{dT}\varphi_r, \tag{1.98}$$

where dT/T is the infinitesimal Carnot cycle factor for TEG and T/dT is the one for TEC. As the Carnot process is a reversible one, the reduced "efficiencies"[25] η_r and φ_r play the role of an "irreversibility factor," which at least measures the distance to reversibility for both TEG and TEC due to a nonperfect TE engine. Such considerations were first published by E. Altenkirch [99, 100, 102].

The reduced efficiency of a TEG η_r is defined as the ratio of the products of conjugated forces and fluxes [351], where we have to pay attention to the fact that the electrical power production in a volume dV is given by the production density $\pi_{\text{el}} = \mathbf{j} \cdot \mathbf{E}$, also denoted as differential electrical power. Note that the net differential power *output* is given by $-\pi_{\text{el}}$, see Section 5.4 and [357–359]. By using Eqs. 1.76 and 1.43, we obtain $\mathbf{j} \cdot \mathbf{E} - T\mathbf{v}_S = 1/T \mathbf{q} \cdot \nabla T = \mathbf{j}_S \cdot \nabla T$, and with $\mathbf{j} \cdot \mathbf{E} = \nabla \cdot \mathbf{q} = \mathbf{j} \cdot \nabla \Phi$ and $\mathbf{j}_S = \Phi \mathbf{j}/T$, we finally obtain

$$\eta_r = \frac{\mathbf{j} \cdot \mathbf{E}}{\mathbf{j}_S \cdot \nabla T} = \frac{\pi_{\text{el}}}{\pi_{\text{el}} - T\mathbf{v}_S} = \frac{1}{1 - \frac{T\mathbf{v}_S}{\pi_{\text{el}}}} = \frac{1}{1 + \frac{T\mathbf{v}_S}{|\pi_{\text{el}}|}} \implies \eta_r = \frac{T}{\Phi}\frac{\nabla\Phi \cdot \mathbf{j}}{\nabla T \cdot \mathbf{j}}, \tag{1.99}$$

which coincides with Clingman's result [273] and corresponds to the reduced variation of the TE potential $\nabla\Phi/\Phi$ when changing the other potential $\nabla T/T$, which is coherent with a general definition of the efficiency of a nonequilibrium thermodynamic process of coupled fluctuating parameters. The reduced efficiency expression can be rewritten from u and Φ expressions, that is, with $u = \frac{1}{\Phi - T\alpha}$,

$$\frac{\nabla\Phi \cdot \mathbf{j}}{\nabla T \cdot \mathbf{j}} = \alpha(1 - \frac{\alpha}{z}u) = \alpha\left(1 - \frac{1}{rzT}\right), \quad u = -\frac{z}{\alpha^2}\frac{\nabla\Phi \cdot \mathbf{j}}{\nabla T \cdot \mathbf{j}} + \frac{z}{\alpha} \quad \text{and} \quad \frac{1}{r} = zT\left(1 - \frac{1}{\alpha}\frac{\nabla\Phi \cdot \mathbf{j}}{\nabla T \cdot \mathbf{j}}\right),$$

respectively, see, for example, [289, 351]. The result is for TEG

$$\eta_r = \frac{1 - \frac{u\alpha}{z}}{1 + \frac{1}{u\alpha T}} = \frac{1 - \frac{\alpha}{z(\Phi - \alpha T)}}{1 + \frac{z(\Phi - \alpha T)}{\alpha zT}} = \frac{\alpha T}{\Phi}\left[1 - \frac{1}{zT(\frac{\Phi}{\alpha T} - 1)}\right] = \frac{1 - \frac{1}{rzT}}{1 + r}. \tag{1.100}$$

---

25) In References [335, 356] reduced efficiencies $\eta_r^{(g)} \equiv \eta_r$, $\eta_r^{(c)} \equiv \varphi_r$ are introduced for both TEG and TEC, respectively.

This classical expression of the reduced efficiency presents a maximum for the compatibility factor $u_{opt} = s$ given as follows. Also, the optimal dissipation ratio can be derived: curve sketching $\frac{d}{dr}\eta_r(r) = 0$ by means of an algebra tool gives

$$r_{opt} = \frac{1 \pm \sqrt{1+zT}}{zT}. \tag{1.101}$$

It is worth noticing that

$$r_{opt} = \frac{1 \pm \sqrt{1+zT}}{zT} = \frac{1}{-1 \pm \sqrt{1+zT}},$$

that is, there is no contradiction to Eq. (5.60) in Section 5.8.

An analogous approach can be found for the reduced coefficient of performance of a TEC, φ_r. As a consequence of the underlying TE effects (which are inverse to each other, and similar are the definitions of the global performance parameters, efficiency η and coefficient of performance φ),[26] the reduced coefficient of performance φ_r is inversely defined:

$$\varphi_r = \frac{\mathbf{j}_S \cdot \nabla T}{\mathbf{j} \cdot \mathbf{E}} = \frac{\pi_{el} - Tv_S}{\pi_{el}} = 1 - \frac{Tv_S}{\pi_{el}} = \frac{\Phi}{T}\frac{\nabla T \cdot \mathbf{j}}{\nabla \Phi \cdot \mathbf{j}}. \tag{1.102}$$

For a direct comparison of TEG and TEC, we recommend a unified 1D model for both generator and cooler single elements [335]. Note that $u(T)$ differs only by sign if TEG and TEC are operated at reversed boundary temperatures, but otherwise in the same working conditions. For this case of directly comparing TEG and TEC, we find formally $\varphi_r = 1/\eta_r$, and the reduced efficiencies present a maximum for $u = s$; where s is the compatibility factor [351], $u_{opt} = s^{(g)} = \frac{\sqrt{1+zT}-1}{\alpha T}$ of a TEG, but $u_{opt} = s^{(c)} = \frac{-\sqrt{1+zT}-1}{\alpha T}$ of a TEC.

The reduced efficiency and local coefficient of performance are defined as functions of u in their ranges of typical use ($0 \leq u \leq 2\,s^{(g)}$ for TEG and $2\,s^{(c)} \leq u \leq 0$ for TEC). Note that the signs of the compatibility factors depend on whether a p-type leg or an n-type leg (or element) is considered.

In the special situation of maximum local TEG efficiency ($u = s^{(g)} > 0$) and maximum local TEC coefficient of performance ($u = s^{(c)} < 0$), the two values are formally equivalent,

$$\eta_{r,opt} = \varphi_{r,opt}^{-1} = \frac{\sqrt{1+zT}-1}{\sqrt{1+zT}+1} < 1,$$

as η_r and φ_r are local irreversibility factors. Note that the COP of a device is allowed to exceed 1, but there is a limit for the local optimum of this quantity as the respective Carnot limit cannot be exceeded by η and φ. The aforementioned relation shows again that zT is a thermodynamic material quantity determining the maximum irreversibility factor that is the same for both interrelated TE effects, that is, Seebeck and Peltier effects.

26) We follow Sherman's notation here and use φ instead of COP in TEC formulae.

1.10 Thermodynamics and Thermoelectric Potential

The equivalent optimal TE potential is given by

$$\Phi_{opt}^{(g/c)} = \alpha T + \frac{1}{s^{(g/c)}} = \alpha T \left[\frac{\sqrt{1+zT}}{\sqrt{1+zT} \mp 1} \right] = (1 + r_{opt})\, \alpha T, \qquad (1.103)$$

where the minus sign applies for TEG, but the plus sign for TEC. Comparing with Eq. (1.92), we find once again that

$$r_{opt} = \frac{1}{-1 \pm \sqrt{1+zT}}.$$

Note that dissipation is negligible for both TEG and TEC in the limit $zT \to \infty$. Furthermore, we find

$$\eta_{r,opt} = \left(2\frac{\Phi_{opt}^{(g)}}{\alpha T} - 1 \right)^{-1} \quad \text{and} \quad \varphi_{r,opt} = 2\frac{\Phi_{opt}^{(c)}}{\alpha T} - 1. \qquad (1.104)$$

Since the maximum reduced efficiency coincides with the minimal entropy production, we conclude that the optimal value of the TE potential, Φ_{opt}, defines the best working conditions for the system. Therefore, it is obvious that an optimal value Φ_{opt} is correlated to an optimal ratio between dissipative and reversible heat flux given by $u_{opt} = s$, or, more precisely, by an optimal dissipation ratio r (see also Section 5.8). A detailed discussion for the TEG is presented in Refs [354, 360].

The total efficiency η and the total coefficient of performance φ, respectively, of a finite generator and a cooler element are obtained by summing up the local contributions based on the reduced efficiency all over the TE element in an integral sense, see Eqs. (5.4) and (5.5). The particular case of maximum performance of an infinitely staged[27] TEG and TEC has been investigated by Sherman et al. [271, 361]. By considering the expressions 1.99, 1.100 and 1.102, we can now plot the reduced coefficient of performance $\varphi_r = \eta_r^{-1}$ as a function of the reduced variable $\xi = \frac{\Phi}{\alpha T} = 1 + r$ (which is identical to Clingman's "dimensionless heat flux" c introduced in Chapter 5, see Eq. (5.44)). The general expression is then

$$\varphi_r = \frac{\xi}{1 - \frac{1}{zT(\xi-1)}}.$$

Note that this formula is equivalent to Eq. (1.100). This expression can now be plotted for all the different working modes of the TE engine, see Figure 1.17. The plot can be separated into three zones, corresponding, respectively, from right to left, to the TEG, TEC, and finally TEH mode. As previously described, the optimal working conditions are obtained for $\varphi_{r,opt} = 2\frac{\Phi_{opt}}{\alpha T} - 1 = 2\xi_{opt} - 1 = 2\, r_{opt} + 1$.

The resulting reduced efficiencies and coefficient of performances are sprayed around this optimal line. As expected, there is an optimal $\varphi_{r,opt}$ value for the TEG

---

27) The device (or TE element) is broken up into an infinite number of stages. Note that the terms "perfectly infinitely staged element" and "self-compatible element" introduced next can be used synonymously.

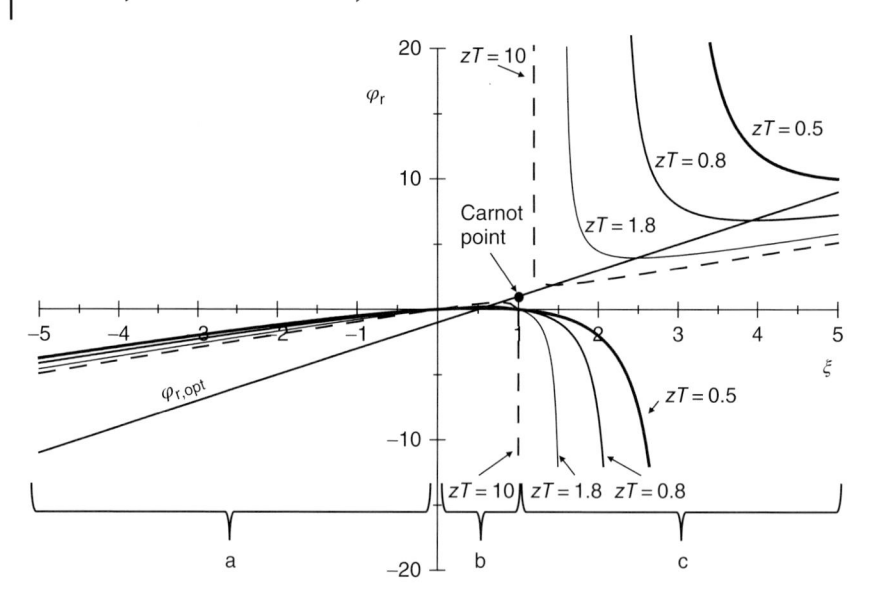

Figure 1.17 ϕ_r plot versus ξ for three different values of zT. Three different modes appear: (a) TEH mode, (b) traditional TEC mode, (c) TEG mode. The optimal working conditions are also given.

and TEC mode, corresponding to the maximal efficiency of a TEG and the maximal coefficient of performance of a TEC, respectively. When zT increases, these two optimal values becomes closer and closer. At the extreme we find the Carnot configuration, given by $zT \rightarrow \infty$, which leads to the condition $\varphi_r = \xi = 1$. In this case, the curve reduces to a single point of coordinates $(1, 1)$. There is a vertical asymptote passing through the "Carnot point," separating the TEG from the TEC mode.[28] The location of the Carnot point, exactly in between the two modes, means that the Carnot engine, as a reversible engine, may work with no distinction, as a generator or a receptor as depicted in Figure 1.18. It is worth noticing that the TEG and TEH modes are separated by a zone where the Peltier flux only counterbalances the conduction flux. This zone can be understood as a *regulation zone* of the conductive flux. The generator and receptor mode can also be considered using the classical, but extended current–voltage plot.

1.10.3
Thermoelectric Potential and Nonequilibrium Thermodynamics

We begin with the discussion of the Gibbs free energy G of an open one-component system, which is characterized by the thermodynamic variables T, p, N, and we recall known facts: Gibbs free energy is proportional to the numbers of particle, $G = \mu_c N$, with μ_c being the chemical potential.

---

28) From a practical point of view, a TEC can be understood as a TEG working at a current value larger than its short-circuit current.

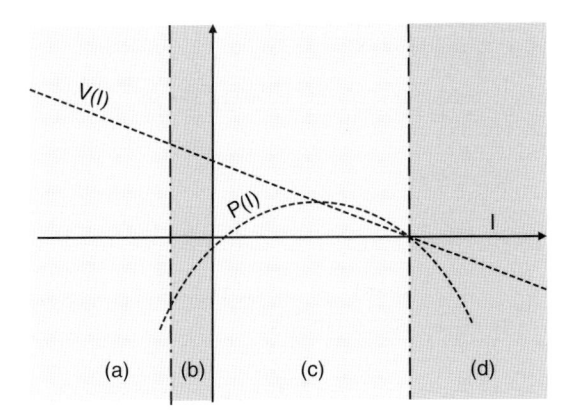

Figure 1.18 Current–voltage response. TEH: (a), Regulation: (b), TEG: (c), TEC: (d).

In the infinitesimal limit, changes of G due to energy and particle exchange are given by

$$dG(T, p, N) = -S\, dT + V\, dp + \mu_c\, dN. \tag{1.105}$$

The thermodynamics of adding one particle to the system was first investigated in Vining's paper in 1997 [328]. Since Vining did not focus on compatibility, he used a capacity matrix formalism by taking into consideration the coupling between thermal and electric transport. Such an approach may be considered obsolete because the relative current density u and the TE potential $\Phi = \alpha T + 1/u$ automatically include this coupling. Moreover, there is no continuity between a thermostatic description and a nonequilibrium thermodynamic description; in a more formal way, there is no continuity between the capacity matrix and the matrix of kinetic coefficients.

We consider now a stationary nonequilibrium state maintained by a total energy flux $\mathbf{j}_E$. For the description of such systems, the fundamental relation of classical equilibrium thermodynamics

$$dS = \frac{1}{T}(dU + p\, dV - \mu_c\, dN) \tag{1.106}$$

is extended to thermodynamically nonequilibrium systems, and it is supposed that the transformation between two different states (characterized by one of the state functions S, U, F, H, G) is performed by a sequence of quasi-stationary states.

Here we show that the internal energy U is also suitable for describing the state of TE systems operating out of equilibrium; G and U are interconnected by Legendre transformations

$$G = U + pV - ST \quad \Rightarrow \quad dU = T\, dS - p\, dV + \mu_c\, dN. \tag{1.107}$$

Note that the relation for dU is of particular importance because it is based on the differentials of the extensive values only.

To specify the changes in U for a TE system, we consider the relationship between the TE potential Φ and the total energy flux or the internal energy.

According to the first law of thermodynamics, the expression of the total energy flux $\mathbf{j}_E$, heat flux $\mathbf{j}_Q$, and particles flux $\mathbf{j}_N$ is given by $\mathbf{j}_E = \mathbf{j}_Q + \mu_e \mathbf{j}_N$, where μ_e is here the electrochemical potential, and $\mathbf{j} = e\, \mathbf{j}_N$. Since $\mathbf{j}_Q = \Phi\, \mathbf{j} = e\, \Phi\, \mathbf{j}_N$, see Eq. (1.92), we obtain for the total energy flux

$$\mathbf{j}_E = \mathbf{j}_Q + \mu_e \mathbf{j}_N = (e\Phi + \mu_e)\mathbf{j}_N = \left(e\alpha T + \mu_e + \frac{e}{u}\right)\mathbf{j}_N$$

$$= (e\alpha T + \mu_e)\, \mathbf{j}_N + \frac{1}{u}\mathbf{j} = (e\alpha T + \mu_e)\, \mathbf{j}_N - \kappa\, \nabla T. \tag{1.108}$$

Note that Eq. (1.108) separates the heat transported by the carriers (first term) and the Fourier heat occurring in any material (second term).

Applying energy and particle conservation, that is, $\nabla \cdot \mathbf{j}_E = 0$ and $\nabla \cdot \mathbf{j}_N = 0$, the nabla calculus directly gives $\nabla(e\, \Phi + \mu_e) \cdot \mathbf{j}_N = 0$. This equation is fulfilled if $\nabla(e\, \Phi + \mu_e) \perp \mathbf{j}_N$ or, alternatively, if

$$\nabla(e\, \Phi + \mu_e) = 0 \quad \Rightarrow \quad e\, \Phi + \mu_e = \text{const.} \quad \text{resp.} \quad \mu_e = -e\,(\Phi - \Phi_0). \tag{1.109}$$

Equation (1.109) has already been discussed in Section 5.8.

Obviously, Eq. (1.108) contains irreversible heat conduction via u as part of the TE potential $\Phi = \alpha T + 1/u$, that is, we have $\Phi \to \alpha T$ in the limit of a reversible process management. By definition of the specific heat under constant volume in a quasistatic description (index "qs")

$$C_V = \frac{\delta Q_{\text{rev}}}{dT} \simeq \frac{\delta Q_{\text{qs}}}{dT}, \tag{1.110}$$

the correct relation within the framework of equilibrium thermodynamics is (with $dV = 0$)

$$dU(T, N) = \delta Q_{\text{rev}} + \delta W = C_V\, dT + (e\alpha T + \mu_e)\, dN, \tag{1.111}$$

where $\left(\frac{\partial U}{\partial N}\right)_T = (e\alpha T + \mu_e)$ is the fraction of the internal energy related to the carrier transport. Note that both the electrical and the thermal balance are influenced when adding a charged particle to the TE system. By comparing Eq. (1.111) with $dU = T\, dS + \mu_e\, dN$, we obtain

$$dS = \frac{C_V}{T}\, dT + e\alpha\, dN, \tag{1.112}$$

which means that any change in entropy is caused by two effects: one is heat accumulation (or restitution by C_V) and the second is due to the transport of entropy by the variation of the number of carriers.

From Eq. (1.111), we find for the difference between two equilibrium states

$$\Delta U = \int dU = \int C_V\, dT + \int (e\alpha T + \mu_e)\, dN = U_2 - U_1, \tag{1.113}$$

irrespective of the "path" between them. Usually, the transformation between these two states is performed by a sequence of quasi-stationary states, but we should keep in mind that such a strategy (lastly a "kinematic" description) cannot give us an answer to the system's driving forces, neither it gives insight into the system's evolution, which is governed by entropy-based principles.

Following Prigogine's principle of minimal entropy production, we clearly observe that there is a specific path that is used by the system, under constraints, to produce less entropy. This had been firstly investigated by Sherman *et al.* [271] and Clingman [273]. By applying the compatibility approach together with thermodynamic arguments, we can state additionally that minimum entropy production is obtained when the TE potential Φ takes a specific optimal value. There is no doubt that Φ is the correct potential for the description of the TE applications TEG, TEC, and TEH by taking into account heat conduction as an irreversible process in TE materials.

References

1. Kaye, J. and Welsh, J.A. (eds) (1960) *Direct Conversion of Heat to Electricity*, John Wiley & Sons, Inc.
2. Angrist, S.W. (1965) *Direct Energy Conversion*, Allyn and Bacon, Inc., Boston, MA.
3. Sutton, G.W. (ed.) (1966) *Direct Energy Conversion*, Inter-University Electronics Series, vol. 3, McGraw-Hill Book Company.
4. Decher, R. (1997) *Direct Energy Conversion*, Oxford University Press.
5. Seebeck, T.J. (1823) Magnetische Polarisation der Metalle und Erze durch Temperatur-Differenz. Technical report, (1822–1825) Reports of the Royal Prussian Academy of Science, Berlin.
6. Velmre, E. (2007) Thomas Johann Seebeck. *Proc. Est. Acad. Sci. Eng.*, **13**, 276–282.
7. Velmre, E. (2010) Thomas Johann Seebeck and his contribution to the modern science and technology. Electronics Conference (BEC), 2010 12th Biennial Baltic, pp. 17–24.
8. Nolas, G.S., Sharp, J., and Goldsmid, H.J. (2001) *Thermoelectrics: Basic Principles and New Materials Development*, Springer Series in Material Science, vol. **45**, Springer-Verlag, Berlin.
9. Riffat, S.D. and Ma, X. (2003) Thermoelectrics: a review of present and potential applications. *Appl. Therm. Eng.*, **23** (8), 913–935.
10. Minnich, A.J., Dresselhaus, M.S., Ren, Z.F., and Chen, G. (2009) Bulk nanostructured thermoelectric materials: current research and future prospects. *Energy Environ. Sci.*, **2** (5), 466–479.
11. Snyder, G.J. and Toberer, Eric.S. (2008) Complex thermoelectric materials. *Nat. Mater.*, 7, 105–114.
12. Sootsman, J.R., Chung, D.Y., and Kanatzidis, M.G. (2009) New and old concepts in thermoelectric materials. *Angew. Chem. Int. Ed.*, **48** (46), 8616–8639.
13. Rowe, D.M. and Bhandari, C.M. (1983) *Modern Thermoelectrics*, Reston Publishing Company, Inc., Reston, VA.
14. Birkholz, U. (1984) Thermoelektrische Bauelemente, in *Amorphe und polykristalline Halbleiter*, Halbleiterelektronik, vol. **18** (ed. W. Heywang), Springer-Verlag.
15. Rowe, D.M. (ed.) (1995) *CRC Handbook of Thermoelectrics*, CRC Press, Boca Raton, FL.
16. Rowe, D.M. (ed.) (2006) *CRC Handbook of Thermoelectrics: Macro to Nano*, CRC Press, Boca Raton, FL.
17. Rowe, D.M. (ed.) (2012) *Materials, Preparation, and Characterization in Thermoelectrics*, Thermoelectrics and its Energy Harvesting, vol. **1**, CRC Press.
18. Rowe, D.M. (ed.) (2012) *Modules, Systems, and Applications in Thermoelectrics*, Thermoelectrics and its Energy Harvesting, vol. **2**, CRC Press.
19. Jansch, D. (ed.) (2011) *Thermoelectrics Goes Automotive: Thermoelektrik, II*, Expert Verlag - IAV.
20. Anatychuk, L.I. (1998) *Physics of Thermoelectricity*, Thermoelectricity, vol. **1**, Institute of Thermoelectricity, Ukraine.
21. Anatychuk, L.I. (2003) *Thermoelectric Power Converters*, Thermoelectricity,

vol. **2**, Institute of Thermoelectricity, Kyiv-Chernivtsi.
22. Tritt, T.M. (1996) Thermoelectrics run hot and cold. *Science*, **272** (5266), 1276–1277.
23. Tritt, T.M. (ed.) (2000) *Recent Trends in Thermoelectric Materials Research I*, Semiconductors and Semimetals, vol. **69**, Academic Press, San Diego, CA,.
24. Tritt, T.M. (ed.) (2001) *Recent Trends in Thermoelectric Materials Research II*, Semiconductors and Semimetals, vol. **70**, Academic Press, San Diego, CA.
25. Tritt, T.M. (ed.) (2001) *Recent Trends in Thermoelectric Materials Research III*, Semiconductors and Semimetals, vol. **71**, Academic Press, San Diego, CA.
26. Martin, J., Tritt, T., and Uher, C. (2010) High temperature Seebeck coefficient metrology. *J. Appl. Phys.*, **108** (12), 121101.
27. Kanatzidis, M.G., Mahanti, S.D., and Hogan, T.P. (2003) *Chemistry, Physics and Materials Science of Thermoelectric Materials: Beyond Bismuth Telluride*, Fundamental Materials Research, Kluwer Academic/Plenum Publishers.
28. Tritt, T.M. (2011) Thermoelectric phenomena, materials, and applications. *Annu. Rev. Mater. Res.*, **41** (1), 433–448.
29. Shakouri, A. (2011) Recent developments in semiconductor thermoelectric physics and materials. *Annu. Rev. Mater. Res.*, **41** (1), 399–431.
30. Shakouri, A. and Zebarjadi, M. (2009) Nanoengineered materials for thermoelectric energy conversion, in *Thermal Nanosystems and Nanomaterials*, Topics in Applied Physics, vol. **118** (ed. S. Volz), Springer-Verlag, Berlin / Heidelberg, pp. 225–299.
31. Vineis, C.J., Shakouri, A., Majumdar, A., and Kanatzidis, M.G. (2010) Nanostructured thermoelectrics: Big efficiency gains from small features. *Adv. Mater.*, **22** (36), 3970–3980.
32. Zebarjadi, M., Esfarjani, K., Dresselhaus, M.S., Ren, Z.F., and Chen, G. (2012) Perspectives on thermoelectrics: from fundamentals to device applications. *Energy Environ. Sci.*, **5** (1), 5147–5162.
33. Goldsmid, H.J. (2010) *Introduction to Thermoelectricity*, Springer-Verlag.
34. Joffe, A.F. (1958) The revival of thermoelectricity. *Sci. Am.*, **199** (5), 31–37.
35. Goldsmid, H.J. and Douglas, R.W. (1954) The use of semiconductors in thermoelectric refrigeration. *Br. J. Appl. Phys.*, **5** (11), 386–390.
36. Ioffe, A.F. (1957) *Semiconductor Thermoelements and Thermoelectric Cooling*, Infosearch, Ltd., London.
37. Goldsmid, H.J. (1960) *Applications of Thermoelectricity*, Butler & Tanner Ltd.
38. Tauc, J. (1957) Generation of an EMF in semiconductors with nonequilibrium current carrier concentrations. *Rev. Mod. Phys.*, **29** (3), 308–324.
39. Joffe, A.F. and Stil'bans, L.S. (1959) Physical problems of thermoelectricity. *Rep. Prog. Phys.*, **22** (1), 167–203.
40. Samoilowitsch, A.G. and Korenblit, L.L. (1953) Gegenwärtiger Stand der Theorie der thermoelektrischen und thermomagnetischen Erscheinungen in Halbleitern. *Fortschr. Phys.*, **1**, 486–554. Unshortened translation of the article published in Uspechi Fiz. Nauk, 49, p. 243 and 337 (1953).
41. Telkes, M. (1954) Power output of thermoelectric generators. *J. Appl. Phys.*, **25** (8), 1058–1059.
42. Joffé, A.F. (1957) *Halbleiter-Thermoelemente*, Akademie-Verlag, Berlin.
43. Peschke, K. (1957) Thermoelemente und Gleichstromthermogeneratoren. *Electr. Eng. (Arch. Elektrotech.)*, **43**, 328–354, doi: 10.1007/BF01407247.
44. Justi, E. (1958) Elektrothermische Kühlung und Heizung - Grundlagen und Möglichkeiten, in *70. Sitzung der Arbeitsgemeinschaft für Forschung des Landes Nordrhein-Westfalen*, Westdeutscher Verlag.
45. Egli, P.H. (1960) *Thermoelectricity*, John Wiley & Sons, Inc., New York.
46. Gray, P.E. (1960) *The Dynamic Behavior of Thermoelectric Devices*, Technology Press of the Massachusetts Institute of Technology, Cambridge, MA.
47. Cadoff, I.B. and Miller, E. (eds) (1960) *Thermoelectric Materials and Devices*, Materials Technology Series, Reinhold Publishing Cooperation, New York.

Lectures presented during the course on Thermoelectric Materials and Devices sponsored by the Department of Metallurgical Engineering in cooperation with the Office of Special Services to Business and Industry, New York, June 1959 and 1960.

48. Birkholz, U. (1960) Zur experimentellen Bestimmung der thermoelektrischen Effektivität von Halbleitern. *Solid-State Electron.*, **1** (1), 34–38.

49. Heikes, R.R. and Ure, R.W. Jr. (1961) *Thermoelectricity: Science and Engineering*, Interscience Publishers, Inc., New York.

50. Zener, C. (1961) Putting electrons to work. *Trans. Am. Soc. Met.*, **53**, 1052–1068.

51. Tauc, J. (1962) *Photo and Thermoelectric Effects in Semiconductors*, Pergamon Press, New York.

52. Finn, B.S. (1963) Developments in Thermoelectricity, 1850–1920. PhD thesis. University of Wisconsin.

53. Jaumot, F.E. Jr. (1963) Thermoelectricity, in *Advances in Electronics and Electron Physics*, Advances in Electronics and Electron Physics, vol. **17** (eds L. Marton and C. Marton), Academic Press, pp. 207–243.

54. Harman, T.C. and Honig, J.M. (1967) *Thermoelectric and Thermomagnetic Effects and Applications*, McGraw-Hill Book Company, New York.

55. Regel', A.R. and Stil'bans, L.S. (1968) Thermoelectric power generation. *Sov. Phys. Semicond.*, **1** (11), 1341–1345. Translated from Fizika i tekhnika poluprovodnikov, **1** (11), 1614–1619, 1967.

56. Finn, B.S. (1967) Thomson's dilemma. *Phys. Today*, **20** (9), 54–59.

57. Finn, B.S. (1980) *Thermoelectricity*, Advances in Electronics and Electron Physics, vol. **50**, Academic Press, pp. 175–240.

58. Peters, F. (1908) *Thermoelemente und Thermosäulen; ihre Herstellung und Anwendung*, Number 30 in Monographie über angewandte Elektrochemie. W. Knapp, Halle a.d.S. A digital version of the book is available at http://digital.bibliothek.uni-halle.de/hd/content/titleinfo/1345813.

59. Lidorenko, N.S. and Terekov, A.Ya. (2007) On the history of thermo-electricity development in Russia. *J. Thermoelectric.*, **2** (8), 32–37.

60. Ulrich, F. and Aepinus, T. (1951) *Theory of Electricity and Magnetism*, The USSR Academy of Sciences Publications.

61. Korzhuyev, M.A. and Katin, I.V. (2011) On the sequence of discovery of the basic thermoelectric effects. *J. Thermoelectric.*, **3**, 79.

62. Anatychuk, L.I. (1994) Seebeck or Volta? *J. Thermoelectric.*, **1994** (1), 9–10.

63. Anatychuk, L.I. (2004) On the discovery of thermoelectricity by Volta. *J. Thermoelectric.*, **2004** (2), 5–10.

64. Anatychuk, L., Stockholm, J., and Pastorino, G. (2010) On the discovery of thermoelectricity by A. Volta. 8th European Conference on Thermoelectrics - ECT 2010.

65. Korzhuyev, M.A. and Katin, I.V. (2009) Formal analysis of Galvani and Volta experiments related to thermoelectricity. *J. Thermoelectric.*, **1**, 11–17.

66. Park, R.M. and Hoersch, H.M. (eds) (1993) *Manual on the Use of Thermocouples in Temperature Measurement*, 4th edn, ASTM International.

67. Valli, E. (1793) *Experiments on Animal Electricity with their Application to Physiology*, Printed for J. Johnson, St. Paul's Church-Yard, London.

68. Kipnis, N. (2003) Changing a theory: the case of Volta's contact electricity. *Nuovo Voltiana*, **5**, 143–162.

69. Pastorino, G. (2009) Alessandro Volta and his role in thermoelectricity. *J. Thermoelectric.*, **1**, 7–10.

70. Khramov, Yu.A. (1983) *Fiziki. Bibliograficheskii Spravochnik (Physicist. Bibliographic Reference Book*, 2nd edn, Nauka, Moscow.

71. Glazunov, V.N. (1990) *Search for Operating Principles of Technical Systems*, Rechnoy Transport Publishers, Moscow.

72. Seebeck, T.J. (1826) Ueber die magnetische Polarisation der Metalle und Erze durch Temperatur-Differenz. *Ann. Phys.*, **82**, 1–20.

73. Oersted, H.C. (1820) *Experimenta Circa Effectum Conflictus Electrici in*

Acum Magneticam, Typis Schultzianis, Hafiniae [Copenhagen].

74. Seebeck, T.J. (1821) Ueber den Magnetismus der galvanischen Kette. Technical report, Reports of the Royal Prussian Academy of Science, Berlin.

75. Justi, E. (1948) *Leitfähigkeit und Leitungsmechanismus fester Stoffe*, Thermoelektrische Effekte, Chapter 3, Vandenhoeck & Ruprecht, pp. 72–119.

76. Meisser, W., Kohler, M., and Reddemann, H. (1935) *Elektronenleitung, galvanomagnetische, thermoelektrische und verwandte Effekte*, Handbuch der Experimentalphysik, vol. **11**, Akademische Verlagsgesellschaft m. b. H..

77. von Yelin, J.C. (1823) Der Thermo-Magnetismus der Metalle, eine neue Entdeckung. *Ann. Phys.*, **73** (4), 415–429.

78. von Yelin, J.C. (1823) Neue Versuche über die magneto - motorische Eigenschaft der bisher so genannten unmagnetischen Metalle. *Ann. Phys.*, **73** (4), 361–364.

79. von Yelin, J.C. (1823) Aus einem Briefe von Yelin's. *Ann. Phys.*, **73** (4), 432–432.

80. Oersted, H.C. (1823) Nouvelles expériences de M. Seebeck sur les actions électro-magnétiques. (Note communiquée par M. Oersted [le 3 mars 1823]). *Ann. Chim. Phys.*, **22**, 199–201.

81. Oersted, H.C. (1823) Notiz von neuen electrisch - magnetischen Versuchen des Herrn Seebeck in Berlin. *Ann. Phys.*, **73** (4), 430–432.

82. Fourier, J.B.J. and Oersted, H.C. (1823) Sur quelques nouvelles expériences thermoélectrique faites par M. le Baron Fourier et M. Oersted. *Ann. Chim. Phys.*, **22**, 375–389.

83. Peltier, F.A. (1847) *Notice sur la vie et les travaux scientifiques de J. C. A. Peltier*, Imprimerie de Édouard Bautruche, Rue de la harpe.

84. Lenz, E. (1838) Einige Versuche im Gebiete des Galvanismus. *Ann. Phys.*, **120** (6), 342–349.

85. Charles, J. and Peltier, A. (1844) *Essai sur la coordination des causes qui précèdent, produisent et accompagnent les phénomènes électriques*, Hayez.

86. Thomson, W. (1851) On the mechanical theory of thermo-electric currents. *Proc. R. Soc. Edinburgh*, **3**, 91–98. also in Philos. Mag., **3** (1852), 529–535–in Thomson, Math. Phys. Pap., **1**, 316–323.

87. Thomson, W. (1848) On an absolute thermometric scale. *Philos. Mag.*, **33**, 313–317.

88. Thomson, W. (1882) *On an Absolute Thermometric Scale*, vol. **1**, Cambridge University Press, pp. 100–106.

89. Thomson, W. (1854) Account of researches in thermo-electricity. *Proc. R. Soc. London*, **7**, 49–58.

90. Thomson, W. (1856–1857) On the electro-dynamic qualities of metals: effects of magnetization on the electric conductivity of nickel and of iron. *Proc. R. Soc. London*, **8**, 546–550.

91. Thomson, W. (1856) On the electrodynamic qualities of metals. *Philos. Trans. R . Soc. London*, **146**, 649–751.

92. Magnus, G. (1851) Ueber thermoelektrische Ströme. *Ann. Phys.*, **159** (8), 469–504.

93. Hunt, L.B. (1964) The early history of the thermocouple. *Platinum Met. Rev.*, **8** (1), 23–28.

94. Kauffman, G.B. (1976) Gustav Magnus and his green salt. *Platinum Met. Rev.*, **20** (1), 21–24.

95. Kant, H. (2002) Ein Lehrer mehrerer Physikergenerationen - Zum 200, Geburtstag des Berliner Physikers Gustav Magnus, http://www2.hu-berlin.de/presse/zeitung/archiv/01_02/num_6/magnus.htm.

96. Rayleigh, L. (1885) On the thermodynamic efficiency of the thermopile. *Philos. Mag. Ser. 5*, **20**, 361–363.

97. Levine, S.N. (ed.) (1961) *Selected Papers on New Techniques for Energy Conversion*, Dover Publications

98. Unger, S. and Schwarz, J. (2010) *Edmund Altenkirch - Pionier der Kältetechnik*, Statusbericht des Deutschen Kälte- und Klimatechnischen Vereins, vol. **23**, DKV, Hannover. (in German).

99. Altenkirch, E. (1909) Über den Nutzeffekt der Thermosäulen. *Phys. Z.*, **10**, 560–580.

100. Altenkirch, E. (1911) Elektrothermische Kälteerzeugung und reversible

elektrische Heizung. *Phys. Z.*, **12**, 920–924.
101. Self, D. (2014) Thermo-electric generators.
102. Willy, W., Altenkirch, E., and Gehlhoff, G.R. (1914) Thermo-electric heating and cooling body.
103. Eucken, A. (1911) Über die Temperaturabhängigkeit der Wärmeleitfähigkeit fester Nichtmetalle. *Ann. Phys.*, **339** (2), 185–221.
104. Eucken, A. (1911) Die Wärmeleitfähigkeit einiger Kristalle bei tiefen Temperaturen. *Phys. Z.*, **12** (22–23), 1005–1008.
105. Eucken, A. (1911) Die Wärmeleitfähigkeit einiger Kristalle bei tiefen Temperaturen. *Abh. Dtsch. Phys. Ges.*, **13** (20), 829–835.
106. Eucken, A. and Kuhn, G. (1928) Results of new measurements of the warmth conductivity of firm crystallized substances in 0(o) and -190(o) C. *Z. Phys. Chem. Stoch. Verwandtschafts.*, **134**, 193–219.
107. Domenicali, C.A. (1954) Irreversible thermodynamics of thermoelectricity. *Rev. Mod. Phys.*, **26** (2), 237–275.
108. Benedicks, C.A.F. (2008) COPYRIGHT 2008 Charles Scribner's Sons.
109. Benedicks, C.A.F. (1904) *Recherches physiques et physico-chimiques sur l'acier au carbone: thèse pour le doctorat*, Librairie de l'Université (C. J. Lundström).
110. Benedicks, C.A.F. (1918) Ein für Thermoelektrizität und metallische Wärmeleitung fundamentaler Effekt. *Annal. Phys.*, **360** (1), 1–80.
111. Benedicks, C.A.F. (1918) Ein für Thermoelektrizität und metallische Wärmeleintung fundamentaler Effekt. *Ann. Phys.*, **360** (2), 103–150.
112. Benedicks, C.A.F. (1920) Galvanometrischer Nachweis der Thermospannung 1. Art beim flüssigen Quecksilber. *Ann. Phys.*, **367** (11), 185–217.
113. Benedicks, C.A.F. (1921) *The Homogeneous Electro-Thermic Effect (Including the Thomson Effect as a Special Case)*, Chapman and Hall, Ltd. A comment on the book in Nature, **109** (2741), 608–(1922), doi: 10.1038/109608b0.
114. Benedicks, C. (1929) Jetziger Stand der grundlegenden Kenntnisse der Thermoelektrizität, in *Ergebnisse der exakten Naturwissenschaften*, Springer Tracts in Modern Physics, vol. **8**, Springer-Verlag, Berlin / Heidelberg, pp. 25–68, doi: 10.1007/BFb0111908.
115. Fuschillo, N. (1952) A critical study of the asymmetrical temperature gradient thermoelectric effect in copper and platinum. *Proc. Phys. Soc. London, Sect. B*, **65** (11), 896.
116. (1922) Book reviews: Carl Benedicks - the homogeneous electro-thermal effect. (Including the Thomson Effect as a Special Case.). *Nature*, **109**, 608.
117. Mahan, G.D. (1991) The Benedicks effect: nonlocal electron transport in metals. *Phys. Rev. B*, **43** (5), 3945–3951. Feb
118. Tauc, J. (1953) The theory of the thermal EMF of semi-conductors. *Czech. J. Phys.*, **3**, 282–302.
119. Tauc, J. (1956) Electronic phenomena in semi-conductors with a temperature gradient. *Czech. J. Phys.*, **6**, 108–122, doi: 10.1007/BF01699870.
120. Trousil, Z. (1956) Proof of the anomalous thermal EMF on germanium. *Czech. J. Phys.*, **6**, 170–172, doi: 10.1007/BF01699880.
121. Grigorenko, A.N., Nikitin, P.I., Jelski, D.A., and George, T.F. (1990) Two-dimensional treatment of nonlinear thermoelectricity in homogeneous metals. *Phys. Rev. B*, **42** (12), 7405–7408.
122. Piotrowski, T., Jung, W., and Sikorski, S. (2007) Application of non-linear theory to analysis of Benedicks effect in semiconductors. *Phys. Status Solidi A*, **204** (4), 1063–1067.
123. The Nobel prize in physics 1946 - Percy Williams Bridgman.
124. Kemble, E.C. and Birch, F. (1970) *Percy Williams Bridgman 1882–1961*, The National Academies Press.
125. McMillan, P.F. (2005) Pressing on: the legacy of Percy W. Bridgman. *Nat. Mater.*, **4**, 715–718.
126. Bridgman, P.W. (1919) A critical thermodynamic discussion of the Volta, thermo-electric and thermionic effects. *Phys. Rev.*, **14** (4), 306–347.

127. Bridgman, P.W. (1926) Thermal conductivity and thermal E.M.F. of single crystals of several non-cubic metals. *Proc. Am. Acad. Arts Sci.*, **61** (4), 101–134.
128. Bridgman, P.W. (1928) Thermoelectric phenomena in crystals and general electrical concepts. *Phys. Rev.*, **31** (2), 221–235.
129. Bridgman, P.W. (1929) On the nature of the transverse thermo-magnetic effect and the transverse thermo-electric effect in crystals. *Proc. Natl. Acad. Sci. U.S.A.*, **15** (10), 768–773.
130. Bridgman, P.W. (1929) On the application of thermodynamics to the thermoelectric circuit. *Proc. Natl. Acad. Sci. U.S.A.*, **15** (10), 765–768.
131. Bridgman, P.W. (1932) Comments on the note by E. H. Kennard on "Entropy, reversible processes and thermocouples". *Proc. Natl. Acad. Sci. U.S.A.*, **18** (3), 242–245.
132. Bridgman, P.W. (1932) A new kind of E.M.F. and other effects thermodynamically connected with the four transverse effects. *Phys. Rev.*, **39** (4), 702–715.
133. Bridgman, P.W. (1940) The second law of thermodynamics and irreversible processes. *Phys. Rev.*, **58** (9), 845.
134. Bridgman, P.W. (1951) The nature of some of our physical concepts. I. *Br. J. Philos. Sci.*, **1** (4), 257–272.
135. Bridgman, P.W. (1951) The nature of some of our physical concepts. II. *Br. J. Philos. Sci.*, **2** (5), 25–44.
136. Bridgman, P.W. (1951) The nature of some of our physical concepts. III. *Br. J. Philos. Sci.*, **2** (6), 142–160.
137. Silk, T.W. and Shofield, A.J. (2008) Thermoelectric effects in anisotropic systems: measurement and applications, arXiv:0808.3526v1.
138. Kikoin, I.K. and Sominskii, M.S. (1961) Abram Fedorovich Ioffe (on his eightieth birthday). *Sov. Phys. Usp.*, **3** (5), 798.
139. Vedernikov, M.V. and Iordanishvili, E.K. (1998) A. F. Ioffe and origin of modern semiconductor thermoelectric energy conversion. 17th International Conference on Thermoelectrics - Proceedings ICT98, IEEE, pp. 37–42.
140. Ioffe, A.F. (1950) *Energeticheskie osnovy termoelektricheskikh bataryei iz poluprovodnikov*, USSR Academy of Sciences.
141. Joffe, A.F. (1956) Heat transfer in semiconductors. *Can. J. Phys.*, **34** (12A), 1342–1355.
142. Joffe, A.F., Joffe, A.V., Ayrapetyans, S.V., Colomiez, N.V., and Stil'bans, L.S. (1956) *Rep. Acad. Sci.*, **106**, 6.
143. Joffe, A.F., Joffe, A.V., Ayrapetyans, S.V., Colomiez, N.V., and Stil'bans, L.S. (1956) *Rep. Acad. Sci.*, **106**, 981.
144. Anatychuk, L.I. (2002) With reference to the history of using semiconductors in thermoelectricity. *J. Thermoelectric.*, **4** (12), 7–10.
145. Slack, G.A. (2011) The thermoelectrics of bulk materials: a history. Online, JulyGlen Slack's Plenary Presentation for the Outstanding Achievement Award at the 30th International Conference on Thermoelectrics.
146. Borrego, J.M. (1963) Carrier concentration optimization in semiconductor thermoelements. *IEEE Trans. Electron Devices*, **10** (6), 364–370.
147. Mahan, G.D. (1991) Inhomogeneous thermoelectrics. *J. Appl. Phys.*, **70** (8), 4551–4554.
148. Telkes, M. (1938) *Westinghouse Research Report R-94264-8*, Westinghouse Corporation, Pittsburgh, PA.
149. Wood, C. (1988) Materials for thermoelectric energy conversion. *Rep. Prog. Phys.*, **51** (4), 459–539.
150. Telkes, M. (1947) The efficiency of thermoelectric generators. I. *J. Appl. Phys.*, **18** (12), 1116–1127.
151. Telkes, M. (1954) Solar thermoelectric generators. *J. Appl. Phys.*, **25** (6), 765–777.
152. Goldsmid, H.J. (1964) *Thermoelectric Refrigeration*, Plenum Publishing Corporation.
153. Slack, G.A. (1995) New materials and performance limits for thermoelectric cooling, in *CRC Handbook of Thermoelectrics* (ed. Rowe, D.M.), CRC Press, pp. 407–440.
154. Vining, C.B., Rowe, D.M., Stockholm, J., and Rao, K.R. (2005) Appendix I: History of the international thermoelectric

society, in *Thermoelectric Handbook - Macro to Nano*, CRC Press, Taylor & Francis Group, pp. AI1–AI8.
155. Dresselhaus, M.S., Chen, G., Tang, M.Y., Yang, R.G., Lee, H., Wang, D.Z., Ren, Z.F., Fleurial, J.-P., and Gogna, P. (2007) New directions for low-dimensional thermoelectric materials. *Adv. Mater.*, **19** (8), 1043–1053.
156. Kanatzidis, M.G. (2009) Nanostructured thermoelectrics: the new paradigm? *Chem. Mater.*, **22** (3), 648–659.
157. Bell, L.E. (2008) Cooling, heating, generating power, and recovering waste heat with thermoelectric systems. *Science*, **321** (5895), 1457–1461.
158. Korzhuyev, M.A. and Temyakov, V.V. (2012) O vklade akademika f. epinusa (1724–1802) v uchenyie o termoelektrichetvo.
159. Volta, A. (1800) On the electricity excited by the mere contact of conducting substances of different kind. *Philos. Trans. R. Soc. London*, **90**, 403–431.
160. Fourier, J.B.J. (1822) *Théorie Analytique de la Chaleur*, F. Didot.
161. Narasimhan, T.N. (1999) Fourier's heat conduction equation: history, influence, and connections. *Rev. Geophys.*, **37** (1), 151–172.
162. Narasimhan, T. (1999) Fourier's heat conduction equation: history, influence, and connections. *J. Earth Syst. Sci.*, **108**, 117–148, doi: 10.1007/BF02842327.
163. Léonard, N. and Carnot, S. (1824) *Réflexions sur la puissance motrice du feu et sur les machines propres à développer cette puissance*, Bachelier, Paris.
164. Léonard, N. and Carnot, S. (1892) *Betrachtungen über die bewegende Kraft des Feuers und die zur Entwickelung dieser Kraft geeigneten Maschinen (1824)*, Wilhelm Engelmann-Verlag. Translated and published by W. Ostwald.
165. Ohm, G.S. (1827) *Die galvanische Kette, Mathematisch Bearbeitet*, T. H. Riemann, Berlin.
166. Klinger, W. (1993) Leben und Werk von Georg Simon Ohm, in *Wege in der Physikdidaktik - Rückblick und Perspektive*, vol. **3**, Verlag Palm & Enke, Erlangen, pp. 18–36.
167. Charles, J. and Peltier, A. (1834) Nouvelles expériences sur la caloricité des courants électrique. *Ann. Chim. Phys.*, **56**, 371–386.
168. Joule, J.P. (1843) On the calorific effects of magneto-electricity, and on the mechanical value of heat. *Philos. Mag. Ser. 3*, **23** (153), 347–355.
169. Becquerel, E. (1866) Memoire sur les pouvoir thermo-électriques des corps et sur les piles thermo-électriques. *Ann. Chim. Phys., Quatrième Sér.*, **8**, 389–436.
170. Haken, W. (1910) Beitrag zur Kenntnis der thermoelektrischen Eigenschaften der Metallegierungen. *Ann. Phys.*, **337** (7), 291–336.
171. Debye, P. (1914) Zustandsgleichung und Quantenhypothese mit einem Anhang über Wärmeleitung, in *Vorträge über die kinetische Theorie der Materie und der Elektrizität*, Teubner, Berlin, pp. 19–60.
172. Julius, M. and Strutt, O. (1928) Zur Wellenmechanik des Atomgitters. *Ann. Phys.*, **86**, 319–324.
173. Gurevich, L. (1945) Thermoelectric properties of conductors I. *J. Phys. (U.S.S.R)*, **9**, 477.
174. Gurevich, L. (1946) Thermoelectric properties of conductors II. *J. Phys. (U.S.S.R)*, **10**, 67.
175. Gurevich, L. (1946) Thermomagnetic and galvanomagnetic properties of conductors III. *J. Phys. (U.S.S.R)*, **10**, 174.
176. Dudkin, L.D. and Abrikosov, N.Kh. (1957) Effect of Ni on the properties of the semiconducting compound $CoSb_3$. *Z. Neorg. Khim.*, **2**, 212–221.
177. Dudkin, L.D. and Abrikosov, N.Kh. (1959) On the doping of the semiconductor compound $CoSb_3$. *Sov. Phys.-Solid State*, **1**, 126–133.
178. Skrabek, E.A. and Trimmer, D.S. (1972) Thermoelectric device including an alloy of GeTe and AgSbTe as the P-type element. US Patent 3,945,855, filed March 8, 1976.
179. Skrabek, E.A. and Trimmer, D.S. (1995) Properties of the general TAGS

system, in *CRC Handbook of Thermoelectrics*, Chapter 22, CRC Press LLC, pp. 267–275.

180. Slack, G.A. (1979) The thermal conductivity of nonmetallic crystals. *Solid State Phys.*, **34**, 1–71.
181. Ross, R.G. and Andersson, P. (1982) Clathrate and other solid phases in tetrahydrofuran-water system: thermal conductivity and heat capacity under pressure. *Can. J. Chem.*, **60**, 881–892.
182. Hicks, L.D. and Dresselhaus, M.S. (1993) Effect of quantum-well structures on the thermoelectric figure of merit. *Phys. Rev. B: Condens. Matter Mater. Phys.*, **47**, 12727.
183. Hicks, L.D. and Dresselhaus, M.S. (1993) Thermoelectric figure of merit of a one-dimensional conductor. *Phys. Rev. B: Condens. Matter Mater. Phys.*, **47**, 16631(R).
184. Fleurial, J.-P., Caillat, T., and Borshchevsky, A. (1994) Skutterudites: a new class of promising thermoelectric material, in *Proceedings of the 13th International Conference on Thermoelectrics*, AIP Conference Proceedings, vol. 316 (ed. B. Mathiprakasam), American Institute of Physics, New York, pp. 40–44.
185. Sales, B.C., Mandrus, D., and Williams, R.K. (1996) Filled skutterudite antimonides: a new class of thermoelectric materials. *Science*, **272** (5266), 1325–1328.
186. Chen, B., Xu, J.-H., Uher, C., Morelli, D.T., Meisner, G.P., Fleurial, J.-P., Caillat, T., and Borshchevsky, A. (1997) Low-temperature transport properties of the filled skutterudites $CeFe_{4-x}Co_xSb_{12}$s. *Phys. Rev. B*, **55**, 1476–1480.
187. Sato, Y., Terasaki, I., and Tajima, S. (1997) Seebeck effect in the mixed state of the 60-k phase of single-crystal $YBa_2Cu_3O_y$. *Phys. Rev. B*, **55**, R14749–R14752.
188. Terasaki, I., Sasago, Y., and Uchinokura, K. (1997) Large thermoelectric power in $NaCo_2O_4$ single crystals. *Phys. Rev. B*, **56**, R12685–R12687.
189. Yakabe, H., Kikuchi, K., Terasaki, I., Sasago, Y., and Uchinokura, K. (1997) Thermoelectric properties of transition-metal oxide $NaCo_2O_4$ system. Proceedings of the 16th International Conference on Thermoelectrics, IEEE, pp. 523–529.
190. Venkatasubramanian, R., Siivola, E., Colpitts, T., and O'Quinn, B. (2001) Thin-film thermoelectric devices with high room-temperature figures of merit. *Nature*, **413**, 597–602.
191. Harman, T.C., Taylor, P.J., Walsh, M.P., and LaForge, B.E. (2002) Quantum dot superlattice thermoelectric materials and devices. *Science*, **297** (5590), 2229–2232.
192. Hsu, K.F., Loo, S., Guo, F., Chen, W., Dyck, J.S., Uher, C., Hogan, T., Polychroniadis, E.K., and Kanatzidis, M.G. (2004) Cubic $AgPb_mSbTe_{2+m}$: Bulk thermoelectric materials with high figure of merit. *Science*, **303** (5659), 818–821.
193. Tang, X.F., Xie, W.J., Li, H., Zhao, W.Y., Zhang, Q.J., and Niino, M. (2007) Preparation and thermoelectric transport properties of high-performance p-type Bi_2Te_3 with layered nanostructure. *Appl. Phys. Lett.*, **90**, 012102.
194. Poudel, B., Hao, Q., Ma, Y., Lan, Y., Minnich, A., Yu, B., Yan, X., Wang, D., Muto, A., Vashaee, D., Chen, X., Liu, J., Dresselhaus, M.S., Chen, G., and Ren, Z. (2008) High-thermoelectric performance of nanostructured bismuth antimony telluride bulk alloys. *Science*, **320** (5876), 634–638.
195. Heremans, J.P., Jovovic, V., Toberer, E.S., Saramat, A., Kurosaki, K., Charoenphakdee, A., Yamanaka, S., and Snyder, G.J. (2008) Enhancement of thermoelectric efficiency in Pb Te by distortion of the electronic density of states. *Science*, **321** (5888), 554–557.
196. Coleman, B.D. and Dill, E.H. (1971) Thermodynamic restrictions on the constitutive equations of electromagnetic theory. *Z. Angew. Math. Phys. ZAMP*, **22** (4), 691–702.
197. Coleman, B.D. and Dill, E.H. (1971) On the thermodynamics of electromagnetic fields in materials with memory. *Arch. Ration. Mech. Anal.*, **41** (2), 132–162.
198. Anatychuk, L.I. (1998) Generalization in physics of thermoelectric energy conversion and new trends of thermoelectricity development.

17th International Conference on Thermoelectrics, pp. 9–17.
199. Anatychuk, L.I. (2003) The law of thermoelectric induction and its application for extending the capabilities of thermoelectricity. 22nd International Conference on Thermoelectrics, pp. 472–475.
200. Amendola, G. and Manes, A. (2005) Minimum free energy in linear thermomagnetism. *Q. Appl. Math.*, **LXIII** (4), 645–672.
201. Amendola, G. and Manes, A. (2006) Maximum recoverable work in linear thermoelectromagnetism. *Nonlinear Oscil.*, **9** (3), 281–311.
202. Anatychuk, L.I. (2012) Thermoelectric induction in power generation: prospects and proposals, in *CRC Handbook of Thermoelectrics: Thermoelectrics and its Energy Harvesting*, Chapter 2 (ed. D.M. Rowe), CRC Press, Boca Raton, FL.
203. Brechet, S.D., Reuse, F.A., and Ansermet, J.-P. (2012) Thermodynamics of continuous media with electromagnetic fields. *Eur. Phys. J. B*, **85** (12), 412–(20 pages).
204. Fieschi, R., de Groot, S.R., Mazur, P., and Vlieger, J. (1954) Thermodynamical theory of galvanomagnetic and thermomagnetic phenomena II: reciprocal relations for moving anisotropic mixtures. *Physica*, **20** (1-6), 245–258.
205. Fieschi, R., de Groot, S.R., and Mazur, P. (1954) Thermodynamical theory of galvanomagnetic and thermomagnetic phenomena. I: reciprocal relations in anisotropic metals. *Physica*, **20** (1-6), 67–76.
206. Fieschi, R., de Groot, S.R., and Mazur, P. (1954) Thermodynamical theory of galvanomagnetic and thermomagnetic phenomena III: explicit expressions for the measurable effects in isotropic metals. *Physica*, **20** (1-6), 259–273.
207. Fieschi, R. (1955) Thermodynamical theory of galvanomagnetic and thermomagnetic phenomena. *Il Nuovo Cimento Ser. 10*, **1** (1), 1–47.
208. Campbell, L.L. (1923) *Galvanomagnetic and Thermomagnetic Effects – The Hall and Allied Phenomena*, Longmans, Grenn and Co.
209. Akulov, N. (1934) Zur Theorie der Hall-, Nernst-, Ettingshausen- und Righi-Leduc-Effekte. *Z. Phys.*, **87** (11-12), 768–777.
210. Callen, H.B. (1947) On the theory of irreversible processes. PhD thesis. Massachusetts Institute of Technology - (M.I.T.), Cambridge, MA.
211. Putley, E.H. (1960) *The Hall Effect and Related Phenomena*, Butterworths, London.
212. Delves, R.T. (1965) Thermomagnetic effects in semiconductors and semimetals. *Rep. Prog. Phys.*, **28** (1), 249.
213. Goldsmid, H.J. (1986) *Electronic Refrigeration*, Pion, London.
214. Newnham, R.E. (2005) Properties of materials: anisotropy, symmetry, structure, in *Galvanomagnetic and Thermomagnetic Phenomena*, Chapter 20, Oxford University Press, pp. 223–233.
215. Newnham, R.E. (2005) Properties of materials: anisotropy, symmetry, structure, in *Thermoelectricity*, Chapter 21, Oxford University Press, pp. 234–242.
216. Askerov, B.M. and Figarova, S.R. (2010) *Thermodynamics, Gibbs Method and Statistical Physics of Electron Gases*, Springer Series on Atomic, Optical, and Plasma Physics, vol. **57**, Springer-Verlag.
217. Hering, E., Martin, R., and Stohrer, M. (2012) *Physik für Ingenieure*, Springer-Verlag.
218. Maggi, G.A. (1850) Thermal conductivity of iron in a magnetic field. *Arch. d. Genève*, **14**, 132.
219. Hall, E.H. (1879) On a new action of the magnet on electric currents. *Am. J. Math.*, **2** (3), 287–292.
220. von Ettingshausen, A. and Nernst, W. (1886) Ueber das Auftreten electromotorischer Kräfte in Metallplatten, welche von einem Wärmestrome durchflossen werden und sich im magnetischen Felde befinden. *Ann. Phys.*, **265** (10), 343–347.
221. Leduc, S.A. (1887) Sur la conductibilité du bismuth dans un champ magnétique et la déviation des lignes isothermes. *C. R. Hebd. Séances Acad. Sci.*, **104**, 1783.

222. Nernst, W. (1887) Ueber die electromotorischen Kräfte, welche durch den Magnetismus in von einem Wärmestrome durchflossenen Metallplatten geweckt werden. *Ann. Phys.*, **267** (8), 760–789.

223. von Ettingshausen, A. (1887) Ueber eine neue polare Wirkung des Magnetismus auf die galvanische Wärme in gewissen Substanzen. *Ann. Phys.*, **267** (8), 737–759.

224. von Ettingshausen, A. (1887) Bemerkungen zu dem Aufsatze: "Ueber eine neue polare Wirkung des Magnetismus auf die galvanische Wärme in gewissen Substanzen". *Ann. Phys.*, **269** (1), 126–128.

225. von Ettingshausen, A. (1887) Ueber den Einfluss magnetischer Kräfte auf die Art der Wärmeleitung im Wismuth. *Ann. Phys.*, **269** (1), 129–136.

226. Righi, A. (1887) Sur la conductibilité calorifique du bismuth dans un champ magnétique. *C. R. Hebd. Séances Acad. Sci.*, **105**, 168.

227. von Ettingshausen, A. and Nernst, W. (1888) Ueber das thermische und galvanische Verhalten einiger Wismuth-Zinn-Legirungen im magnetischen Felde. *Ann. Phys.*, **269** (3), 474–492.

228. Leduc, S.A. (1888) Modifications de la conductibilité calorifique du bismuth dans un champ magnétique. *J. Phys. Theor. Appl.*, **7** (1), 519–525.

229. Angrist, S.W. (1961) Galvanomagnetic and thermomagnetic effects. *Sci. Am.*, **205**, 124–136.

230. Wolfe, R. (1964) Magnetothermoelectricity. *Sci. Am.*, **210**, 70–82.

231. Zawadzki, W. (1962) Thermomagnetic effects in semiconductors. *Phys. Status Solidi B*, **2** (4), 385–410.

232. Angrist, S.W. (1961) The direct conversion of heat to electricity by means of a Nernst effect thermomagnetic generator. PhD thesis. The Ohio State University.

233. O'Brien, B.J. and Wallace, C.S. (1958) Ettingshausen effect and thermomagnetic cooling. *J. Appl. Phys.*, **29** (7), 1010–1012.

234. El-Saden, M.R. (1962) Theory of the Ettingshausen cooler. *J. Appl. Phys.*, **33** (5), 1800–1803.

235. Harman, T.C. and Honig, J.M. (1962) Theory of galvano-thermomagnetic energy conversion devices. I. Generators. *J. Appl. Phys.*, **33** (11), 3178–3188.

236. Harman, T.C. and Honig, J.M. (1962) Theory of galvano-thermomagnetic energy conversion devices. II. Refrigerators and heat pumps. *J. Appl. Phys.*, **33** (11), 3188–3194.

237. Harman, T.C. and Honig, J.M. (1962) Operating characteristics of transverse (Nernst) anisotropic galvano-thermomagnetic generators. *Appl. Phys. Lett.*, **1** (2), 31–32.

238. Cuff, K.F., Horst, R.B., Weaver, J.L., Hawkins, S.R., Kooi, C.F., and Enslow, G.M. (1963) The thermomagnetic figure of merit and Ettingshausen cooling in Bi-Sb. *Appl. Phys. Lett.*, **2** (8), 145–146.

239. Goldsmid, H.J. (1963) The Ettingshausen figure of merit of bismuth and bismuth - antimony alloys. *Br. J. Appl. Phys.*, **14** (5), 271.

240. Harman, T.C. (1963) Theory of the infinite stage Nernst-Ettingshausen refrigerator. *Adv. Energy Convers.*, **3** (4), 667–676.

241. Harman, T.C. and Honig, J.M. (1963) Nernst and Seebeck coefficients in bismuth at high magnetic fields. *Adv. Energy Convers.*, **3** (3), 525–528.

242. Harman, T.C. and Honig, J.M. (1963) Operating characteristics of Nernst refrigerators for anisotropic materials. *J. Appl. Phys.*, **34** (1), 239–240.

243. Harman, T.C. and Honig, J.M. (1963) Erratum: operating characteristics of transvere (Nernst) anisotropic galvano-thermomagnetic generators. *Appl. Phys. Lett.*, **2** (2), 44–44.

244. Harman, T.C. and Honig, J.M. (1963) Theory of galvano-thermomagnetic energy conversion devices. III. Generators constructed from anisotropic materials. *J. Appl. Phys.*, **34** (1), 189–194.

245. Harman, T.C., Honig, J.M., and Tarmy, B.M. (1963) Theory of galvano-thermomagnetic energy conversion devices. V. Devices constructed from anisotropic materials. *J. Appl. Phys.*, **34** (8), 2225–2229.

246. Harman, T.C., Honig, J.M., and Tarmy, B.M. (1963) Galvano-thermomagnetic

phenomena. IV. Application to anisotropic adiabatic nernst generators. *J. Appl. Phys.*, **34** (8), 2215–2224.

247. Horst, R.B. (1963) Thermomagnetic figure of merit: Bismuth. *J. Appl. Phys.*, **34** (11), 3246–3254.

248. Kooi, C.F., Horst, R.B., Cuff, K.F., and Hawkins, S.R. (1963) Theory of the longitudinally isothermal Ettingshausen cooler. *J. Appl. Phys.*, **34** (6), 1735–1742.

249. Norwood, M.H. (1963) Theory of Nernst generators and refrigerators. *J. Appl. Phys.*, **34** (3), 594–599.

250. Ure, R.W. (1963) Theory of materials for thermoelectric and thermomagnetic devices. *Proc. IEEE*, **51** (5), 699–713.

251. Varga, B., Reich, A.D., and Madigan, J.R. (1963) Thermoelectric and thermomagnetic heat pumps. *J. Appl. Phys.*, **34** (12), 3430–3441.

252. Wright, D.A. (1963) Ettingshausen cooling in pyrolytic graphite. *Br. J. Appl. Phys.*, **14** (6), 329.

253. Delves, R.T. (1964) Figure of merit for Ettingshausen cooling. *Br. J. Appl. Phys.*, **15** (1), 105.

254. Simon, R. (1964) Maximum thermomagnetic figure of merit of two-band, multivalley semiconductors. *Solid-State Electron.*, **7** (6), 397–412.

255. Planck, M. (1879) Über den zweiten Hauptsatz der mechanischen Wärmetheorie. PhD thesis. Könglichen Universität München.

256. Onsager, L. (1931) Reciprocal relations in irreversible processes. I. *Phys. Rev.*, **37** (4), 405–426.

257. Onsager, L. (1931) Reciprocal relations in irreversible processes. II. *Phys. Rev.*, **38** (12), 2265–2279.

258. Onsager, L. (1945) Theories and problems of liquid diffusion. *Ann. N.Y. Acad. Sci.*, **46**, 241–265.

259. Callen, H.B. (1948) The application of Onsager's reciprocal relations to thermoelectric, thermomagnetic, and galvanomagnetic effects. *Phys. Rev.*, **73** (11), 1349–1358.

260. de Groot, S.R. (1963) *Thermodynamics of Irreversible Processes*, North-Holland Publishing Company, Amsterdam.

261. Tolman, R.C. and Fine, P.C. (1948) On the irreversible production of entropy. *Rev. Mod. Phys.*, **20** (1), 51–77.

262. Sommerfeld, A. and Frank, N.H. (1931) The statistical theory of thermoelectric, galvano- and thermomagnetic phenomena in metals. *Rev. Mod. Phys.*, **3** (1), 1–42.

263. Darrow, K.K. (1929) Statistical theories of matter, radiation and electricity. *Rev. Mod. Phys.*, **1** (1), 90.

264. Haase, R. (1952) Thermodynamisch-phänomenologische Theorie der irreversiblen Prozesse, in *Ergebnisse der Exakten Naturwissenschaften*, Springer Tracts in Modern Physics, vol. **26**, Springer-Verlag, Berlin/Heidelberg, pp. 56–164, doi: 10.1007/BFb0109311.

265. Domenicali, C.A. (1953) Irreversible thermodynamics of thermoelectric effects in inhomogeneous, anisotropic media. *Phys. Rev.*, **92** (4), 877–881.

266. Domenicali, C.A. (1954) Stationary temperature distribution in an electrically heated conductor. *J. Appl. Phys.*, **25** (10), 1310–1311.

267. de Groot, S.R. and van Kampen, N.G. (1954) On the derivation of reciprocal relations between irreversible processes. *Physica*, **21** (1-5), 39–47.

268. Bernard, W. and Callen, H.B. (1959) Irreversible thermodynamics of non-linear processes and noise in driven systems. *Rev. Mod. Phys.*, **31** (4), 1017–1044.

269. Tykodi, R.J. (1959) Thermodynamics, stationary states, and steady-rate processes. III. The thermocouple revisited. *J. Chem. Phys.*, **31** (6), 1517–1521.

270. Miller, D.G. (1960) Thermodynamics of irreversible processes. The experimental verification of the Onsager reciprocal relations. *Chem. Rev.*, **60** (1), 15–37.

271. Sherman, B., Heikes, R.R., and Ure, R.W. Jr. (1960) Calculation of efficiency of thermoelectric devices. *J. Appl. Phys.*, **31** (1), 1–16.

272. Littman, H. and Davidson, B. (1961) Theoretical bound on the thermoelectric figure of merit from irreversible thermodynamics. *J. Appl. Phys.*, **32** (2), 217–219.

273. Clingman, W.H. (1961) Entropy production and optimum device design. *Adv. Energy Convers.*, **1**, 61–79.
274. Clingman, W.H. (1961) New concepts in thermoelectric device design. *Proc. IRE*, **49** (7), 1155–1160.
275. Littman, H. (1962) A clarification of the theoretical upper bound on the thermoelectric "figure of merit" derived from irreversible thermodynamics. *J. Appl. Phys.*, **33** (8), 2655–2656.
276. El-Saden, M.R. (1962) Irreversible thermodynamics and the theoretical bound on the thermomagnetic figure of merit. *J. Appl. Phys.*, **33** (10), 3145–3146.
277. Nourbehecht, B. (1963) Irreversible Thermodynamic Effects in Inhomogeneous Media and their Applications in Certain Geoelectric Problems. Technical report, AD 409622, Geophysics Laboratory - Massachusetts Institute of Technology, Cambridge, MA.
278. Borrego, J.M. (1964) Zener's maximum efficiency derived from irreversible thermodynamics. *Proc. IEEE*, **52** (1), 95.
279. Osterle, J. (1964) A unified treatment of the thermodynamics of steady-state energy conversion. *Appl. Sci. Res.*, **12**, 425–434.
280. Curzon, F.L. and Ahlborn, B. (1975) Efficiency of a Carnot engine at maximum power output. *Am. J. Phys.*, **43**, 22–24.
281. Andresen, B., Salamon, P., and Berry, R.S. (1977) Thermodynamics in finite time: extremals for imperfect heat engines. *J. Chem. Phys.*, **66** (4), 1571–1577.
282. Hatsopoulos, G.N. and Keenan, J.H. (1981) *Thermoelectricity*, John Wiley & Sons, Inc., New York, pp. 675–683.
283. Salamon, P. and Nitzan, A. (1981) Finite time optimizations of a Newton's law Carnot cycle. *J. Chem. Phys.*, **74** (6), 3546–3560.
284. Ondrechen, M.J., Rubin, M.H., and Band, Y.B. (1983) The generalized carnot cycle: a working fluid operating in finite time between finite heat sources and sinks. *J. Chem. Phys.*, **78** (7), 4721–4727.
285. Gupta, V.K., Shanker, G., Saraf, B., and Sharma, N.K. (1984) Experiment to verify the second law of thermodynamics using a thermoelectric device. *Am. J. Phys.*, **52** (7), 625–628.
286. Rockwood, A.L. (1984) Relationship of thermoelectricity to electronic entropy. *Phys. Rev. A*, **30** (5), 2843–2844.
287. de Vos, A. (1988) Thermodynamics of radiation energy conversion in one and in three physical dimensions. *J. Phys. Chem. Solids*, **49** (6), 725–730.
288. Mackey, M.C. (1989) The dynamic origin of increasing entropy. *Rev. Mod. Phys.*, **61** (4), 981.
289. Snyder, G.J. (2006) Thermoelectric power generation: efficiency and compatibility, in *CRC Handbook of Thermoelectrics: Macro to Nano*, Chapter 9 (ed. D.M. Rowe), Taylor and Francis, Boca Raton, FL.
290. Gordon, J.M. (1991) Generalized power versus efficiency characteristics of heat engines: the thermoelectric generator as an instructive illustration. *Am. J. Phys.*, **59** (6), 551–555.
291. Lampinen, M.J. (1991) Thermodynamic analysis of thermoelectric generator. *J. Appl. Phys.*, **69** (8), 4318–4323.
292. Wu, C. (1993) Heat transfer effect on the specific power availability of heat engines. *Energy Convers. Manage.*, **34** (12), 1239–1247.
293. Özkaynak, S., Gökun, S., and Yavuz, H. (1994) Finite-time thermodynamic analysis of a radiative heat engine with internal irreversibility. *J. Phys. D: Appl. Phys.*, **27** (6), 1139.
294. Bejan, A. (1996) Entropy generation minimization: the new thermodynamics of finite-size devices and finite-time processes. *J. Appl. Phys.*, **79** (3), 1191–1218.
295. Bejan, A. (1996) Method of entropy generation minimization, or modeling and optimization based on combined heat transfer and thermodynamics. *Rev. Gale Therm.*, **35** (418–419), 637–646.
296. Parrott, J.E. (1996) Thermodynamic theory of transport processes in semiconductors. *IEEE Trans. Electron Devices*, **43** (5), 809–826.
297. Agrawal, D.C. and Menon, V.J. (1997) The thermoelectric generator as an

endoreversible carnot engine. *J. Phys. D: Appl. Phys.*, **30** (3), 357.
298. Chen, J., Yan, Z., and Wu, L. (1997) Nonequilibrium thermodynamic analysis of a thermoelectric device. *Energy*, **22** (10), 979–985.
299. Hoffmann, K.H., Burzler, J.M., and Schubert, S. (1997) Endoreversible thermodynamics. *J. Non-Equilib. Thermodyn.*, **22**, 311–355.
300. Chen, L., Wu, C., and Sun, F. (1999) Finite time thermodynamic optimization or entropy generation minimization of energy systems. *J. Non-Equilib. Thermodyn.*, **24** (4), 327–359.
301. Arenas, A., Vázquez, J., Sanz-Bobi, M.A., and Palacios, R. (2000) Performance of a thermoelectric module using the thermodynamic relationship temperature-entropy $(T-S)$. 19th International Conference on Thermoelectrics, Cardiff, Wales, August 20–24, 2000, Barbow Press, Church Village, Mid Glamorgan.
302. Nuwayhid, R.Y., Moukalled, F., and Noueihed, N. (2000) On entropy generation in thermoelectric devices. *Energy Convers. Manage.*, **41** (9), 891–914.
303. Chua, H.T., Ng, K.C., Xuan, X.C., Yap, C., and Gordon, J.M. (2002) Temperature-entropy formulation of thermoelectric thermodynamic cycles. *Phys. Rev. E*, **65** (5), 056111.
304. Garrido, J. (2002) Observable variables in thermoelectric phenomena. *J. Phys. Chem. B*, **106** (41), 10722–10724.
305. Xuan, X.C. (2002) Optimum design of a thermoelectric device. *Semicond. Sci. Technol.*, **17** (2), 114.
306. Bell, L.E. (2003) Alternate thermoelectric thermodynamic cycles with improved power generation efficiencies, in *Thermoelectrics, 2003 22nd International Conference on - ICT, La Grande-Motte, France, August 17–21, 2003*, IEEE (Institute of Electrical and Electronics Engineers), Piscataway, NJ, pp. 558–562.
307. Chakraborty, A. (2005) Thermoelectric cooling devices - thermodynamic modelling and their application in adsorption. PhD thesis. Department of Mechanical Engineering - National University of Singapore, Singapore.
308. Chakraborty, A., Saha, B.B., Koyama, S., and Ng, K.C. (2006) Thermodynamic modelling of a solid state thermoelectric cooling device: temperature-entropy analysis. *Int. J. Heat Mass Transfer*, **49** (19-20), 3547–3554.
309. Chakraborty, A. and Ng, K.C. (2006) Thermodynamic formulation of temperature-entropy diagram for the transient operation of a pulsed thermoelectric cooler. *Int. J. Heat Mass Transfer*, **49** (11–12), 1845–1850.
310. Pramanick, A.K. and Das, P.K. (2006) Constructal design of a thermoelectric device. *Int. J. Heat Mass Transfer*, **49** (7–8), 1420–1429.
311. Xuan, X. and Li, D. (2006) Thermodynamic analysis of electrokinetic energy conversion. *J. Power Sources*, **156** (2), 677–684.
312. Chen, M., Rosendahl, L., Bach, I., Condra, T., and Pedersen, J. (2007) Irreversible transfer processes of thermoelectric generators. *Am. J. Phys.*, **75** (9), 815–820.
313. Fronczak, A., Fronczak, P., and Hołyst, J.A. (2007) Thermodynamic forces, flows, and Onsager coefficients in complex networks. *Phys. Rev. E*, **76** (6), 061106.
314. Izumida, Y. and Okuda, K. (2009) Onsager coefficients of a finite-time carnot cycle. *Phys. Rev. E*, **80** (2), 021121.
315. Goddard, J. (2011) On the thermoelectricity of W. Thomson: towards a theory of thermoelastic conductors. *J. Elast.*, **104** (1–2), 267–280, doi: 10.1007/s10659-011-9309-6.
316. Pearson, K. (1924) Historical note on the origin of the normal curve of errors. *Biometrika*, **16** (3/4), 402–404.
317. Le Cam, L. (1986) The central limit theorem around 1935. *Stat. Sci.*, **1** (1), 78–91.
318. Dutka, J. (1991) The early history of the factorial function. *Arch. Hist. Exact Sci.*, **43** (3), 225–249.
319. Huang, B.-L. and Kaviany, M. (2008) Ab initio and molecular dynamics predictions for electron and phonon transport in bismuth telluride. *Phys.*

Rev. B: Condens. Matter Mater. Phys., **77**, 125209 (19 pages).
320. Hinsche, N., Mertig, I., and Zahn, P. (2011) Effect of strain on the thermoelectric properties of silicon: an ab initio study. *J. Phys.: Condens. Matter*, **23**, 295502.
321. Pottier, N. (2007) *Physique statisitique hors équilibre, processus irréversibles linéaires*, Savoirs Actuels EDP Sciences/CNRS Editions, Paris.
322. Rocard, Y. (1967) *Thermodynamique*, 2nd edn, Masson, Paris.
323. Callen, H.B. and Greene, R.F. (1952) On a theorem of irreversible thermodynamics. *Phys. Rev.*, **86** (5), 702–710.
324. Callen, H.B. (1960) *Thermodynamics*, John Wiley & Sons, Inc.
325. Drude, P. (1900) Zur Elektronentheorie der Metalle. *Ann. Phys.*, **306** (3), 566–613.
326. Drude, P. (1900) Zur Ionentheorie der Metalle. *Phys. Z.*, **1**, 161–165.
327. Drude, P. (1902) Zur Elektronentheorie der Metalle. *Ann. Phys.*, **312** (3), 687–692.
328. Vining, C.B. (1997) The thermoelectric process, in *Materials Research Society Symposium Proceedings: Thermoelectric Materials - New Directions and Approaches* (eds T.M. Tritt, M.G. Kanatzidis Jr.,, H.B. Lyon, and G.D. Mahan), Materials Research Society, pp. 3–13.
329. Ball, C.A.B., Jesser, W.A., and Maddux, J.R. (1995) The distributed Peltier effect and its influence on cooling devices. Proceedings of the 14th International Conference on Thermoelectrics, pp. 305–309.
330. Buist, R.J. (1995) The extrinsic Thomson effect. Proceedings of the 14th International Conference on Thermoelectrics, A.F. Ioffe Physical-Technical Institute, St. Petersburg, pp. 301–304.
331. Belov, I.M., Volkov, V.P., and Manyakin, O. (1999) Optimization of Peltier thermocouple using distributed Peltier effect. 18th International Conference on Thermoelectrics, IEEE (Institute of Electrical and Electronics Engineers), Piscataway, NJ, pp. 316–318.
332. Landau, L.D. and Lifshitz, E.M. (1984) *Electrodynamics of Continuous Media*, 2nd edn, Butterworth Heinemann, Oxford.
333. Carslaw, H.S. and Jaeger, J.C. (1959) *Conduction of Heat in Solids*, 2nd edn, Oxford University Press.
334. Seifert, W., Ueltzen, M., and Müller, E. (2002) One-dimensional modelling of thermoelectric cooling. *Phys. Status Solidi A*, **1** (194), 277–290.
335. Seifert, W., Müller, E., and Walczak, S. (2006) Generalized analytic description of one-dimensional non-homogeneous TE cooler and generator elements based on the compatibility approach, in *25th International Conference on Thermoelectrics, Vienna, Austria, August 06–10, 2006* (ed. P. Rogl), IEEE, Piscataway, NJ, pp. 714–719.
336. Snyder, G.J., Fleurial, J.-P., Caillat, T., Yang, R., and Chen, G. (2002) Supercooling of Peltier cooler using a current pulse. *J. Appl. Phys.*, **92** (3), 1564–1569.
337. Snyder, G.J., Fleurial, J.-P., Caillat, T., Chen, G., and Yang, R.G. (2004) Current Pulses Momentarily Enhance Thermoelectric Cooling. Technical report, NPO-30553, NASA Tech Briefs.
338. Stil'bans, L.S. and Fedorovitch, N.A. (1958) The operation of refrigerating thermoelectric elements in nonstationary conditions. *Sov. Phys.-Tech. Phys.*, **3** (1), 460–463.
339. Parrott, J.E. (1960) The interpretation of the stationary and transient behaviour of refrigerating thermocouples. *Solid-State Electron.*, **1** (2), 135–143.
340. Reich, A.D. and Madigan, J.R. (1961) Transient response of a thermocouple circuit under steady currents. *J. Appl. Phys.*, **32** (2), 294–301.
341. Landecker, K. and Findlay, A.W. (1961) Study of the fast transient behaviour of Peltier junctions. *Solid-State Electron.*, **3** (3-4), 239–260.
342. Parrott, J.E. (1962) The stationary and transient characteristics of refrigerating thermocouples. *Adv. Energy Convers.*, **2**, 141–152.
343. Naer, V.A. (1965) Transient regimes of thermoelectric cooling and heating units. *J. Eng. Phys. Thermophys.*, **8** (4), 340–344.

344. Iordanishvili, E.K. and Malkovich, B.E.Sh. (1972) Experimental study of transient thermoelectric cooling. *J. Eng. Phys. Thermophys.*, **23** (3), 1158–1163.
345. Miner, A., Majumdar, A., and Ghoshal, U. (1999) Thermoelectromechanical refrigeration based on transient thermoelectric effects. *Appl. Phys. Lett.*, **75** (8), 1176–1178.
346. Thonhauser, T., Mahan, G.D., Zikatanov, L., and Roe, J. (2004) Improved supercooling in transient thermoelectrics. *Appl. Phys. Lett.*, **85** (15), 3247–3249.
347. Bechtold, T., Rudnyi, E.B., and Korvink, J.G. (2005) Dynamic electro-thermal simulation of microsystems - a review. *J. Micromech. Microeng.*, **15** (11), R17–R31.
348. Hussein, M., Hanim, S., and Zamri, Y.M. (2007) The transient response for different types of erodible surface thermocouples using finite element analysis. *Therm. Sci.*, **11** (4), 49–64.
349. Zhou, Q., Bian, Z., and Shakouri, A. (2007) Pulsed cooling of inhomogeneous thermoelectric materials. *J. Phys. D: Appl. Phys.*, **40** (14), 4376–4381.
350. Ursell, T.S. and Snyder, G.J. (2002) Compatibility of segmented thermoelectric generators. 21st International Conference on Thermoelectrics, Piscataway, NJ, September 25–29, 2002, IEEE (Institute of Electrical and Electronics Engineers), pp. 412–417.
351. Snyder, G.J. and Ursell, T.S. (2003) Thermoelectric efficiency and compatibility. *Phys. Rev. Lett.*, **91** (14), 148301.
352. Kedem, O. and Caplan, S.R. (1965) Degree of coupling and its relation to efficiency of energy conversion. *Trans. Faraday Soc.*, **61**, 1897–1911.
353. Apertet, Y., Ouerdane, H., Goupil, C., and Lecœur, Ph. (2012) Internal convection in thermoelectric generator models. *J. Phys. Conf. Ser.*, **395**, 012103–(8 pages). 6th European Thermal Sciences Conference (Eurotherm 2012).
354. Goupil, C., Seifert, W., Zabrocki, K., Müller, E., and Snyder, G.J. (2011) Thermodynamics of thermoelectric phenomena and applications. *Entropy*, **13** (8), 1481–1517.
355. Buda, I.S., Lutsyak, V.S., Khamets, U.M., and Shcherbina, L.A. (1991) Thermodynamic definition of the thermoelectric figure of merit of an anisotropic medium. *Phys. Status Solidi A*, **123** (2), K139–K143.
356. Seifert, W., Zabrocki, K., Snyder, G.J., and Müller, E. (2010) The compatibility approach in the classical theory of thermoelectricity seen from the perspective of variational calculus. *Phys. Status Solidi A*, **207** (3), 760–765.
357. Schilz, J., Müller, E., Helmers, L., Kang, Y.S., Noda, Y., and Niino, M. (1999) On the composition function of graded thermoelectric materials. *Mater. Sci. Forum*, **308–311**, 647–652.
358. Müller, E., Drašar, v.C., Schilz, J., and Kaysser, W.A. (2003) Functionally graded materials for sensor and energy applications. *Mater. Sci. Eng., A*, **362** (1-2), 17–39. Papers from the German Priority Programme (Functionally Graded Materials).
359. Seifert, W., Zabrocki, K., Müller, E., and Snyder, G.J. (2010) Power-related compatibility and maximum electrical power output of a thermogenerator. *Phys. Status Solidi A*, **207** (10), 2399–2406.
360. Goupil, C. (2009) Thermodynamics of the thermoelectric potential. *J. Appl. Phys.*, **106**, 104907.
361. Sherman, B., Heikes, R.R., and Ure, R.W. Jr. (1960) Calculation of efficiency of thermoelectric devices, in *Thermoelectric Materials and Devices*, Materials Technology Series, Chapter 15 (eds I.B. Cadoff and E. Miller), Reinhold Publishing Cooperation, New York, pp. 199–226.

2
Continuum Theory of TE Elements

Knud Zabrocki, Christophe Goupil, Henni Ouerdane, Yann Apertet, Wolfgang Seifert[†], and Eckhard Müller

2.1
Domenicali's Heat Balance Equation

2.1.1
Tensorial Character of Material Properties

In general, thermoelectric material properties can be expressed mathematically as tensors displaying their anisotropic character due to a noncubic crystal lattice structure or other orientation-dependent effects.

As $\dot{\mathbf{q}}$ and ∇T are vectors, it is totally clear that the thermal conductivity $\hat{\kappa}$ in the general form as written in Eq. (2.1) is a tensor of second rank

$$\dot{\mathbf{q}}(\mathbf{r}, t) = -\hat{\kappa}_{\mathbf{j}} \nabla T(\mathbf{r}, t). \tag{2.1}$$

So, in Cartesian coordinates, the thermal conductivity tensor can be shown as a matrix of its components

$$\hat{\kappa} = \begin{pmatrix} \kappa_{xx} & \kappa_{xy} & \kappa_{xz} \\ \kappa_{yx} & \kappa_{yy} & \kappa_{yz} \\ \kappa_{zx} & \kappa_{zy} & \kappa_{zz} \end{pmatrix}.$$

In Reference [1], Miller provided some statements on the tensor calculus in the framework of thermodynamics of irreversible processes. It is totally clear that the entries of the material coefficients' tensors depend on the choice of the axes. The tensors may take simpler forms, if the axes coincide with the crystal axes, insofar as this is possible. By using the corresponding symmetries, our coordinate system can be rotated. For a fourfold axis of symmetry (C_4) with the z axis as crystallographic principal axis, a tensor for the thermal conductivity is obtained with the following form:

$$\hat{\kappa} = \begin{pmatrix} \kappa_{xx} & \kappa_{xy} & 0 \\ -\kappa_{xy} & \kappa_{xx} & 0 \\ 0 & 0 & \kappa_{zz} \end{pmatrix}. \tag{2.2}$$

Continuum Theory and Modelling of Thermoelectric Elements, First Edition. Edited by Christophe Goupil.
© 2016 Wiley-VCH Verlag GmbH & Co. KGaA. Published 2016 by Wiley-VCH Verlag GmbH & Co. KGaA.

This result can be found in Ref. [1], and the fact that all the orthorhombic (C_{2v}, D_2, D_{2h}), all the cubic (T, T_d, T_h, O, O_h) and certain triclinic (D_3, C_{3v}, D_{3d}), hexagonal ($D_6, C_{6v}, D_{3h}, D_{6h}$), and tetragonal ($D_4, C_{4v}, D_{2d}, D_{4h}$) crystal classes satisfy

$$\kappa_{ik} = \kappa_{ki},$$

for *all* i and k by *geometrical considerations alone*. The remaining classes have to be tested whether the Onsager reciprocal relations hold or not. Such classes are the triclinic (C_1, C_i), monoclinic (C_2, C_3, C_{2h}), and trigonal (C_3, C_{3i}), tetragonal (C_4, S_4, C_{4h}), and hexagonal (C_6, C_{3h}, C_{6h}) classes. The tensor in Eq. (2.2) corresponds to the simplest nontrivial cases concerning trigonal, tetragonal, and hexagonal systems. Obviously, it can be symmetric only if $\kappa_{xy} \equiv 0$, which has to be shown by the experiments, for details see [1].

In Reference [2], it is stated that there is an important difference between transport tensors and thermoelectric tensors. The Onsager's principle is valid only for electrical and thermal conductivity, that is, the corresponding transport tensors are supposed to be symmetric. This is not the case for the Seebeck and Peltier coefficients as they correlate two different flows. From this fact, it follows that in the most general case (triclinic), $\alpha_{mn} \neq \alpha_{nm}$ as well as $\Pi_{mn} \neq \Pi_{nm}$, where nine coefficients instead of six have to be determined.

In Table 2.1, the thermoelectric coefficients for other symmetry groups are given. Neumann's principle states that *"the symmetry of any physical property of a crystal must include the symmetry elements of the point group of the crystal,"* [2, p. 34].

Although in this work the focus lies on the isotropic case, where all material properties are scalars instead of tensors, the heat balance can be easily generalized for the anisotropic case, which is shown in the following subsection. For more details on anisotropic thermoelements and their physical background, see [3–10] and references therein.

2.1.2
Heat Balance and Source Terms

The properties can depend on both temperature T and spatial coordinates x_m with ($m = 1, 2, 3$), for example, Cartesian coordinates x, y, z. For the Seebeck coefficient, this means, for example, $\alpha = \alpha_{mn}(x_m, T)$. This dependence gives two distinct parts of the spatial derivative of the material properties, for example, for the Seebeck coefficient

$$\frac{\partial \alpha_{mn}}{\partial x_m} = \left(\frac{\partial \alpha_{mn}}{\partial x_m}\right)_T + \left(\frac{\partial \alpha_{mn}}{\partial T}\right)_{x_m} \frac{\partial T}{\partial x_m}, \qquad (2.3)$$

where α_{mn} denote the components of the Seebeck coefficient tensor $\hat{\alpha}$.

The general expression for the production of heat in an inhomogeneous, anisotropic medium, crystalline or otherwise (steady state), is given by

Table 2.1 Seebeck and Peltier coefficients for the 32 crystallographic point groups and the 7 Curie groups for textured solids (Table 21.4 in Ref. [2]).

Point groups and Tensor terms

Point groups	Tensor
$1, \bar{1}$	$\begin{pmatrix} \alpha_{11} & \alpha_{12} & \alpha_{13} \\ \alpha_{21} & \alpha_{22} & \alpha_{23} \\ \alpha_{31} & \alpha_{32} & \alpha_{33} \end{pmatrix}$
$2, m, 2/m$	$\begin{pmatrix} \alpha_{11} & 0 & \alpha_{13} \\ 0 & \alpha_{22} & 0 \\ \alpha_{31} & 0 & \alpha_{33} \end{pmatrix}$
$222, mm2, mmm$	$\begin{pmatrix} \alpha_{11} & 0 & 0 \\ 0 & \alpha_{22} & 0 \\ 0 & 0 & \alpha_{33} \end{pmatrix}$
$3, \bar{3}, 4, \bar{4}, 4/m, 6, \bar{6}, 6/m, \infty, \infty/m$	$\begin{pmatrix} \alpha_{11} & \alpha_{12} & 0 \\ -\alpha_{12} & \alpha_{11} & 0 \\ 0 & 0 & \alpha_{33} \end{pmatrix}$
$32, 3m, \bar{3}m, 422, 4mm, \bar{4}2m, 4/mmm, 622, 6mm, \bar{6}m2, 6/mmm, \infty m, \infty 2, \infty/mm$	$\begin{pmatrix} \alpha_{11} & 0 & 0 \\ 0 & \alpha_{11} & 0 \\ 0 & 0 & \alpha_{33} \end{pmatrix}$
$23, m3, 432, \bar{4}3m, m3m, \infty\infty m, \infty\infty$	$\begin{pmatrix} \alpha_{11} & 0 & 0 \\ 0 & \alpha_{11} & 0 \\ 0 & 0 & \alpha_{11} \end{pmatrix}$

Domenicali's thermal energy balance equation [3, 11, 12]

$$\frac{\partial}{\partial x_m}\left(\kappa_{mn}\frac{\partial T}{\partial x_n}\right) = -\varrho_{mn} j_m j_n + T\frac{\partial(\alpha_{mn} j_n)}{\partial x_m}, \tag{2.4}$$

where κ_{mn} and ϱ_{mn} are analogously the components of the tensor of the thermal conductivity $\hat{\kappa}$ and the electrical resistivity $\hat{\varrho}$. Note that Eq. (2.4) is an analogous version of the equations discussed in Section 1.8 in tensor description.

Since $\alpha_{mn} = \alpha_{mn}(T, x_m)$, the last term on the right-hand side denotes three contributions. Furthermore, the Fourier contribution on the left-hand side v_q^F balances with the Joule source term v_q^J, the Peltier source term v_q^P, the Thomson source term v_q^T, and the Bridgman source term v_q^B, that is, Eq. (2.4) can be written in short $v_q^F = v_q^J + v_1^P + v_1^T + v_q^B$. The different volumetric source density contributions are listed in Table 2.2. The split up of the so-called Peltier–Thomson term due to the chain rule corresponds to different physical heat source densities, the (distributed) *Peltier* heat v_q^P and the *Thomson* heat v_q^T, as it has been already discussed in Section 1.8. Buist called these terms the *extrinsic* and the *intrinsic* Thomson effect, respectively, see [13]. In the 3D case, there occurs a third term if the electrical current density depends on the position, representing the *Bridgman* heat [3, 14] and references in Section 1.1.8. Silk and Schofield pointed out that "the thermoelectric anisotropy has important consequences in the form

Table 2.2 Volumetric source density terms.

Fourier v_q^F	Joule v_q^J	Peltier v_q^P	Thomson v_q^T	Bridgman v_q^B
$\partial_m(\kappa_{mn}\partial_n T)$	$-\varrho_{mn}j_m j_n$	$Tj_n(\partial_m \alpha_{mn})_T$	$Tj_n[\alpha'_{mn}(T)]_{x_m}\partial_m T$	$T\alpha_{mn}\partial_m j_n$

The Abbreviation $\frac{\partial}{\partial x_m} \equiv \partial_m$ is used and the prime denotes the derivative with respect to T.

of thermoelectric eddy currents and the Bridgman effect," see [8]. The special role of eddy currents and its prospects to find new kinds of thermoelements is discussed in Ref. [10]. Based on a generalized Maxwell theory combining Maxwell's equation of electromagnetic induction with thermoelectric effects, an inverse problem of thermoelectricity, where the temperature distribution and thermoelement configuration are sought from given material parameters and current configurations, which result in the induction of thermoelectric currents, is introduced and solved in this work. Only if the Seebeck coefficient is anisotropic, the Bridgman effect can be observed. It describes the compensation (heating/cooling) due to current direction changes in a material with anisotropic Seebeck coefficient. This is what distinguishes it from the Peltier and Thomson effects, which appear in isotropic materials as well. Note that the Bridgman term reduces to $-T\alpha \partial j_m/\partial x_m$ for an isotropic Seebeck coefficient $\alpha_{mn} = \alpha\delta_{mn}$ and that this term vanishes in the steady state. As it is easily verified in the 1D case, you can't find the heat due to the steady-state condition $\nabla \cdot j = 0$ is not detected, which leads to $\partial j(x)/\partial x = 0$, that is, $j = $ const. In an anisotropic material, different directions are linked to different amounts of flowing Peltier heat.

A detailed discussion of this effect and some examples are given in Domenicali's review article [3] and in Ref. [8].

For isotropic (chemically homogeneous) material and steady-state condition, the thermal energy balance is in vector notation

$$\nabla \cdot (-\kappa \nabla T) = \varrho j^2 - \tau \mathbf{j} \cdot \nabla T, \qquad (2.5)$$

with the Joule heat per volume ϱj^2 and Thomson coefficient $\tau = T\frac{d\alpha}{dT}$.

Let us consider an infinitesimal section of a thermoelectric leg in a temperature gradient and an electric field. The temperature gradient will induce a heat flux ($\kappa \nabla T$) across this segment, according to Fourier's law. The divergence of this heat is equal to the source terms: irreversible Joule heating (ϱj^2) and reversible Thomson heat ($T\frac{d\alpha}{dT}\mathbf{j} \cdot \nabla T$) both of which depend on the electric current density.

In thermoelectric systems, Eq. (2.5) is typically discussed in one dimension by assuming that the heat flux and electrical current are in parallel [15]. Remember that the material properties α, σ, and κ are in general temperature- and position-dependent. Here we want to concentrate on decoupled dependencies, that is, either a temperature **or** spatial dependence of the material coefficients, to gain analytical results for the performance parameters of a TE element. A numerical investigation of a coupled material's local as well as temperature dependence is performed by Kaliazin et al. [16].

Within the framework of a 1D model [17–19] and with temperature-dependent material properties, Eq. (2.5) reduces in the steady state to

$$\kappa(T)\frac{\partial^2 T}{\partial x^2} + \frac{d\kappa}{dT}\left(\frac{\partial T}{\partial x}\right)^2 - jT\frac{d\alpha}{dT}\frac{\partial T}{\partial x} = -\frac{j^2}{\sigma(T)}, \qquad (2.6)$$

where a constant current density j is supposed to flow in x-direction. Equation (2.6) is a nonlinear differential equation in T (with nonconstant coefficients), which has to be supplemented by boundary conditions, see, for example, [18].

It is important to note that if there is a "one-to-one correspondence" of temperature T and position x, the temperature profile $T(x)$ is a continuous and strictly monotonous one and thus a bijective function. This especially applies to the maximum efficiency and maximum power output of a thermogenerator (TEG) and maximum coefficient of performance (COP) of a Peltier cooler (TEC), and all configurations with a current lower than the optimal current if constant or real, temperature-dependent, material properties are considered.[1] Then, there exists an inverse function $x(T)$ corresponding to $T(x)$ and vice versa, and Eq. (2.6) can be transformed into

$$\frac{d}{dx}\left(-\kappa(x)\frac{dT(x)}{dx}\right) + jT(x)\frac{d\alpha(x)}{dx} = \frac{j^2}{\sigma(x)} \qquad (2.7)$$

with $\sigma(x) = 1/\varrho(x)$. Identical temperature profiles $T(x)$ are calculated from Eqs. (2.6) and (2.7) if spatial material profiles over the length of the TE element are given by $\alpha(x) = \alpha[T(x)]$, $\sigma(x) = \sigma[T(x)]$, and $\kappa(x) = \kappa[T(x)]$.

2.1.3
Spatial and Temperature Averaging of the Material Properties

By taking the Seebeck coefficient as an example, the interrelation between the temperature dependence of the material parameters and their spatial dependence is given by

$$\bar{\alpha} = \frac{1}{T_s - T_a}\int_{T_a}^{T_s}\alpha(T)dT = \frac{1}{L}\int_0^L \alpha(x)\frac{T'(x)}{(T_s - T_a)/L}dx. \qquad (2.8)$$

As the function $LT'(x)/(T_s - T_a)$ varies around unity, both the temperature average and the spatial average turn out to be close for moderate gradients, where the spatial average shall be defined as $\alpha_{av} = 1/L\int_0^L \alpha(x)dx$ with $\bar{\alpha} \approx \alpha_{av}$ where the index "av" denotes the spatial average. Note that $\bar{\alpha} \approx \alpha_{av}$ holds exactly for linear temperature profiles and in the CPM case, where the material properties are supposed to be constant and thus independent of the temperature and the position in the TE element.

---

1) Within the constant properties model (CPM), we find for the slope of the temperature profile at the sink side $T'(L) = 0$ for the maximum temperature difference ($\varphi = 0$), and $T'(L) > 0$ for $\varphi > 0$. Note that the CPM is a suitable reference for moderately temperature-dependent material properties.

Naturally, Eq. (2.7) can also be used as an independent differential equation when the spatial dependence of the material parameters $\alpha(x), \sigma(x)$, and $\kappa(x)$ is investigated. Such investigations where the coefficients explicitly depend on the position but are independent of T had been performed first by Ybarrondo [20, 21].

Let us emphasize that, in contrast to Eq. (2.6), the differential equation (2.7) with purely spatially dependent material profiles is linear in T. This fact enables us to search for analytical solutions based on the principle of superposition, see, for example, [22].

For arbitrary continuous monotonic spatial functions of the material profiles $\alpha(x), \sigma(x)$, and $\kappa(x)$, the calculation of the temperature profile can be performed numerically by a 1D finite element method code (1D TE FEM) or the algorithm of multisegmented elements as well as other approaches, see for example, [16, 19, 22–36]. The basis of the algorithm of multisegmented elements is shown in Section 3.1.2. In Reference [37], Anatychuk shows interrelations between the material properties in case of functionally graded materials (FGM). An easy approach to investigate FGM beyond the classical CPM is using the linear spatial profiles of the material properties [22, 38]. Although such profiles do not display the behavior of a real material in all its depth, it is possible to gain information about the general trends.

2.2
Transferred Heat Balance

Since the efficiency of a TEG and the COP of a TEC are derived from thermal and electrical quantities (see next section), it is not surprising that the authors have used combined terms of TE quantities to describe TE systems efficiently.

Clingman used in Ref. [39] a "dimensionless heat flux" c for performance optimization of TE devices; he defined c by the ratio of (total) heat flow $\dot{Q}$ to Peltier heat flow $I\alpha T$. For more information, see [40] and Section 5.8 in this book.

Nearly at the same time, new concepts were published that use reduced variables (ratios) to transcribe the constitutive relations of thermoelectric coupling and the thermal energy balance. Kedem and Caplan [41] described two coupled flows by their "degree of coupling," an approach that reduces both the thermoelectric flows and forces to dimensionless numbers. Burshtein pointed out in 1957 [42] (as also later, similarly, Miozhes [43] and Sherman et al. [23]) that the 1D thermal energy balance (2.7) can be transformed by means of a relative Fourier heat flux y:

$$y(T) = -\frac{\kappa(T)}{jx'(T)} \Rightarrow \frac{d}{dT}y(T) = -T\frac{d\alpha(T)}{dT} - \frac{\varrho(T)\kappa(T)}{y(T)}. \tag{2.9}$$

The usage of y simplifies the TE potential, see Eq. (1.92), but becomes ill-defined for an open-circuit generator, when $j = 0$.

A more detailed analysis was performed by Snyder and coworkers, who showed that the compatibility approach (see Chapter 5) to functionally grading thermoelectric materials for power generation yields a single device that achieves the performance expected from a staged generator with an infinite number of stages without losses from the interfaces [15, 44, 45]. They introduced the relative current density u as the ratio of electric and thermal fluxes, see also Eq. (1.90),

$$u = \frac{-j^2}{\kappa \nabla T \cdot \mathbf{j}}.$$

This approach can be used to analyze the local performance of a TE material and its optimization, see Section 1.10.2 and Figure 1.17. A detailed discussion is given in Chapter 5, especially in Section 5.8.

With $u(T) = 1/y(T)$, Eq. (2.9), see [45], translates into

$$\frac{d}{dT}\left(\frac{1}{u}\right) = -T\frac{d\alpha}{dT} - u\rho\kappa \quad \text{or, alternatively,} \quad \frac{du}{dT} = u^2\left(\tau + \frac{\alpha^2}{z}u\right). \quad (2.10)$$

For the derivation of Eq. (2.10), see also Section 6.5.1.

An alternative method was proposed by Drabkin and Ershova [46], who suggested to transform the thermal energy balance into a system of first-order differential equations for the temperature profile $T(x)$ and the relative heat flux difference $\frac{1}{j}[\alpha j T(x) + \kappa T'(x)] = \frac{1}{j}(q_\pi - q_\kappa)$ leading to a Hamilton-like optimization procedure (Pontryagin's maximum method [47]).

Before we go on with the optimization process of TE devices and elements in this chapter, let us first have a look at a descriptive method that is often used in the corresponding calculation, the so-called constant properties model (CPM).

2.3
Ioffe's Description and Performance Parameters of CPM Devices

The CPM is a widespread and adequate reference for comparing both thermoelectric materials and devices. It has been introduced by A. Ioffe [48, 49] and is therefore sometimes denoted as Ioffe approximation or Ioffe CPM approximation.

The aim of the CPM is an analytical calculation that gives the temperature profile $T(x)$ as a result, which is the basis for the determination of all performance and device parameters. The CPM can be used to compare and value other methods concerning their quality and quantity. For TE devices being not necessarily homogeneous, the *device* or *extrinsic* figure of merit introduced by Goldsmid [50]

$$Z = \frac{S^2}{RK}, \quad (2.11)$$

is observed, where S is the device effective Seebeck coefficient, R is the isothermal (and internal) Ohmic resistance, and K is the thermal conductance of the device or thermocouple [50]. Z represents a global value of the TE device, whereas the intrinsic figure of merit z was introduced in Chapter 1 and covers the local properties of the material. For a single "ideal CPM element" with constant cross-sectional area A_c, there is a relation between the local and global values

$$R = \frac{L}{A_c \sigma} = \frac{L}{A_c}\rho \quad \text{and} \quad K = \frac{A_c}{L}\kappa, \quad (2.12)$$

where L is the length of the element. From Eq. (2.12), we find that $RK = \kappa\varrho$, such that for the single element under ideal conditions, that is, parallel current and heat flux, no contact resistances, the extrinsic and intrinsic figure of merit coincide. It is noteworthy that in general for efficiency calculations, neither the resistivity ϱ nor the thermal conductivity κ has an isolated specific influence, but both always appear together as a product, see for example, [36, 51]. Specific separate effects are discussed in later sections of this chapter.

As the simplest unit of a TE device, usually a TE couple consisting of two TE elements is discussed, one being a p-type semiconductor and the other a n-type semiconductor element. Of course, every element of this p–n couple has an individual intrinsic figure of merit, that is, $z_p = Z_p$ and $z_n = Z_p$, but it is totally clear that in general there is no simple relation between the device Z and Z_p, Z_n, as pointed out in Refs [50, 52]. A principle handwaving rule, however, is that the larger the figures of merit of the individual elements are, the larger is the Z of a couple. For a couple with two distinct elements, which have naturally different material properties, that is, α_p, α_n, ϱ_p, ϱ_n, (or σ_p, σ_n), κ_p, and κ_n, these exhibit different optimal working conditions, that is, optimal current densities. Because a unique current flows through the couple, geometric optimization of the elements has to be performed to achieve maximal performance.

The device Z is equal to the material's figure of merit z, not only for an "ideal CPM element" but also within the symmetric CPM model where the n-leg and p-leg of a couple have identical properties with opposite sign of the Seebeck coefficient.

2.3.1
Single-Element Device

Clearly distinguishing from the design optimization of real TE devices, we will restrict initially to an "ideal" thermogenerator or Peltier single-element device. Such an *ideal CPM element* is supposed to exhibit no thermal losses due to radiation or thermal bypasses, no thermal or electric contact resistance, and only one-dimensional flow is assumed, that is, electrical current and heat flux are parallel, see also [53]. Every nonparallel arrangement without a magnetic field, where there is an arbitrary angle between the temperature gradient and the electrical current, leads to a reduction of the performance, which is deduced from a generalized figure of merit shown by Gryaznov et al. [54]. For the sake of simplicity, a single but representative (segmented or continuously graded) TE generator or cooler element (p-type or n-type, of element length L and constant cross-sectional area A_c) is often considered as part of a TE device or as a single-element device instead of addressing an entire module or device.[2] Thus, further considerations are based on fixed parameters L and fixed boundary temperatures T_a and T_s within

---

2) In a 1D approach, the segmentation or grading is clearly in the same direction as the electrical current and heat flux are. In a quasi-1D or in a 2D/3D approach, the direction of the electrical current and the heat flow in comparison to the grading direction have to be taken into account, see [55–59].

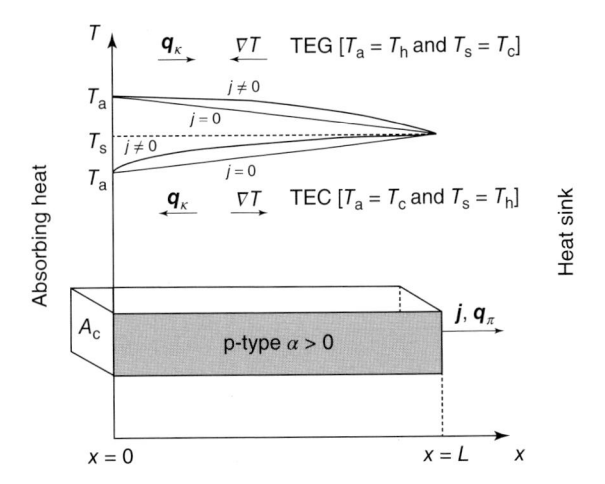

Figure 2.1 Unified 1D model of a thermoelectric element: lower temperature profile: cooling operation (= TEC with $T_a < T_s$); upper temperature profile: thermoelectric generator (= TEG with $T_a > T_s$). Bowing of the profile is indicated qualitatively. Typically, the bowing in the TEG case is weak due to relatively lower current in efficient operation, compared to the TEC. The arrows for ∇T and q_κ indicate the actual gradient and flow direction of positive magnitude. Analytically all vectors are counted positive in x-direction, whereas the magnitude of the flow vectors with left-headed arrows adopts negative numerical values in the 1D formulae.

the framework of a unified 1D model for both TEG and TEC (see Figure 2.1 and [31, 36]), where T_a is the temperature at the heat absorbing side (hot side T_h for TEG, but cold side T_c for TEC), and T_s denotes the heat sink temperature, which is in many cases fixed not far from room temperature. Alternatively, mixed boundary conditions such as fixing the temperature of the sink side T_s and the heat flux at the absorbing side q_a can be used in the corresponding optimization processes.

The total heat flux and its components in Figure 2.1 are indicated by the symbols $\mathbf{q} = -\kappa\nabla T + \alpha T \mathbf{j} =: \mathbf{q}_\kappa + \mathbf{q}_\pi$ with Fourier and Peltier heat fluxes $\mathbf{q}_\kappa$ and $\mathbf{q}_\pi$, respectively (see also [19]).[3] Note that all flows are counted positive according to right-headed arrows. Further note that $T(x)$ peaks in the interior of the TE element only above the optimal current for maximum efficiency and maximum COP, respectively.

All calculations in this chapter refer to an ideal CPM element with its 1D arrangement. Other shapes of elements need more detailed consideration, for example, circular shape [60–65], where the formulae for the power output may change while the efficiency remains shape-independent. Details on this can be found in Section 2.8.

---

3) Take care of the terms and the notation used. Normally, Q denotes the **heat** in units of 1 J, whereas with $\dot{Q} = \frac{\partial Q}{\partial t}$, the **heat transfer rate, heat flow,** or **thermal power** in units of 1 W is denoted. For the sake of simplicity, the dot is sometimes omitted, as we do here. The **heat flux** is often used, which is the heat transfer rate per cross-sectional area $q = \frac{1}{A_c}\dot{Q} = \frac{1}{A_c}\frac{\partial Q}{\partial t}$ in units of 1 W/m².

2.3.2
Performance Parameters of a Thermoelectric Element with Constant Material Properties

Within CPM, the material properties are supposed to be constant in both ways, chemically from the spatial sense and physically due to the temperature, which results in a quite simple thermal energy equation

$$\frac{d^2 T}{dx^2} = -\frac{j^2}{\sigma \kappa} = \text{const.} = -c_o, \qquad (2.13)$$

where the abbreviation $c_o = j^2/(\sigma \kappa)$ is used and which balances only Fourier and Joule heat. It can be solved analytically by integration for both Dirichlet conditions (boundary temperature) and mixed boundary conditions (given heat flux and temperature at the boundaries) resulting in a parabolic temperature profile, that is, for the Dirichlet boundary conditions:

$$T(x) = T_a + (T_s - T_a)\frac{x}{L} + \frac{j^2}{2\sigma \kappa} x(L-x) \quad \text{with} \quad T_a = T(x=0),\ T_s = T(x=L). \qquad (2.14)$$

For an open circuit, that is, $j = 0$, a linear temperature profile results between the two heat reservoirs at T_s and T_a. The curvature of the temperature profile is due to resistive heating (Joule heat). Once having calculated the temperature profile $T(x)$, we can determine all performance parameters as a function of the electrical current density j, for example,

- TEG: net power output density[4] (electrical power output per cross-sectional area A_c):

$$p_\text{net}(j) = -p(j) = -\frac{P_\text{el}(j)}{A_c}, \qquad (2.15a)$$

where the electrical power output is

$$P_\text{el} = \iiint_V \mathbf{j} \cdot \mathbf{E}\, dV, \qquad (2.15b)$$

with the assumption of a constant A_c and considerations from Section 1.8.4, it follows

$$P_\text{el} = A_c \int_0^L \left(\frac{j^2}{\sigma} + j\alpha T'(x)\right) dx, \qquad (2.15c)$$

where P_el is the electrical power output, p is the corresponding density.
- TEC: cooling power density (absorbed heat per time and cross-sectional area):

$$q_a(j) = \frac{\dot{Q}_a(j)}{A_c} = [-\kappa(x)T'(x) + j\alpha(x)T(x)]_{x=0}, \qquad (2.16)$$

where $\dot{Q}_a$ is the cooling power (absorbed heat per time unit).

---
[4] Power output is defined here according to thermodynamic rules: quantities put into the system are positive.

- TEG: efficiency

$$\eta(j) = \frac{P_{\text{net}}(j)}{q_a(j)}, \tag{2.17}$$

- TEC: coefficient of performance

$$\varphi(j) = \frac{\dot{Q}_a(j)}{P_{\text{el}}(j)} = \frac{q_a(j)}{p(j)}. \tag{2.18}$$

For more information, we refer to [36].

Within CPM, we find the temperature gradients at both element sides from Eqs. (2.13) and (2.14):

$$\left(\frac{dT}{dx}\right)_{x=0} = \frac{T_s - T_a}{L} + \frac{c_o}{2}L, \quad \left(\frac{dT}{dx}\right)_{x=L} = \frac{T_s - T_a}{L} - \frac{c_o}{2}L, \tag{2.19}$$

with c_o as defined in Eq. (2.13). Note that $j^2 \rho L = \kappa c_o L$ represents the density of Joule heat because we have within the framework of CPM: $I^2 R/A_c = j^2 R A_c = j^2 L/\sigma = \kappa c_o L$ with the isothermal (and internal) Ohmic resistance $R = L/(\sigma A_c)$. Since we have $\mathbf{q} = \mathbf{q}_\kappa + \mathbf{q}_\pi$, the conductive heat fluxes at the boundaries can be written as

$$\text{absorbed:} \quad q(0) - \alpha T_a j = -\kappa \frac{T_s - T_a}{L} - j^2 \frac{L}{2\sigma}, \tag{2.20a}$$

$$\text{flowing out:} \quad q(L) - \alpha T_s j = -\kappa \frac{T_s - T_a}{L} + j^2 \frac{L}{2\sigma}. \tag{2.20b}$$

Obviously, the mean temperature gradient $\Delta T/L$ is overlaid by Joule heat, which is symmetrically distributed[5] over the length of the TE element, see also Figure 2.2 where a single-element device is shown, and Figure 1.15 in Chapter 1, where a couple of one p-type and one n-type element is illustrated.

Written with global values ($I = j A_c$, $K = \kappa A_c/L$ etc.), we find from Eq. (2.20a) for the thermal power input $\dot{Q}_a = q(0) A_c$ at the absorbing side $x = 0$ (index a)

$$\dot{Q}_a = \alpha T_a I - \frac{1}{2} I^2 R - K(T_s - T_a). \tag{2.21}$$

A description using global variables has been firstly introduced by Altenkirch [66] and often used in technological applications, see for example, [49, 67].

In the next subsection, we sum up the expressions for the three configurations TEG, TEC, and TEH within the framework of the Ioffe CPM approximation, see Figure 2.2. For the sake of simplicity, we use the load ratio $m = R_L/R$ for TEG. Then, as we will see, the expressions for the maximal power and maximal efficiencies lead to different values of the m parameter in the following summary of formulae, that is, $m_{\text{opt},P} = M_P$ and $m_{\text{opt},\eta} = M_\eta$, respectively, which can also be found in Ref. [40].

---

5) The Joule effect induced by the circulation of the electrical current has been often considered to be spread over the hot and cold sources in equal contributions $1/2 R I^2$. Obviously, this is valid in general only under the CPM condition, see Eqs. (2.20).

86 | *2 Continuum Theory of TE Elements*

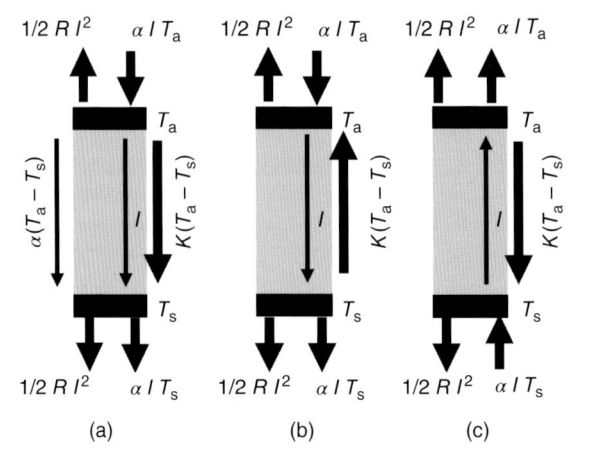

Figure 2.2 Single element device consisting of one thermoelectric leg. The three working modes are: a) TEG, b) Cooler, c) Heater.

2.3.2.1 Thermoelectric Generator (TEG)

We are using here $T_h = T_a$, $T_c = T_s$ and $\dot{Q}_{in} = \dot{Q}_a$, $\dot{Q}_{out} = \dot{Q}_s$, respectively (Figure 2.3).

- Incoming thermal power:

$$\dot{Q}_{in} = \alpha T_h I - \frac{1}{2} R I^2 + K(T_h - T_c).$$

- Outgoing thermal power:

$$\dot{Q}_{out} = \alpha T_c I + \frac{1}{2} R I^2 + K(T_h - T_c).$$

- Electrical power produced:

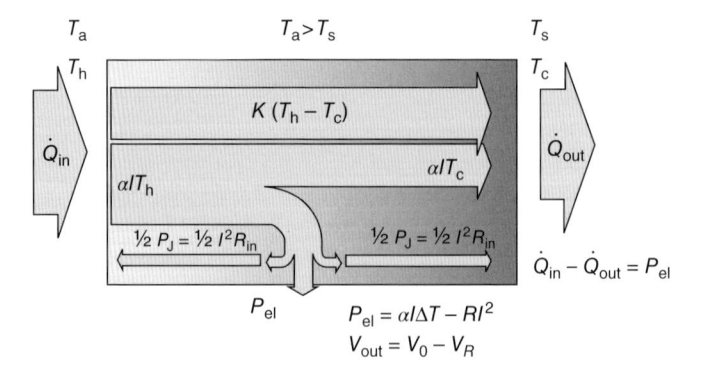

$$P_{el} = \dot{Q}_{in} - \dot{Q}_{out} = \alpha I (T_h - T_c) - R I^2.$$

Figure 2.3 Contributions in a TEG (CPM).

2.3 Ioffe's Description and Performance Parameters of CPM Devices

- Open voltage:

$$V_0 = \alpha(T_h - T_c)$$

By considering a resistor of resistance R_L connected to the TEG, we now define the load ratio $m = R_L/R$. Then we obtain the expressions of the output voltage and current

$$V_{out} = V_0 \frac{R_L}{R + R_L} = V_0 \frac{m}{1+m}, \quad (2.22a)$$

$$I_{out} = \frac{V_0}{R + R_L} = \frac{V_0}{R(1+m)}, \quad (2.22b)$$

and

$$P_{el} = \frac{V_0^2}{R} \frac{m}{(m+1)^2}. \quad (2.22c)$$

In the following figures 2.4, the output parameters, see Eqs. (2.22), dependent on the load ratio m are illustrated.

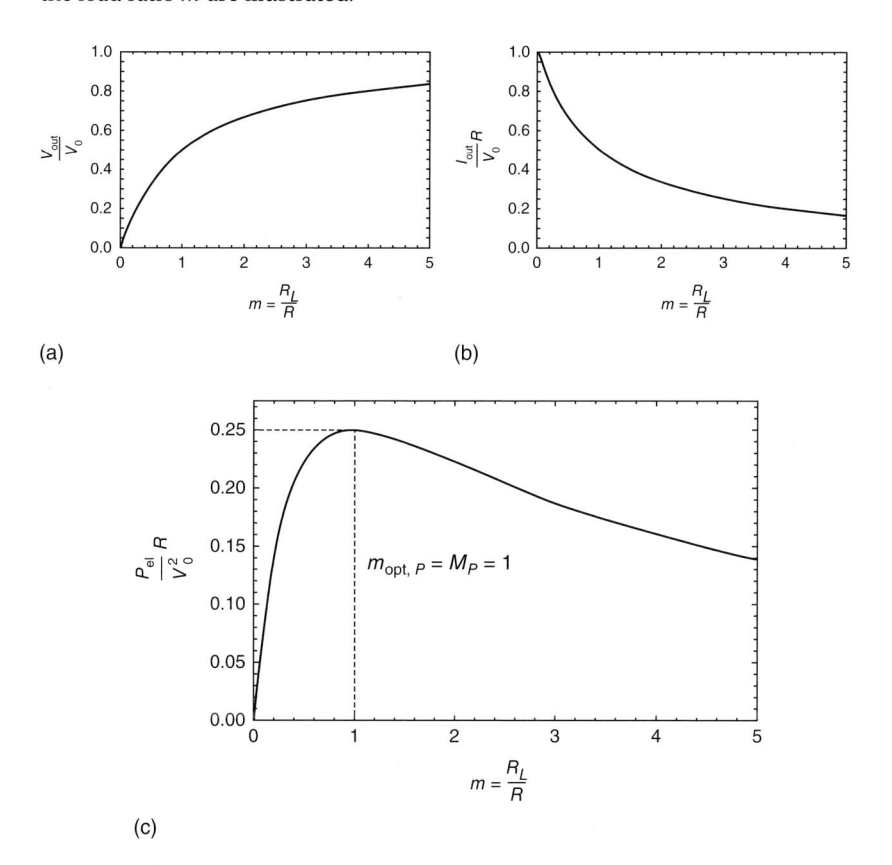

Figure 2.4 Performance parameters of a TEG in dependence on the load ratio m. (a) Output voltage; (b) Output current; (c) Output power.

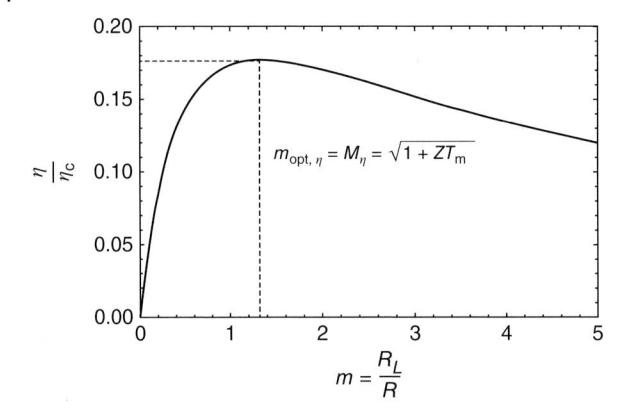

Figure 2.5 The efficiency of a TEG in dependence on the load ratio. $T_h = 600K$ and $T_c = 300K$ results in a temperature difference of $\Delta T = T_h - T_c = 300K$, a Carnot efficiency $\eta_C = \frac{\Delta T}{T_h} = \frac{1}{2}$, and an average temperature $T_m = (T_h+T_c)/2 = 450K$. The numerical example shown in the graph is for $ZT_h = 1$ ($ZT_m = 3/4$).

Then the efficiency can be expressed in a compact form,

- Efficiency:

$$\eta = \frac{P_{el}}{Q_{in}} = \frac{\Delta T}{T_h} \frac{m}{m+1+\frac{(m+1)^2}{ZT_h} - \frac{1}{2}\frac{\Delta T}{T_h}} \quad \text{with} \quad \frac{1}{Z} = \frac{KR}{\alpha^2}. \quad (2.23)$$

The efficiency of a TEG η in dependence on the load ratio m is shown in Figure 2.5 for a particular example.

As we can notice, the efficiency is the product of the reversible Carnot efficiency $\eta_C = \frac{\Delta T}{T_h} = \frac{T_h - T_c}{T_h}$ and the irreversible factor denoted as reduced efficiency η_r, which equals

$$m/\left(m+1+\frac{(m+1)^2}{ZT_h} - \frac{1}{2}\frac{\Delta T}{T_h}\right).$$

- Maximum efficiency:
From the derivation $\frac{\partial \eta}{\partial m} = 0$, we obtain after a few algebra steps, see example, [68, p. 141 ff]

$$\eta_{max} = \eta_C \frac{M_\eta - 1}{M_\eta + \frac{T_c}{T_h}} \quad (2.24)$$

with $M_\eta = \sqrt{1+ZT_m}$ where $T_m = (T_h+T_c)/2$.

Note that ZT is the only material parameter affecting the maximum efficiency.

- Maximum electrical power output:

$$\frac{\partial P_{el}}{\partial I} = 0 = \alpha(T_h - T_c) - 2RI,$$

$$I_{max,P} = \frac{\alpha(T_h - T_c)}{2R} = \frac{V_0}{2R}.$$

The last equation tells us that the maximal power output is obtained when the electrical load resistance R_L is equal to the internal resistance of the TEG. Then the maximal output power is obtained for $M_P = m_{opt,P} = 1$. The maximal power expression reduces to

$$P_{el,max} = \frac{\alpha^2 \Delta T^2}{4R} = \frac{1}{4}\frac{V_0^2}{R}. \tag{2.25}$$

The maximal producible power does not depend directly on ZT itself but independent of ΔT on the term α^2/R called power factor of a device. One can notice that the conditions for maximal output power ($M_P = 1$) and maximal efficiency ($M_\eta = \sqrt{1 + ZT_m}$) are different. This means that given a fabricated TE device where the geometric lengths and areas are fixed, more power will be produced if additional heat is supplied and higher current drawn than in the maximum efficiency configuration. However, when designing an optimal device for a particular application, the optimum design will have the geometry and current for maximum efficiency because this will provide more power with the same designed heat input. In practical cases, both conditions are quite close to each other in a sense that the output power will decrease only slightly when switching from maximum power to maximum efficiency case, whereas the similar applies for the efficiency, when switching back to the maximum power case.

2.3.2.2 Thermoelectric Cooler (TEC)

We are using here $T_c = T_a, T_h = T_s$, and $\dot{Q}_{in} = \dot{Q}_a, \dot{Q}_{out} = \dot{Q}_s$, respectively (Figure 2.6). This is well imaginable as long as the TEC operates in its regular mode $T_a < T_s$, whereas practical applications occur with $T_a > T_s$.

- Absorbed thermal power:

$$\dot{Q}_{in} = \alpha T_c I - \frac{1}{2}RI^2 - K(T_h - T_c). \tag{2.26}$$

- Outgoing thermal power:

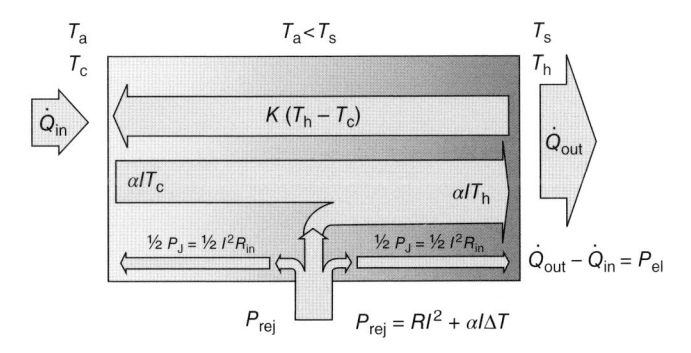

$$\dot{Q}_{out} = \alpha T_h I + \frac{1}{2}RI^2 - K(T_h - T_c). \tag{2.27}$$

Figure 2.6 Contributions in a TEC (CPM).

- Consumed electrical power:
$$P_{rec} = \dot{Q}_{out} - \dot{Q}_{in} = \alpha(T_h - T_c)I + RI^2.$$
- Maximum cooling power:
$$I_{Q_c^{max}} = \frac{\alpha T_c}{R}, \tag{2.28}$$

$$Q_c^{max} = \frac{1}{2}\frac{\alpha^2 T_c^2}{R} - K(T_h - T_c) = K\left[\frac{1}{2}ZT_c^2 - (T_h - T_c)\right]. \tag{2.29}$$

One can notice that the maximal cooling power is substantially driven by the figure of merit of the material and of the device, respectively, but not exclusively. The weighting of the influence of the individual material parameters depends on the temperature difference. In the heat pump operation ($\Delta T = 0$), the power factor is relevant, as for the TEG power generation.

- Coefficient of performance φ_c [6]:
$$\varphi_c = \frac{\dot{Q}_{in}}{P_{rec}} = \frac{\alpha T_c I - \frac{1}{2}RI^2 - K(T_h - T_c)}{\alpha(T_h - T_c)I + RI^2}. \tag{2.30}$$

- Maximum φ_c:
$$\frac{\partial \varphi_c}{\partial V_0} = 0 \text{ with } V_0 = \alpha(T_h - T_c) \text{ gives:}$$

$$\varphi_c^{max} = \frac{T_c}{(T_h - T_c)} \frac{\sqrt{1 + ZT_m} - T_h/T_c}{\sqrt{1 + ZT_m} + 1}. \tag{2.31}$$

V_0 is the open voltage, a specific value of the electric voltage V. The Carnot factor $\varphi_C = T_c/(T_h - T_c)$ is the reversible term of φ_c^{max}. The second factor contains the irreversible contributions.

It should be noticed that Eq. (2.31) is similar to the expression obtained for the TEG configuration, see Eq. (2.24). Both formulae are well known and often written as [49, 67, 69, 70]

$$\eta_{max} = \frac{T_h - T_c}{T_h} \frac{\sqrt{1 + ZT_m} - 1}{\sqrt{1 + ZT_m} + T_c/T_h} \tag{2.32}$$

and

$$\frac{1}{\varphi_c^{max}} = \frac{T_h - T_c}{T_c} \frac{\sqrt{1 + ZT_m} + 1}{\sqrt{1 + ZT_m} - T_h/T_c}, \tag{2.33}$$

offering the possibility for non-CPM devices to compare their device figure of merit ZT with the corresponding effective device figure of merit based on CPM (the exact value of which will depend on the temperature range). Moreover, maximum efficiency or φ is only achieved under specific ideal working conditions, and practical applications usually do not exactly fulfill these conditions!

---

6) We follow here Sherman's notation and use φ instead of COP in the formulae.

- Maximum temperature difference: The maximum temperature difference is achieved for $Q_c^{max} = 0$ and hence for $\varphi_c = 0$. In this case, we obtain from Eq. (2.29)

$$\Delta T_{max} = (T_h - T_c)_{max} = \frac{1}{2} Z T_c^2, \tag{2.34}$$

depending only on Z, from a material point of view.

2.3.2.3 Thermoelectric Heater (TEH)

TEH is a device that operates similarly to TEC, whereas the rejected power is counted. Since the absorbing side may be colder or warmer than the rejecting side, using T_a and T_s might be most adequate here

- Incoming thermal power:

$$\dot{Q}_a = \alpha T_a I - \frac{1}{2} R I^2 - K(T_a - T_s). \tag{2.35}$$

- Outgoing thermal power:

$$\dot{Q}_s = \alpha T_s I + \frac{1}{2} R I^2 - K(T_a - T_s). \tag{2.36}$$

- Consumed electrical power:

$$P_{rec} = \dot{Q}_s - \dot{Q}_a = -\alpha(T_a - T_s)I + R I^2. \tag{2.37}$$

- Coefficient of performance:

$$\varphi_w = \frac{Q_s}{P_{rec}} = \frac{\alpha T_a I + \frac{1}{2} R I^2 - K(T_a - T_s)}{\alpha(T_a - T_s)I + R I^2}. \tag{2.38}$$

- Maximum φ_w:

$$\varphi_w^{max} = \frac{T_a}{(T_a - T_s)} \left(1 - 2 \frac{\sqrt{1 + Z T_m} - 1}{Z T_a}\right) \tag{2.39}$$

The Carnot factor for φ is here $\varphi_C = T_a/(T_a - T_s)$, whereas the irreversible contribution is given by the second term.

Note that all terms in the maximum power expressions are proportional to $1/R$ or K and thus to $1/L$, where the efficiency is independent of the element geometry. With the reduction of the length, the maximum power will rise above any limit.

2.3.3
Inverse Performance Equations and Effective Device Figure of Merit

In the previous subsection, it has been stated that the formulae for the maximum efficiency, see Eq. (2.32), and for the maximum φ, see Eq. (2.33), are used to define conveniently a device's figure of merit ZT. Given a maximum performance value

(η_{max} or φ_c^{max}), one can subsequently solve Eqs. (2.32), (2.33) for the device figure of merit Z. The results of a straightforward calculation are:

$$\text{TEG:} \quad ZT_m = \left(\frac{\Delta T + \eta_{max} T_c}{\Delta T - \eta_{max} T_h}\right)^2 - 1, \tag{2.40}$$

$$\text{TEC:} \quad ZT_m = \left(\frac{T_h + \Delta T \varphi_c^{max}}{T_c - \Delta T \varphi_c^{max}}\right)^2 - 1 \tag{2.41}$$

$$= \left(\frac{\Delta T + T_h/\varphi_c^{max}}{\Delta T - T_c/\varphi_c^{max}}\right)^2 - 1. \tag{2.42}$$

Note that the Eqs. (2.40) and (2.42) are equivalent when replacing $T_c \Leftrightarrow T_h$ and $\eta_{max} \Leftrightarrow 1/\varphi_c^{max}$. Further, note that a comparison of the resulting effective device Z with the material's figure of merit z can also be of interest. (For ideal CPM devices $Z = z$ holds.)

Let us particularly consider the limit $\Delta T \to dT$, that is, an infinitesimal segment with $T_c, T_h \to T_m$, and $T_c/T_h \to 1$. In this case, we have locally $\eta_{max} = \frac{dT}{T_h} \eta_{r,max}$ with reduced efficiency η_r [45, 71], see also Section 1.10.2, gives

$$ZT_m|_{dT} = \left(\frac{dT + dT \frac{T_c}{T_h} \eta_{r,max}}{dT - dT \eta_{r,max}}\right)^2 - 1 \bigg|_{T_c/T_h \to 1} = \left(\frac{1 + \eta_{r,max}}{1 - \eta_{r,max}}\right)^2 - 1. \tag{2.43}$$

By referring to the optimal material distribution in a single stage device, the relation between $\eta_{r,max}$ and the optimal local efficiency $\eta_{r,opt}$ is given by

$$\eta_{r,max} \le \eta_{r,opt} = \frac{\sqrt{1 + zT_m} - 1}{\sqrt{1 + zT_m} + 1}. \tag{2.44}$$

The optimal local efficiency $\eta_{r,opt}$ is found for the so-called *self-compatible* elements. Self-compatibility is the consideration of compatibility in the same material; for more information, see Chapter 5. This is achieved in $u = s$ material with optimal "reduced efficiencies" $\eta_{r,opt} = \varphi_{r,opt} = \frac{\sqrt{1+zT}-1}{\sqrt{1+zT}+1}$, where z is the material's figure of merit. A straightforward calculation yields

$$\frac{1 + \eta_{r,opt}}{1 - \eta_{r,opt}} = \frac{1 + \frac{\sqrt{1+zT_m}-1}{\sqrt{1+zT_m}+1}}{1 - \frac{\sqrt{1+zT_m}-1}{\sqrt{1+zT_m}+1}} = \frac{(\sqrt{1+zT_m}+1) + (\sqrt{1+zT_m}-1)}{2} = \sqrt{1+zT_m}.$$

Hence we obtain with the limit $\Delta T \to dT$:

$$ZT_m \le (\sqrt{1+zT_m})^2 - 1 = zT_m. \tag{2.45}$$

An analogous proof with the same result can be made for TEC by using the relation $\frac{1}{\varphi_{max}} = \frac{dT}{T_c} \frac{1}{\varphi_{r,max}}$.

The maximum efficiency of a TEG and the maximum φ of a TEC had already been published by Sherman *et al.* [23, 24] and more explicitly by Ybarrondo [72] for an infinitely staged device with optimum load (and without losses from the interfaces) and given figure of merit $Z(T)\,T$:

$$\eta_{\max} = 1 - \exp\left(-\int_{T_s}^{T_a} \frac{1}{T} \frac{\sqrt{1+Z(T)\,T}-1}{\sqrt{1+Z(T)\,T}+1}\,dT\right), \qquad (2.46a)$$

$$\varphi_{\max} = \left[\exp\left(\int_{T_a}^{T_s} \frac{1}{T} \frac{\sqrt{1+Z(T)\,T}+1}{\sqrt{1+Z(T)\,T}-1}\,dT\right) - 1\right]^{-1}. \qquad (2.46b)$$

In 2002, Ursell and Snyder [44] showed that the compatibility approach to functionally grading thermoelectric materials for power generation yields a single device that achieves the same performance as expected from an infinitely staged generator (see also Section 5.2). In Section 5.3.1, it is reviewed that the method of local optimization of reduced efficiencies yields the same maximum performance in self-compatible single elements as given in Eqs. (2.46). Since the maximum performance of any real, single-element device cannot exceed the maximum performance value obtained for an element with optimally graded material, we can conclude from our proof that the effective figure of merit of such a device is limited by its material's zT

$$Z T_m \leq z T(T_m). \qquad (2.47)$$

More roughly, we can state that any device cannot be better than its material. The relation between the material and the device's figure of merit for TEC had already been discussed by Norwood in 1961 [73]. Note that Eq. (2.47) is not valid for the general case, but for CPM, it is correct.

In the 1990s, Bergman et al. [74–77] showed that the effective figure of merit of a composite can never exceed the largest Z value of any of its components (Bergman's theorem). Note that the Bergman–Levy theorem has been challenged recently, see [78].

2.4
Maximum Power and Efficiency of a Thermogenerator Element

We propose now a more detailed analysis of the maximum power (density) p_{out} and efficiency η of a thermogenerator element. The electrical power output can be determined by

$$P_{\text{el}} = \iiint_V \mathbf{j} \cdot \mathbf{E}\,dV = A_c \int_0^L \left(\frac{j^2}{\sigma} + j\alpha\,T'(x)\right)dx, \qquad (2.48)$$

where we assume a 1D planar setup with (constant) cross-sectional area A_c. Then, the net electrical power output density

$$p_{\text{out}}(j) = -\frac{P_{\text{el}}}{A_c}, \qquad (2.49)$$

is given by the difference between the open voltage and the Ohmic voltage drop. The heat flux or heat transfer rate per cross-sectional area

$$q = \frac{\dot{Q}}{A_c} = -\kappa \nabla T + j\alpha T \rightarrow q(x) = -\kappa \frac{dT}{dx} + j\alpha T(x), \qquad (2.50)$$

and the efficiency of a TEG is usually derived from the ratio of the gained electrical power (power output density) to the thermal input (the absorbed heat)

$$\eta(j) = \frac{p_{out}}{q_a}. \qquad (2.51)$$

In our setup, the hot side (heat absorbing side) of the thermogenerator element is at position $x = 0$. Within CPM, the power output density is obtained from Eq. (2.48)

$$p_{out}(j) = j\alpha(T_a - T_s) - \frac{j^2 L}{\sigma}, \qquad (2.52)$$

where L is the element's length. T_a and T_s are the boundary temperatures at $x = 0$ and $x = L$, respectively. Finally, we obtain for the efficiency

$$\eta(j) = \frac{j\alpha(T_a - T_s) - \frac{j^2 L}{\sigma}}{j\alpha T_a - \kappa \frac{dT}{dx}\big|_{x=0}} = \frac{2\alpha\sigma(T_a - T_s)Lj - 2L^2 j^2}{2\alpha\sigma T_a Lj + 2\kappa\sigma(T_a - T_s) - L^2 j^2}. \qquad (2.53)$$

Given the performance parameters as a function of the electrical current (load), the optimal performance can be found by solving a classical extremum problem [22, 38]. The result for the optimal current density is

$$\frac{dp_{out}}{dj} = 0 \Rightarrow j_{opt,p} = \frac{\sigma\alpha}{2L}(T_a - T_s) = \frac{\sigma\alpha}{2}\frac{\Delta T}{L}. \qquad (2.54)$$

Using this result in Eq. (2.52) gives

$$p_{max} \equiv p_{out}(j_{opt,p}) = \alpha^2 \sigma \frac{(\Delta T)^2}{4L}. \qquad (2.55)$$

Analogously, the optimal current density for the efficiency is determined

$$\frac{d\eta}{dj} = 0 \Rightarrow j_{opt,\eta} = \frac{\kappa}{\alpha L}\frac{\Delta T}{T_m}(-1 + \sqrt{1 + zT_m}), \qquad (2.56)$$

leading to the maximum efficiency within CPM

$$\eta_{max} = \frac{T_a - T_s}{T_a}\frac{\sqrt{1 + zT_m} - 1}{\sqrt{1 + zT_m} + T_s/T_a} = \eta_C \eta_r, \qquad (2.57)$$

which can be divided into the Carnot efficiency $\eta_C = \Delta T/T_a = (T_a - T_s)/T_a$ and the reduced efficiency η_r. The Carnot efficiency is only built up by the boundary temperatures, whereas the reduced efficiency contains material properties in form of the intrinsic figure of merit.

The performance of a TEG element in dependence on the electrical current density is illustrated in Figure 2.7. On the one hand, the power output density $p_{out}(j)$, see Eq. (2.52), and on the other hand, the efficiency $\eta(j)$, see Eq. (2.53), is shown

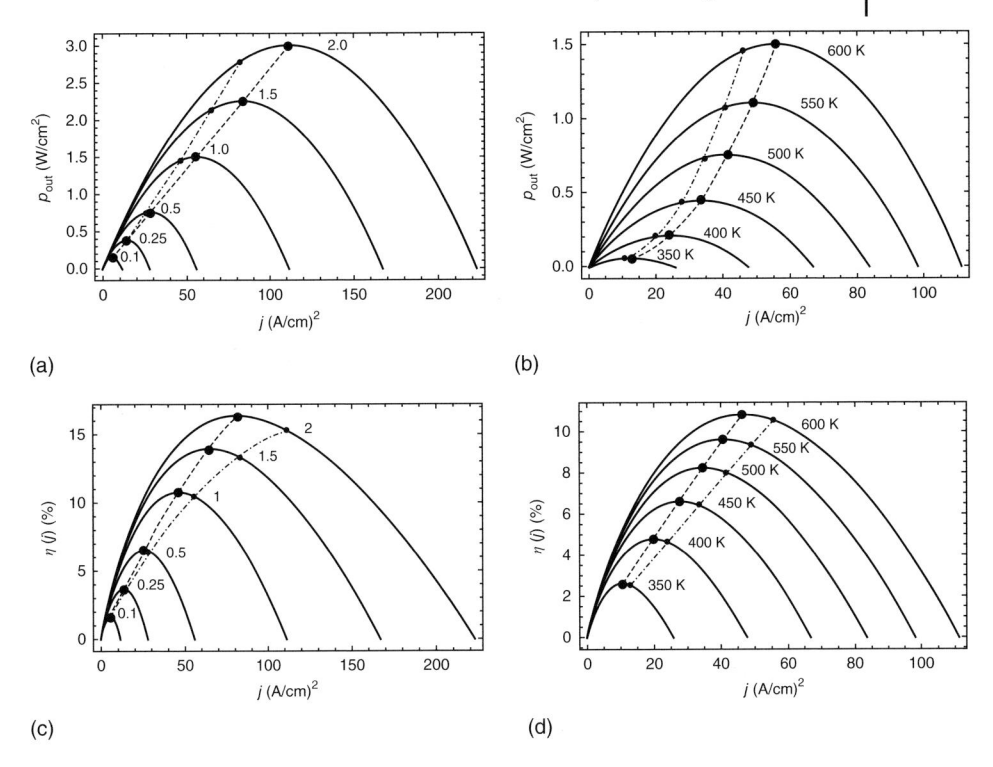

Figure 2.7 Performance of a TEG element in dependence on the electrical current density for the variation of T_a and zT_m. Maximum values are shown with large dots. As there are two optimum current densities, the small dots represent the performance value for the "other" optimum current density, that is, $p_{out}(j_{opt,\eta})$ or $\eta(j_{opt,p})$, respectively. The dashed and dotted–dashed lines serve as guidelines for the corresponding values. (a) Power output density – variation of zT_m; (b) Power output density – variation of T_a; (c) Efficiency – variation of zT_m; (d) Efficiency – variation of T_a.

and varied due to the figure of merit zT_m and the absorbing side temperature T_a. The maximum values of the performance parameters are highlighted by dots and the dashed lines as well as dotted–dashed lines represent the performance parameters corresponding to the contrary optimum current densities , that is, $p_{out}(j_{opt,\eta})$ or $\eta(j_{opt,p})$, respectively.

The values of maximum values for the two variations are summarized in Tables 2.3 and 2.4.

The Ioffe description for the performance parameters of a TEG element presented in this section is only valid for CPM. Moreover, it is a sufficient approximation for a small temperature difference. However, this approximation is often used for configurations in a larger temperature difference where the temperature dependence of the material properties cannot be neglected any more. Then, the calculation will provide qualitative results with increasing inaccuracy, and a straightforward question is how good the approximation is and if there are strict methods to get valid quantitative results for maximum performance.

Table 2.3 Maximum performance values and corresponding electrical current densities for the variation of $zT_m = 0.1, \cdots, 2.0$ for fixed $T_a = 600\text{K}$, $T_s = 300\text{K}$, $\alpha = 180\mu\text{V/K}$, and $\kappa = 1.5\text{W/m K}$.

zT_m (1)	σ (S/m)	$j_{opt,p}$ (A/cm²)	p_{max} (W/cm²)	$\eta(j_{opt,p})$ (%)	$j_{opt,\eta}$ (A/cm²)	η_{max} (%)	$P_{out}(j_{opt,\eta})$ (W/cm²)
0.1	10288.1	5.56	0.15	1.57	5.42	1.58	0.1499
0.25	25720.2	13.89	0.375	3.64	13.11	3.65	0.374
0.5	51440.3	27.78	0.75	6.45	24.97	6.52	0.742
1.0	102881	55.56	1.5	10.53	46.02	10.82	1.456
1.5	154321	83.33	2.25	13.33	64.75	19.96	2.136
2.0	205761	111.11	3	15.38	81.34	16.40	2.785

Table 2.4 Maximum performance values and corresponding electrical current densities for the variation of $T_a = 350\text{K}, \cdots, 600\text{K}$ for fixed $zT_m = 1.0$, $\alpha = 180\mu\text{V/K}$, $\kappa = 1.5\text{W/m K}$, and $T_s = 300\text{K}$.

T_a (K)	T_m (K)	σ (S/m)	$j_{opt,p}$ (A/cm²)	p_{max} (W/cm²)	$\eta(j_{opt,p})$ (%)	$j_{opt,\eta}$ (A/cm²)	η_{max} (%)	$P_{out}(j_{opt,\eta})$ (W/cm²)
600	450	102881	55.56	1.500	10.53	46.02	10.82	1.456
550	425	108932	49.02	1.103	9.35	40.61	9.61	1.070
500	400	115741	41.67	0.750	8.00	34.52	8.23	0.728
450	375	123457	33.33	0.450	6.45	27.61	6.64	0.437
400	350	132275	23.81	0.214	4.65	19.72	4.78	0.208
350	325	142450	12.82	0.058	2.53	10.62	2.61	0.056

2.4.1
Load Resistance as Design Parameter for a Thermogenerator

As the optimization procedure is performed by finding the optimal current densities, it has to be pointed out that in general for a TEG, the electrical current density *is a state parameter determined uniquely by internal design parameters of the given generator and its external operating conditions and cannot be designed to a fixed value as the current in thermoelectric coolers*, as pointed out by Min Chen and coworkers [79, 80].

This is especially the case where a nonideal TEG is investigated, that is, by taking into account finite heat transfer from the heat source and to the heat sink, resulting in an additional irreversibility. Here in the ideal case (infinite heat transfer from the source and to the sink), the electrical current density can be directly substituted by the real design parameter, the load resistance R_L:

$$I = jA_c = \frac{U_\alpha}{R_i + R_L} \Rightarrow j(R_L) = \frac{\alpha \Delta T}{A_c(R_i + R_L)}, \qquad (2.58)$$

where, for the sake of clarity, the internal resistance R_i for the TE element as defined by Eq. (2.12) is introduced. By using Eq. (2.58) in Eq. (2.52), the power output density in dependence on the load resistance is

$$p_{\text{out}}(R_L) = \frac{(\alpha \Delta T)^2}{A_c R_i} \frac{1}{1 + \frac{R_L}{R_i}} \left[1 - \frac{1}{1 + \frac{R_L}{R_i}}\right]. \tag{2.59}$$

By substituting $\xi = [1 + R_L/R_i]^{-1}$, the power output density has the functional form $p_{\text{out}}(\xi) \propto \xi[1 - \xi] = f[\xi]$, from which it can be calculated or directly deduced that for $\xi = 1/2$, that is, $R_i = R_L$, the power output density shows a maximum, see Figure 2.8(b). Note that a similar formula has already been defined in Eq. (2.22c), that is, $\xi = [1 + m]^{-1}$, and after some algebraic steps, the equivalence between Eqs. (2.59) and (2.22c) can be found. The efficiency dependence can be analogously transformed from the electrical current density to the load resistance description by using Eq. (2.58) in Eq. (2.53). After a straightforward calculation, it is found that the ratio

$$M_\eta = \frac{R_L}{R_i} = \sqrt{1 + zT_m} \tag{2.60}$$

delivers the best efficiency value, see also Eq. (2.24). With this abbreviation, the maximum efficiency formula in the Ioffe-CPM approximation, see Eq. (2.57), can be written as

$$\eta_{\max} = \eta_C \eta_r = \eta_C \frac{M_\eta - 1}{M_\eta + v} = \eta_C \frac{M_\eta - 1}{M_\eta + 1 - \eta_C} \quad \text{with} \quad v = \frac{T_s}{T_a}, \tag{2.61}$$

where the maximum efficiency is subdivided into the Carnot efficiency η_C and the reduced efficiency η_r, which contains the irreversibility of the thermoelectric process [40]. The ratio of the (absolute) boundary temperatures $v = \frac{T_s}{T_a}$ is introduced. With that, the Carnot efficiency can be expressed as $\eta_C = 1 - v$.

2.4.2
Efficiency versus Power Approach

Actually, Eq. (2.57) can be used as the starting point of performance calculations for TE devices. It should be noticed that the performance of TE applications had

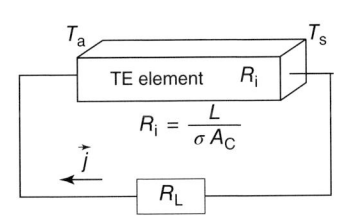

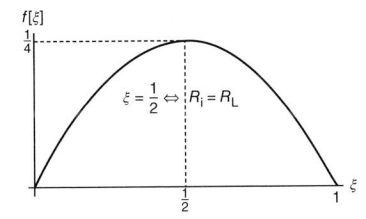

Figure 2.8 Correspondence between the electrical current density j and the load resistance R_L under ideal conditions in the CPM-Ioffe approximation. Left Correlation between current and load; Right Functional form of the power output density.

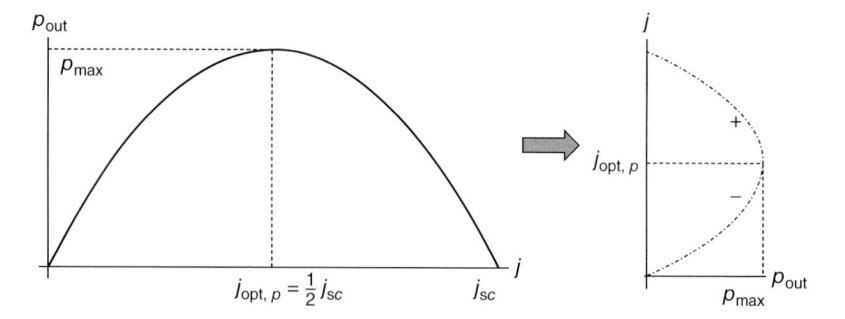

Figure 2.9 Electrical power output density p_{out} as a function of the current density j and vice versa.

been discussed by Altenkirch 100 years ago [66, 81] in a qualitative manner but without naming the figure of merit. The CPM formulae, which Ioffe [49] had established, were used in a variety of articles in the 1950s and 1960s, see for example, [69, 70, 82]. In Eq. (2.57), the material enters via the dimensionless figure of merit zT_m while the "working conditions" are given by the (hot and cold sides) boundary temperatures primarily. In contradiction to the maximal electrical power output (density), the geometric shape and size of the thermoelectric element do not influence the maximal efficiency.

Obviously, both optimal current densities $j_{opt,p}$ and $j_{opt,\eta}$ are different in general, see Eqs. (2.54) and (2.56), which means that both maxima cannot be reached in one configuration obtained by fixed conditions [temperature difference (ΔT), geometry (L, A_c), and material (α, σ, κ)]. This can be illustrated when presenting the efficiency as a function of the net power output, $\eta(p_{out})$, see Figure 2.10. Such an approach was introduced by Gordon [83–85].

He expressed the current density j by means of Eqs. (2.52) and (2.55) as a function of p_{out}, that is,

$$j^{\pm} = j_{opt,p}[1 \pm \sqrt{1 - \tilde{p}}], \tag{2.62}$$

where the power output $\tilde{p} = p_{out}/p_{out}^{max}$ is normalized to its maximal value (Figure 2.9).

After the elimination, there are two branches for the current density j^+ and j^-, see Eq. (2.62) and Figure 2.9, this division into two branches is found in dependence on the efficiency on the power output as well, that is,

$$\eta^{\pm} = \eta^{\pm}(\tilde{p}) = \frac{\tilde{p}}{\frac{\tilde{p}}{2} + \frac{4}{z\Delta T} + \frac{2T_m}{\Delta T}(1 \pm \sqrt{1 - \tilde{p}})}. \tag{2.63}$$

For $\tilde{p} = 1$ and $p_{out} = p_{out}^{max}$, we find with the help of Eq. (2.63)

$$\eta^{\pm}(\tilde{p} = 1) = \frac{2z\Delta T}{8 + z(\Delta T + 4T_m)}. \tag{2.64}$$

The value of maximal power output is the point where the nontrivial convergence of the positive and negative branches of the efficiency takes place. The power output density, where the efficiency is maximal, $\tilde{p}_{\eta,max}$, can be calculated with help of

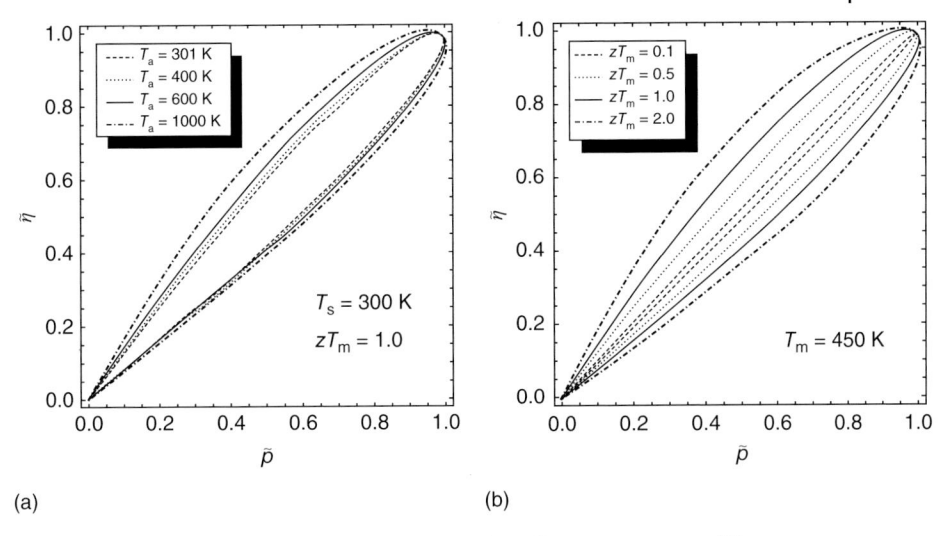

2.4 Maximum Power and Efficiency of a Thermogenerator Element

Figure 2.10 Efficiency versus power output density for different temperature differences ΔT and figure of merits zT_m with $T_c = 300K$. (a) Variation of ΔT; (b) Variation of zT_m.

a classical extreme value task $\frac{d\bar{\eta}}{d\bar{p}} = 0$, which results in

$$\tilde{p}_{\eta,\max}(zT_m) = \frac{2}{\zeta(zT_m)}\left[\sqrt{1+\zeta(zT_m)} - 1\right] \text{ with } \zeta(x) = \frac{x^2}{4(1+x)}. \quad (2.65)$$

The difference between both maximum currents can be displayed easily by building the ratio using Eqs. (2.54) and (2.56)

$$\frac{j_{\text{opt},p}}{j_{\text{opt},\eta}} = \frac{1}{2}zT_m \frac{1}{-1+\sqrt{1+zT_m}} \quad (2.66)$$

Obviously, the ratio is a function of zT_m only, which is illustrated in Figure 2.11.

For small values of the figure of merit, the ratio is close to 1. This can be shown by using the Taylor expansion of $\sqrt{1+zT_m} \approx 1 + \frac{1}{2}zT_m$ for $zT_m \ll 1$ and setting this in Eq. (2.66).

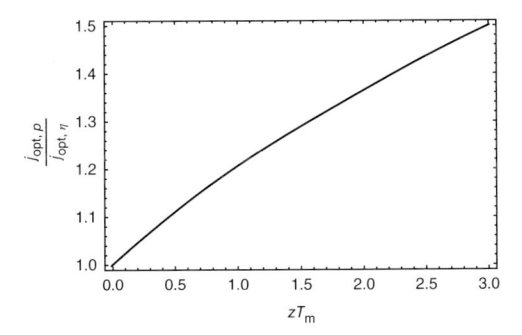

Figure 2.11 Ratio of optimal current densities in dependence on the figure of merit.

2.4.3
Constant Heat Input (CHI) Model

Usually, the performance of a TEG is calculated for constant temperature operation (CTO), that is, from a mathematical point of view, Dirichlet boundary conditions are taken into account and the junction temperatures are supposed to be fixed, see Eq. (2.14). This operation means that there is no change in the junction temperature when the current or, better, the load (or the ratio of the load to the internal resistance m) is varied, that is, the heat source and heat sink are supposed to be infinite. In References [86, 87], another method of operation of the TEG is shown. Here it is supposed that the hot junction receives a constant heat flux from the external source, that is, the heat power absorbed is constant. Such an operation can take place if the heat source of the TEG is a nuclear power source or the heat comes directly from the sun. Both situations can be found in the power source of space vehicles. Mathematically, this means that von Neumann conditions are assumed. Of course, the performance behavior of a TEG changes in a manner analogous to the way as we compare the constant current operation with the constant voltage operation of an electric network. The operation under constant heat flux condition is especially of interest in hybrid systems such as RTG and STEG. The CHI model is described now in more detail. The heat balance at the hot junction is

$$\dot{Q}_a = \dot{Q}_F + \dot{Q}_P - \frac{1}{2}\dot{Q}_J, \tag{2.67}$$

where $\dot{Q}_a$ is the incoming thermal power (at the absorbing side), which is subdivided into three parts:[7]

1) Fourier heat conduction part

$$\dot{Q}_F = \kappa \frac{A_c}{L}(T_a - T_s) = K \Delta T. \tag{2.68}$$

2) Peltier part at the absorbing side

$$\dot{Q}_P = I \alpha T_a. \tag{2.69}$$

3) Joule part at the absorbing side

$$\dot{Q}_J = I^2 \varrho \frac{L}{A_c} = I^2 R_i. \tag{2.70}$$

The electrical power is given by

$$P_{el} = I^2 R_L, \tag{2.71}$$

and the electrical current can be expressed as

$$I = \frac{\alpha(T_a - T_s)}{R_i + R_L} = \frac{\alpha \Delta T}{R_{total}}. \tag{2.72}$$

7) A constant cross-sectional area (A_c) of the TE element is assumed here.

2.4 Maximum Power and Efficiency of a Thermogenerator Element

By setting Eqs. (2.68), (2.69), and (2.70) in Eq. (2.67) and using Eqs. (2.71) and (2.72), the thermal power at the absorbing side can be expressed in terms of the electrical power output.[8]

$$\dot{Q}_a = P_{el}\left(\frac{2m+1}{2m}\right) + \sqrt{\frac{P_{el}}{R_L}}\left[\frac{\alpha(1+m)}{Z} + \alpha T_s\right], \quad (2.73a)$$

$$\dot{Q}_a = P_{el}\left(\frac{2m+1}{2m}\right) + \frac{\alpha}{Z}\sqrt{\frac{P_{el}}{R_L}}[m + \zeta_s + 1], \quad (2.73)$$

where m is as in Eq. (2.60) defined as the ratio between the load and the internal resistance and $\zeta_s = ZT_s$ is the normalized sink junction temperature. To find the optimum load resistance in the case where the internal resistance of the TEG R_i is assumed to be fixed, Eq. (2.73) is differentiated with respect to R_L. Note that only m and P_{el} depend on R_L. By setting dP_{el}/dR_L, the optimum load resistance must fulfill the following relation:

$$M_{R_L}^{3/2} - (1+\zeta_s)M_{R_L}^{1/2} - \frac{Z}{\alpha}\sqrt{P_{el}R_i} = 0, \quad (2.74)$$

where M_{R_L} denotes the optimum value of R_L/R_i (and thus the optimum value of R_L as R_i is fixed). The optimum value M_{R_L} always exceeds $1 + \zeta_s$. The dependence of M_{R_L} on ζ_s is illustrated in Figure 2.12.

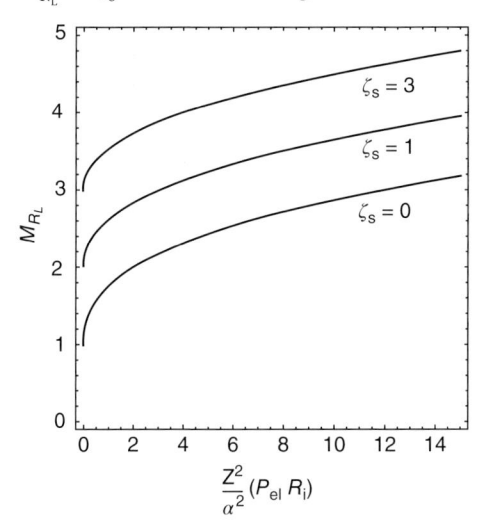

Figure 2.12 Constant heat flux operation. Optimum value of m for fixed internal resistance for different ζ_s.

8) In the original work [86], a thermocouple (two connected TE elements) is investigated. For this, there is a difference between the material's figure of merit z and the device's figure of merit Z (a maximum is calculated and used by adjusting the areas of the element). For a single element $z = \frac{\alpha^2}{\kappa\varrho} = \alpha^2\left(\kappa\frac{A_c}{L}\varrho\frac{L}{A_c}\right)^{-1} = \alpha^2/(KR_i) = Z$ is valid and to prevent the confusion mixing up the material's figure of merit with the z-coordinate, the device Z is used.

Expression (2.74) is only useful if the electrical power output P_{el} is known beforehand, which, of course, in practice is seldom the case. Therefore, a combination of Eqs. (2.73) and (2.74)

$$[M_{R_L} - 1 - \zeta_s][2M_{R_L}^2 + (1 - 2\zeta_s)M_{R_L} + \zeta_s + 1] - \frac{2Z^2}{\alpha^2}(\dot{Q}_a R_i) = 0 \quad (2.75)$$

is helpful as M_{R_L} is given in terms of the presumed known absorbing thermal power. Then, by combining Eqs. (2.74) and (2.75), the optimum efficiency becomes

$$\eta_{R_L} = \frac{P_{el}}{\dot{Q}_a} = \frac{2M_{R_L}(M_{R_L} - 1 - \zeta_s)}{2M_{R_L}^2 + (1 - 2\zeta_s)M_{R_L} + \zeta_s + 1} \quad (2.76)$$

Note that maximizing the power output also maximizes the efficiency as the absorbing thermal power is assumed constant. The normalized temperature difference can be determined to be

$$\zeta_a - \zeta_s = Z\Delta T = (M_{R_L} + 1)(M_{R_L} - 1 - \zeta_s). \quad (2.77)$$

The previous relations are shown in Figures 2.13, where for a fixed internal resistance R_i, the efficiency and (normalized) temperature difference are calculated under optimal operating conditions.

There is further calculation for a fixed load and variable internal resistance found in Ref. [86].

2.5
Temperature-Dependent Materials – Analytic Calculations

In the research of thermoelectrics, the focus lies on the investigation, determination, and measurement of the temperature dependence of the material properties. The physical inhomogeneity of a particular material property, that is, the temperature dependence, results in the temperature dependence of the performance key parameters, power factor f, and figure of merit zT. From a mathematical point of view, the problem gets more involved, because the coefficients in the thermal balance PDE are the state-dependent. A general analytical solution for the temperature distribution can only be found in special cases. This section is meant for reviewing the most important of the special cases, which are solvable analytically. The solutions of those models that are determined can help to evaluate numerical methods and procedures used to calculate the performance in the cases where an analytical solution is not possible. Furthermore, these models give the possibility to identify the general impacts of the temperature dependence of the material properties on the performance, especially in comparison to the classical materials with constant properties.

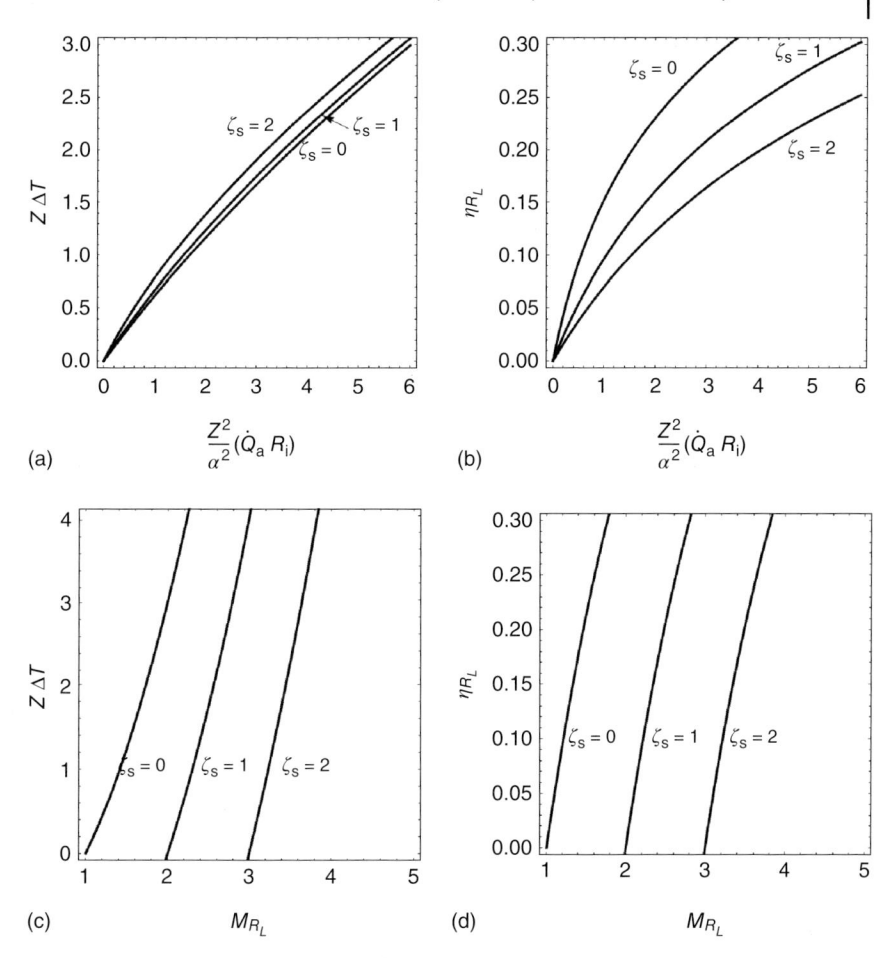

Figure 2.13 Optimum operating conditions for fixed internal resistance. (a) $Z\Delta T$; (b) η_{R_L}; (c) $Z\Delta T$; (d) η_{R_L}.

2.5.1
Inverse Temperature Dependence of the Thermal Conductivity

A simple example of temperature dependence of the material properties was discussed by Mahan [25], where the thermal conductivity is supposed to be inversely dependent on the temperature, that is, $\kappa(T) = k_1/T$, with the parameter k_1 controlling the decay of the thermal conductivity with increasing temperature. This behavior of the thermal conductivity is supposed to be physically realistic, because it is known that the thermal conductivity usually decreases with higher temperatures. The other two material parameters, the Seebeck coefficient and the electrical conductivity, are taken as a constant. The calculation of the performance

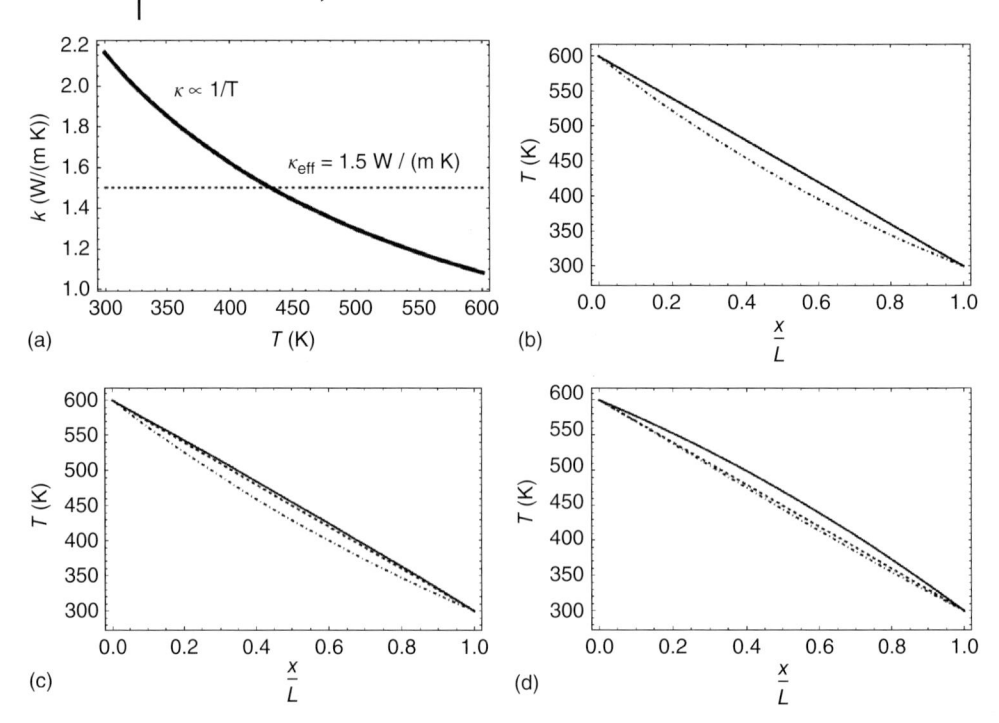

Figure 2.14 Comparison of the temperature profiles for the Ioffe CPM solution (solid line) and Mahan's solution for a temperature-dependent thermal conductivity (dotted–dashed line). The profiles are shown for three different current densities representing the currentless case $j = 0\,\text{A/cm}^2$, the current density close to the optimum values $j = 50\,\text{A/cm}^2$, and the current density close to the short-circuit value with $j = 100\,\text{A/cm}^2$. The dashed line is the linear temperature profile between the fixed boundary temperatures $T_a = 600\,\text{K}$ and $T_s = 300\,\text{K}$. The material properties are $\alpha = 180\,\mu\text{V/K}$, $\sigma = 102881\,\text{S/m}$, and $\kappa_{\text{eff}} = \overline{\kappa} = 1.5\,\text{W/m K} \Rightarrow k_1 = 649.213\,\text{W/m}$. (a) $\kappa(T)$; (b) $j = 0\,\text{A/cm}^2$; (c) $j = 50\,\text{A/cm}^2$; (d) $j = 100\,\text{A/cm}^2$.

parameters is similar to that of the CPM. A simple integration of the heat balance results in a temperature profile

$$T(x) = T_a \exp\left(\frac{x}{L} \ln \frac{T_s}{T_a} + \frac{j^2}{2\sigma k_1} x(L-x)\right), \quad (2.78)$$

where its functional form is totally different from the parabolic profile of the CPM case, see Eq. (2.14). When Eqs. (2.78) and (2.14) are compared, it can be seen that the parabolic CPM temperature profile has negative curvature and the result here with the exponential behavior has a positive curvature, as already pointed out by Mahan [25]. Figure 2.14 shows a comparison between the CPM solution, that is, the parabolic temperature profile, see Eq. (2.14), and temperature-dependent thermal conductivity discussed in this subsection, that is, Eq. (2.78).

The result for the power output is the same as in the case of CPM since the power output does not depend on the thermal conductivity, that is, it is found in Eq. (2.55). For the efficiency, it is easy to verify that the functional form for the maximum value stays the same as for CPM, but the figure of merit z has to be substituted by an effective value z_eff due to the fact that the effective thermal conductivity is identical to the temperature average of $\kappa(T)$

$$\kappa_\text{eff} = \frac{k_1 \ln\left(\frac{T_a}{T_s}\right)}{T_a - T_s} = \frac{1}{T_a - T_s} \int_{T_s}^{T_a} \frac{k_1}{T} dT = \frac{1}{\Delta T} \int_{T_s}^{T_a} \kappa(T) dT = \overline{\kappa(T)} \quad (2.79)$$

The value of the effective figure of merit is then

$$z_\text{eff} = \frac{\alpha^2 \sigma}{\kappa_\text{eff}} = \frac{\alpha^2 \sigma \Delta T}{k_1 \ln\left(\frac{T_a}{T_s}\right)}. \quad (2.80)$$

As you can easily see, the effective values depend on the boundary temperatures T_a and T_s displaying the temperature dependence of the thermal conductivity.

2.5.2
Inverse Temperature Dependence of the Thermal Conductivity and a Variable Electrical Resistivity

In Reference [88], Lee compared the influence of variable electrical resistivity and variable thermal conductivity on the performance of a TEG with the CPM theory for a single-element device. The material properties used in Lee's work are expressed as:

$$\alpha = \alpha_{(m)}, \varrho = \varrho_{(m)} \left(1 + \varrho_1^2 \ln \frac{T}{T_m}\right) \quad \text{and} \quad \kappa = \frac{\kappa_{(m)} T_m}{T}, \quad (2.81)$$

where the subscript "(m)" describes the corresponding value of the material property at $T = T_m = (T_a+T_s)/2$. Note that there is in general a difference between the averages and the latter values, that is, $\alpha_m = (\alpha_a+\alpha_s)/2$ and $\alpha_{(m)} = \alpha(T = T_m)$. The parameter ϱ_1^2 is an adjustable one controlling the temperature dependence of the electrical resistivity. As the Seebeck coefficient is supposed to be constant, the Thomson coefficient is zero in accordance to the second Kelvin relation.

The differential equation and boundary conditions have been transformed into a nondimensional form:[9]

$$\Theta = \frac{T}{T_m}, \xi = \frac{x}{L}, \text{ and } Y = \frac{\alpha_{(m)}}{\kappa_{(m)}} \frac{L}{A_c} I, \quad (2.82)$$

which leads to the following heat balance

$$\frac{d^2(\ln \Theta)}{d\xi^2} + \gamma_{(m)}^2 \varrho_1^2 Y^2 \ln \Theta = -\gamma_{(m)}^2 Y^2. \quad (2.83)$$

9) Note that the nomenclature is changed in comparison to Lee's work [88]. Especially the hot and cold sides are interchanged in this work, which has of course no influence on the physical behavior of the system.

The boundary conditions transform to $\Theta(\xi = 0) = \Theta_a$ and $\Theta(\xi = 1) = \Theta_s$. With the relation $\Theta_a + \Theta_s = 2$, one of these boundary conditions (BCs) can be eliminated. From the transformation

$$\gamma_{(m)}^2 = \frac{1}{z_{(m)} T_m} \tag{2.84}$$

results as a further substitution, where $z_{(m)}$ is calculated from the material properties at the average temperature T_m. Note that this is not an averaged figure of merit. After straightforward calculation, the following solution of Eq. (2.83) with the given BC can be found for the natural logarithm of the nondimensional temperature

$$\ln \Theta = \frac{\csc(\tilde{\gamma} Y)}{\tilde{\gamma}} \{B_a \sin[\tilde{\gamma}(1-\xi)] + B_s \sin(\tilde{\gamma} Y \xi)\} - \frac{1}{\varrho_1^2} \tag{2.85}$$

with

$$\tilde{\gamma} = \gamma_{(m)} \varrho_1, \quad B_a = \tilde{\gamma}\left(\ln \Theta_a + \frac{1}{\varrho_1^2}\right), \quad \text{and} \quad B_s = \tilde{\gamma}\left(\ln \Theta_s + \frac{1}{\varrho_1^2}\right).$$

Clearly, the dimensionless temperature can be calculated using the relation $\Theta = \exp[\ln \Theta]$ with Eq. (2.85). Lee pointed out in Ref. [88] two limiting cases: first is the limit $\varrho_1 \to 0$ giving

$$\frac{d^2(\ln \Theta)}{d\xi^2} = -\gamma_{(m)}^2 Y^2, \Theta(\xi = 0) = \Theta_a \quad \text{and} \quad \Theta(\xi = 1) = \Theta_s, \tag{2.86}$$

but this is not exactly the CPM case but the case discussed in the Section 2.5.1 with an inverse temperature dependence of the thermal conductivity given by Mahan [25]. The solution of Eq. (2.86) is given by

$$\ln \Theta = \ln \Theta_a + \left[\frac{\gamma_{(m)}^2 Y^2}{2} + \ln\left(\frac{\Theta_s}{\Theta_a}\right)\right]\xi - \frac{\gamma_{(m)}^2 Y^2}{2}\xi^2, \tag{2.87}$$

which can be verified by taking the same limit in Eq. (2.85). The classical Ioffe-CPM result can be observed in Lee's nomenclature by

$$\Theta_{CPM} = \Theta_a + \left[\Theta_a - \Theta_s + \frac{\gamma_{(m)}^2 Y^2}{2}\right]\xi - \frac{\gamma_{(m)}^2 Y^2}{2}\xi^2. \tag{2.88}$$

As usual, this case is used as a kind of "standard" to value the effects of variable electrical resistivity and thermal conductivity.

2.5.2.1 Performance Equations

To determine the maximum performance parameters, that is, for the TEG, the maximum power output (density) and the maximum efficiency, the energy balances at the absorbing (hot) side and the sink (cold) side have to be transformed into the dimensionless form

$$\dot{Q}_a = -\kappa A_c \left.\frac{dT}{dx}\right|_{x=0} + \alpha_{(m)} T_a I \Rightarrow \dot{Q}_a = -\left.\frac{d(\ln \Theta)}{d\xi}\right|_{\xi=0} + Y\Theta_a \tag{2.89a}$$

$$\dot{Q}_s = -\kappa A_c \frac{dT}{dx}\bigg|_{x=L} + \alpha_{(m)} T_s I \Rightarrow \dot{\tilde{Q}}_s = -\frac{d(\ln \Theta)}{d\xi}\bigg|_{\xi=1} + Y\Theta_s, \quad (2.89b)$$

$$P_{el} = \dot{Q}_a - \dot{Q}_s \Rightarrow \tilde{P}_{el} = \dot{\tilde{Q}}_a - \dot{\tilde{Q}}_s \quad \text{and} \quad \eta = \frac{P_{el}}{\dot{Q}_a} \Rightarrow \eta = \frac{\tilde{P}_{el}}{\dot{\tilde{Q}}_a} \quad (2.89c)$$

leading to the following transformation relations

$$\dot{\tilde{Q}} = \frac{L}{\kappa_{(m)} T_m A_c} \dot{Q}, \quad \tilde{P}_{el} = \frac{L}{\kappa_{(m)} T_m A_c} P_{el}, \quad \text{and} \quad \tilde{\eta} = \eta. \quad (2.89d)$$

Setting Eq. (2.85) in Eqs. (2.89a)–(2.89c) results in

$$\dot{\tilde{Q}}_a = Y[B_a \cot(\tilde{\gamma} Y) - B_s \csc(\tilde{\gamma} Y) + \Theta_a], \quad (2.90a)$$

$$\dot{\tilde{Q}}_s = Y[B_a \csc(\tilde{\gamma} Y) - B_s \cot(\tilde{\gamma} Y) + \Theta_s], \quad (2.90b)$$

$$\tilde{P}_{el}(Y) = Y\left[\Theta_a - \Theta_s - (B_a + B_s)\tan\left(\frac{\tilde{\gamma} Y}{2}\right)\right], \quad (2.90c)$$

$$\eta(Y) = \frac{\Theta_a - \Theta_s - (B_a + B_s)\tan(\tilde{\gamma} Y/2)}{\Theta_a + B_a \cot(\tilde{\gamma} Y) - B_s \csc(\tilde{\gamma} Y)}. \quad (2.90d)$$

Clearly, the solutions of the performance parameters contain trigonometric, oscillatory functions for which special care has to be taken to obtain physically reasonable results. The values Θ_a, Θ_s, and $\tilde{\gamma}$ are fixed and arbitrary currents Y can be selected although it has to be clarified if the TEG can deliver these currents. In the CPM case, the assumption that the electrical power output has to be positive and the second law of thermodynamics has to be fulfilled allows to uniquely physically obtain the permissible values of the current. This is not the case here and can lead to interpretational dangers. Suppose that $\tilde{\gamma} Y_0$ is found in the range $0 < \tilde{\gamma} Y_0 < \pi$, the corresponding power is $\tilde{P}_{el,0}$, whereas the efficiency η_0 should clearly be less than the Carnot efficiency η_C. The periodic character of the trigonometric functions in Eqs. (2.90d) and (2.90c) represents for higher currents

$$Y_n = Y_0 + \frac{2\pi n}{\tilde{\gamma}} \quad (n \in \mathbb{N}) \quad (2.91)$$

higher power output with the same efficiency values

$$\tilde{P}_{el,n} = \frac{Y_n}{Y_0} \tilde{P}_{el,0} \quad \text{and} \quad \eta_n = \eta_0. \quad (2.92)$$

For large n, it becomes clear that $\tilde{P}_{el,n}$ diverges. For a TEG, a connection between the current and the external load resistance has to be established $\tilde{P}_{el} = I^2 R_L$ with help of Eq. (2.90c):

$$R_L = \frac{1}{Y}\left[\frac{\varrho_{(m)} L}{A_c}\right]\left[\Theta_a - \Theta_s - (B_a + B_s)\tan\left(\frac{\tilde{\gamma} Y}{2}\right)\right]. \quad (2.93)$$

It can be easily verified that for sufficiently low values of R_L, there are multiple solutions of Y, which fulfill Eq. (2.93). From a physical standpoint, it is clear that

Y increases continuously from the open-circuit case $Y = 0$ where $R_L \to \infty$ to the short-circuit case when $R_L = 0$. This range is

$$0 \le Y \le Y_{sc} = \frac{2}{\tilde{\gamma}} \arctan\left(\frac{\Theta_a - \Theta_s}{B_a + B_s}\right), \tag{2.94}$$

where Y_{sc} is the short-circuit current. The performance parameters for the particular cases with $\varrho_1 = 0$ and the CPM case can be expressed with the help of Eqs. (2.87) and (2.88). For the hyperbolic temperature dependence of κ, it is

$$\tilde{P}_{el} = Y(\Theta_a - \Theta_s) - \gamma_{(m)}^2 Y^2, \tag{2.95a}$$

$$\eta = \frac{Y(\Theta_a - \Theta_s) - \gamma_{(m)}^2 Y^2}{Y\Theta_a - \frac{\gamma_{(m)}^2 Y^2}{2} + \ln\left(\frac{\Theta_a}{\Theta_s}\right)}, \tag{2.95b}$$

whereas for the CPM case, we find the performance parameters in the dimensionless form

$$\tilde{P}_{el,cpm} = Y(\Theta_a - \Theta_s) - \gamma_{(m)}^2 Y^2, \tag{2.96}$$

$$\eta_{cpm} = \frac{Y(\Theta_a - \Theta_s) - \gamma_{(m)}^2 Y^2}{Y\Theta_a + \Theta_a - \Theta_s - \frac{\gamma_{(m)}^2 Y^2}{2}}. \tag{2.97}$$

Note that Eqs. (2.95) and (2.96) have the same explicit form and hence it seems that the hyperbolic temperature profile of κ (with constant α and ϱ) has no influence on the performance. For a comparison, similar working conditions have to be chosen. Of course, the maximum performance parameters are of great interest, that is, maximum power output and maximum efficiency. As it can be seen easily, the maximum power output is the same for both cases, whereas the efficiencies differ.

2.5.2.2 Maximum Power Output

As the power output is a function of the current or, better to say, a function of the external load resistance, adjusting this load properly will lead to a value of the current that maximizes the power output. This condition can be calculated by

$$\left.\frac{\partial \tilde{P}_{el}}{\partial Y}\right|_{Y=Y_{max,P}} = 0. \tag{2.98}$$

Doing this in the case of $\varrho_1 \ne 0$ by using Eq. (2.90c), the relation

$$[1 + \tilde{\gamma} Y_{max,P} \csc(\tilde{\gamma} Y_{max,P})] \tan\left(\frac{\tilde{\gamma} Y_{max,P}}{2}\right) = \frac{\Theta_a - \Theta_s}{B_a + B_s}, \tag{2.99}$$

where the values of $Y_{max,P}$ are restricted in the limits given by Eq. (2.94). From Eq. (2.99), $Y_{max,P}$ has to be calculated (numerically) and then $P_{el,max}$ can be determined by using $Y_{max,P}$ in Eq. (2.90c). The value of the efficiency at this current value can be of interest as well, that is, $\eta(Y_{max,P}) = \eta^*$ using Eqs. (2.99) and (2.90d). For the particular case of $\varrho_1 = 0$, the same procedure leads to

$$Y_{max,P} = \frac{\Theta_a - \Theta_s}{2\gamma_{(m)}^2}. \tag{2.100}$$

Substitution of this value into Eq. (2.95) gives the maximum power value

$$\tilde{P}_{el,max} = \frac{(\Theta_a - \Theta_s)^2}{4\gamma_{(m)}^2}. \tag{2.101}$$

By using $Y_{max,P}$ in the equation for the thermal efficiency, see Eq. (2.95), it results in

$$\eta^* = \frac{(\Theta_a - \Theta_s)^2}{(\Theta_a - \Theta_s)(\Theta_a + 1) + 4\gamma_{(m)}^2 \ln\left(\frac{\Theta_a}{\Theta_s}\right)}. \tag{2.102}$$

It is clear that for the CPM case, the values of $Y_{max,P}$ and $\tilde{P}_{el,max}$ are the same as in Eqs. (2.100) and (2.101), respectively. Of course, the value of the efficiency at $Y_{max,P}$ is different compared to the result given in Eq. (2.102), which can be seen if Eq. (2.100) is substituted in Eq. (2.97)

$$\eta^*_{cpm} = \frac{(\Theta_a - \Theta_s)}{\Theta_a + 1 + 4\gamma_{(m)}^2}. \tag{2.103}$$

As $\ln(\Theta_a/\Theta_s) > \Theta_a - \Theta_s$ as long as $\Theta_a \neq \Theta_s$, it can be easily seen that the value of the efficiency in Eq. (2.102) is lower than that of CPM in Eq. (2.103). It is obvious that this difference is due to the hyperbolic temperature difference of the thermal conductivity, that is, $\kappa \propto 1/T$, such that at the hot junction, there is a higher heat absorption in comparison to what is needed if κ is supposed to be constant.

In Figure 2.15, the normalized power output is illustrated. This is the ratio between $\tilde{P}_{el}(Y_{max,P})$ using Eqs. (2.90c) and (2.99) to $\tilde{P}_{el,max}(\varrho_1 = 0)$ defined in Eq. (2.101). Clearly, it can be seen that the (normalized) maximum power output $\hat{P}_{el,max}$ increases with increasing ϱ_1.

2.5.2.3 Maximum Efficiency

The operating mode gaining the maximum thermal efficiency needs to adjust the load and with that the corresponding current differently compared with the case to attain maximum power. The current denoted as $Y^\dagger$, which maximizes the efficiency, can be determined by

$$\left.\frac{\partial \eta}{\partial Y}\right|_{Y=Y^\dagger} = 0. \tag{2.104}$$

A straightforward calculation ends up in a determination equation for $Y^\dagger$

$$C_1 \cot(\tilde{\gamma} Y^\dagger) = C_2 \csc(\tilde{\gamma} Y^\dagger) + C_3$$

with

$$C_1 = B_a \Theta_a + B_s \Theta_s, \quad C_2 = B_a \Theta_s + B_s \Theta_a, \quad \text{and} \quad C_3 = B_a^2 - B_s^2 \tag{2.105}$$

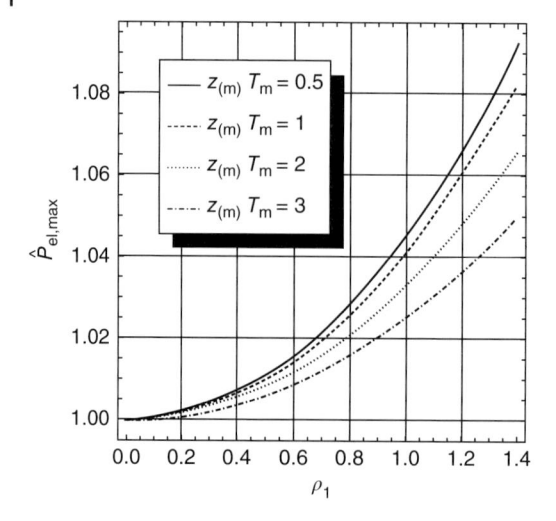

Figure 2.15 Effect of variable electrical resistivity and variable thermal conductivity on maximum power output, $\hat{P}_{el,max} = \tilde{P}_{el}(Y_{max},p)/\tilde{P}_{el,max}(\varrho_1 = 0)$.

In another form, Eq. (2.105) can be written as

$$\csc(\tilde{\gamma} Y^\dagger) = \frac{C_2 \pm C_1 M}{2(\Theta_a - \Theta_s)}, \tag{2.106}$$

where

$$M = \sqrt{1 + \frac{2(\Theta_a - \Theta_s)}{\gamma_{(m)}^2 \ln\left(\frac{\Theta_a}{\Theta_s}\right) [\varrho_1^2 \ln(\Theta_a \Theta_s) + 2]}}. \tag{2.107}$$

From the solution in Eq. (2.106), only the term with the plus sign is relevant. On the one hand, a negative sign would lead to violation of the second law of thermodynamics, and on the other hand, from $C_1 > C_2$ and $M > 1$ using the condition that $\varrho \geq 0$, the plus sign has to be chosen to fulfill the permissible range of $\tilde{\gamma} Y$, see Eq. (2.94). After a straightforward but tedious calculation using Eqs. (2.105), (2.106), (2.90d), and the following relations

$$C_1 - C_2 = (B_a - B_s)(\Theta_a - \Theta_s),$$
$$C_1 + C_2 = 2(B_a + B_s),$$
$$B_a C_1 - B_s C_2 = (B_a^2 - B_s^2)\Theta_a,$$
$$B_a C_2 - B_s C_1 = (B_a^2 - B_s^2)\Theta_s,$$

which yields the result with the physically relevant plus sign

$$\eta_{max} = \eta(Y^\dagger) = \eta_C \frac{M-1}{M + \Theta_s/\Theta_a}. \tag{2.108}$$

Lee pointed out in Ref. [88] that the results found in Eqs. (2.107) and (2.108) had been expressed in a more general form by Cohen and Abeles [89]. In their method, it is not even necessary to solve for the temperature profile. The conclusions that an increasing M leads to an increasing efficiency for (fixed boundary temperatures) can be extracted from the Ioffe-CPM approach. For example, M is increased if $1/\gamma_{(m)} = z_{(m)} T_m$ is increased. It becomes clear from Eq. (2.107) that M increases as ϱ_1^2 increases because $\ln(\Theta_a \Theta_s) \leq 0$. A maximum of M is attained if $\varrho_1^2 = -1/\ln \Theta_s$ as a direct consequence of $\varrho \geq 0$ and Eq. (2.81). Again the particular case can be investigated separately, for example, $\varrho_1 = 0$ for which the calculation is simple and gives

$$Y^\dagger = \frac{\Theta_a - \Theta_s}{\gamma_{(m)}^2 (M+1)} \quad \text{with} \quad M = \sqrt{1 + \frac{\Theta_a - \Theta_s}{\gamma_{(m)}^2 \ln\left(\frac{\Theta_a}{\Theta_s}\right)}}, \qquad (2.109)$$

which can be derived as a limiting case of Eq. (2.107) as well. The efficiency is given by Eq. (2.108) with the corresponding M from Eq. (2.109), whereas

$$P_{el}^\dagger = P_{el}(Y^\dagger) = \frac{(\Theta_a - \Theta_s)^2 M}{\gamma_{(m)}^2 (M+1)^2}. \qquad (2.110)$$

Of course, the Ioffe-CPM values can be determined in this framework

$$Y_{cpm}^\dagger = \frac{\Theta_a - \Theta_s}{\gamma_{(m)}^2 (M_{cpm}+1)}, \quad \eta_{max,cpm} = \eta_C \frac{M_{cpm}-1}{M_{cpm}+\Theta_s/\Theta_a}, \quad \text{and}$$

$$P_{el,cpm}^\dagger = \frac{(\Theta_a - \Theta_s)^2 M_{cpm}}{\gamma_{(m)}^2 (M_{cpm}+1)^2}. \qquad (2.111)$$

The two temperature dependencies of ϱ and κ have counteracting effects as the hyperbolic temperature dependence of the thermal conductivity results in a reduction of the maximum efficiency below the CPM value; this efficiency can be increased if ϱ_1^2 is increased. For all values of $z_{(m)} T_m$, there exists a precise value where the maximum efficiencies are the same. This threshold value is given by

$$(\varrho_1^2)_{thr} = \frac{2}{\ln(\Theta_a \Theta_s)} \left[\frac{\Theta_a - \Theta_s}{\ln\left(\frac{\Theta_a}{\Theta_s}\right)} - 1 \right] \qquad (2.112)$$

The power output at maximum efficiency operation (electrical current density $Y_\dagger$) is always greater than that in the CPM case, and the deviation increases if ϱ_1^2 is increased.

2.5.3
Constant Thomson Coefficient – Logarithmic Behavior of the Seebeck Coefficient

One of the first investigations concerning the temperature dependence of material properties can be found in Burshtein's work [42]. He considered chemically homogeneous (no spatial dependence) but physically inhomogeneous, that is, temperature-dependent, isotropic material properties. Therefore, the heat balance equation (2.4) has the following form[10]

$$\frac{d^2 T}{dx^2} + \frac{d \ln \kappa}{dT}\left(\frac{dT}{dx}\right)^2 - j\frac{\tau}{\kappa}\frac{dT}{dx} + j^2\frac{\varrho}{\kappa} = 0. \qquad (2.113)$$

He already stated that it is in general not possible to find a closed solution for arbitrary $\kappa(T)$, $\varrho(T)$, and $\alpha(T)$ and provided the solution of Ioffe et al. [49, 90], where the material coefficients (here besides $\kappa(T)$ and $\varrho(T)$, the Thomson coefficient $\tau = T\frac{d\alpha}{dT}$) are supposed to be constant or the true temperature-dependent materials replaced by average values in the working temperature range, which results in a slightly different temperature profile compared to Eq. (2.14)

$$T(x) = T_a + \frac{\Delta T - (\varrho j/\tau)L}{1 - e^{\tau j L/\kappa}}(1 - e^{\tau j x/\kappa}) + \frac{\varrho j}{\tau}x. \qquad (2.114)$$

The heat flows

$$-\kappa A_c \left.\frac{dT}{dx}\right|_{x=0} \quad \text{and} \quad -\kappa A_c \left.\frac{dT}{dx}\right|_{x=L}$$

can be calculated with the help of Eq. (2.114). Burshtein expressed the result in terms of "partial" heat flows

$$\dot{Q}_F = \kappa A_c \frac{\Delta T}{L}, \quad \dot{Q}_J = I^2 \varrho \frac{L}{A_c}, \quad \dot{Q}_T = I\tau\Delta T \qquad (2.115)$$

which can be expressed as heat fluxes in our notation, see again footnote 3,

$$q_A^F = \kappa \frac{\Delta T}{L}, \quad q_A^J = j^2 \varrho L, \quad q_A^T = j\tau \Delta T, \qquad (2.116)$$

where the sub-/superscripts F, J, and T denote the parts of Fourier, that is, the dissipative thermal conductivity, Joule, and Thomson contributions. A series expansion in terms of $\xi_T = \dot{Q}_T/\dot{Q}_F = q_A^T/q_A^F$ and $\xi_J = \dot{Q}_J/\dot{Q}_F = q_A^J/q_A^F$ is found then by assuming that in practical cases of a TEG, the ratio $|q_A^T|/q_A^F$ is less than 0.3. The calculation shows that the main contribution to the heat flow/flux is primarily determined by the Fourier contribution. The first correction terms are given as $1/2\xi_J$ and $1/2\xi_T$, and the next set of correction terms are no more than 5% of the fundamental corrections such that for the heat flux at the boundaries, we get

$$q_A|_{x=0} = -\kappa \left.\frac{dT}{dx}\right|_{x=0} = q_A^F - \frac{1}{2}(q_A^J + q_A^T) = \kappa \frac{\Delta T}{L} - \frac{1}{2}j^2\varrho L - \frac{1}{2}j\tau\Delta T \qquad (2.117a)$$

$$q_A|_{x=L} = -\kappa \left.\frac{dT}{dx}\right|_{x=L} = q_A^F + \frac{1}{2}(q_A^J + q_A^T) = \kappa \frac{\Delta T}{L} + \frac{1}{2}j^2\varrho L + \frac{1}{2}j\tau\Delta T \qquad (2.117b)$$

from which we see that the fundamental corrections are entered symmetrically and independently. The interactions of the electric effects are expressed in higher

10) All the expressions are adapted to our notation.

2.5 Temperature-Dependent Materials – Analytic Calculations

terms, which are neglected in this approximation. It is possible to transform the original equation (2.113) by substituting

$$y = -\frac{\kappa}{j}\frac{dT}{dx},$$

which results in

$$\frac{dy}{dT} + \frac{\varrho\kappa}{y} + \tau = 0. \tag{2.118}$$

As for the original equation, this cannot be solved by quadrature for arbitrary temperature-dependent coefficients $\varrho(T), \kappa(T)$, and $\tau(T)$. There are several special cases given by Burshtein [42]. The magnitude of the order of y is controlled by the working conditions and the design of the thermoelectric elements such as $j, \Delta T, L$, and A_c. There are approximations for large y, which means a small resistivity, that is, $\varrho \approx 0$, and a small Joule effect in comparison to the Thomson effect, and for small y, where $\tau \approx 0$, the opposite is true, that is, the Thomson effect is negligible in comparison to the Joule effect. Both limiting cases are of certain interest because generators work at high y, whereas for the operation of heaters and coolers, a low y is found. In the approximation of $y \gg \varrho\kappa/\tau$, the influence of the Thomson effect is highlighted, whereas for $y \ll \varrho\kappa/\tau$, the influence of the Joule effect is investigated. As a final result, Burshtein found

$$-\kappa\frac{dT}{dx}\bigg|_{x=0} = \frac{1}{L}\int_{T_s}^{T_a}\kappa\,dT - j^2 L\frac{\int_{T_s}^{T_a}\left[\varrho\kappa\int_{T_s}^{T}\kappa(T')dT'\right]dT}{\left(\int_{T_s}^{T_a}\kappa\,dT\right)^2}$$

$$-j\frac{\int_{T_s}^{T_a}\left[\tau\int_{T_s}^{T}\kappa(T')dT'\right]dT}{\int_{T_s}^{T_a}\kappa\,dT}, \tag{2.119a}$$

$$-\kappa\frac{dT}{dx}\bigg|_{x=L} = \frac{1}{L}\int_{T_s}^{T_a}\kappa\,dT + j^2 L\frac{\int_{T_s}^{T_a}\left[\kappa\int_{T_s}^{T}\varrho(T')\kappa(T')dT'\right]dT}{\left(\int_{T_s}^{T_a}\kappa\,dT\right)^2}$$

$$+j\frac{\int_{T_s}^{T_a}\left[\kappa\int_{T_s}^{T}\tau(T')dT'\right]dT}{\int_{T_s}^{T_a}\kappa\,dT}. \tag{2.119b}$$

It can be shown that q_A^F and q_A^T are defined as before in Eq. (2.116), whereas for the Joule contribution we get

$$q_A^J = j^2 L\frac{\int_{T_s}^{T_a}\varrho\kappa\,dT}{\int_{T_s}^{T_a}\kappa\,dT} = j^2\bar{\varrho}L. \tag{2.120}$$

The averaging of ϱ as shown in the previous equation is strictly valid in the limit $j \to 0$ but is used for nonzero values as a first approximation, see [68, p. 148].

2.5.3.1 Methods of Averaged Coefficients and Geometric Optimization of a TE Couple

In the book *Thermoelectric and thermomagnetic effects and its applications* [52], the elementary theory of the efficiency of thermoelectric devices is described in Chapter 6. The "method of averaged coefficients" as proposed by Harman and Honig [52] can be seen as a semianalytical method. There it is pointed out that the material properties such as α, $\sigma(\varrho)$, and κ have to be determined in dependence on the spatial coordinates (and/or on T). Nevertheless, the calculation is performed approximately for the sake of simplicity. Doing so, averages of the material coefficients are used. It is not stated which average is considered (temperature average or volume/spatial average) to be more general.

In 1957, Ioffe considered that there are two possibilities for averaging TE material properties [49]: On the one hand, the average over space, which is denoted here with the subscript "av", and on the other hand, the average over temperature, which is denoted with an overbar. If the material properties show only one kind of dependence, either temperature dependence, for example, $\alpha = \alpha[T(x)]$, or position dependence, for example, $\alpha = \alpha(x)$, then there is the possibility to transform them into each other if the temperature profile is known and if it is unique (monotonous), because a one-to-one correspondence can be observed, that is, $T(x) \leftrightarrow x(T)$. For example, the averages for the Seebeck coefficient can be related to each other as follows:

$$\bar{\alpha} = \frac{1}{T_s - T_a} \int_{T_a}^{T_s} \alpha(T)\,dT = \frac{1}{L} \int_0^L \alpha(x) \frac{T'(x)}{(T_s - T_a)/L}\,dx$$

$$\approx \frac{1}{L} \int_0^L \alpha(x)\,dx = \alpha_{av},$$

if $\alpha(x) = \alpha[T(x)]$, that is, if there is no explicit spatial dependence.

Straightforwardly, it can be found that $\int_0^L T'(x)\,dx = \int_{T_a}^{T_s} dT = T_s - T_a$ and that the function $L T'(x)/(T_s - T_a)$ varies around unity. Here it can be verified that both averages are close together for moderate gradients. Further, note that both averages coincide in general for a linear temperature profile, but also clearly for a nonlinear, parabolic temperature profile, which is found in the Ioffe-CPM approximation. For a temperature-dependent material Ioffe suggested

$$\varrho_{av} = \ln\left(\frac{T_a}{T_s}\right) \int_{T_a}^{T_s} \varrho(T) \frac{1}{T}\,dT.$$

as an averaging formula for the resistivity. He stated that the majority of the semiconductors have $\rho T^{-n} = $ const. as the temperature dependence of the resistivity, for which, we obtain

$$\varrho_{av} = \frac{1}{n} \ln\left(\frac{T_a}{T_s}\right) (\varrho_a - \varrho_s),$$

see in [49, p. 62].

The focus in the *method of averaged coefficients* introduced by Harman and Honig, see [52, ch. 6], lies on averaging of κ, ϱ (σ), and τ instead of averaging of the Seebeck coefficient α, where the latter results in a parabolic temperature

2.5 Temperature-Dependent Materials – Analytic Calculations

profile. As the average is a constant, this procedure ends up in the same form of heat equation as described by Burshtein [42]. For the temperature distribution, Eq. (2.114) is verified. For finding the maximum performance values, the calculation in Ref. [52] is similar to that given in Section 2.3.2. The figure of merit of a TE couple is defined as

$$Z \equiv \frac{\overline{\alpha_{pn}}^2}{\overline{KR}}, \tag{2.121}$$

where the overline denotes the corresponding average and $\alpha_{pn} = \alpha_p - \alpha_n$. A similar procedure for a TE couple as in Ref. [52] had been provided by Goldsmid before [50]. In both works, it is pointed out that Eq. (2.121) has been defined for a p–n couple but that it is convenient to define the corresponding quantity for each branch as

$$z_\zeta = \frac{\alpha_\zeta^2}{\overline{\kappa_\zeta \varrho_\zeta}} = \frac{\alpha_\zeta^2}{\overline{K_\zeta R_\zeta}} \qquad \zeta = p \text{ or } n. \tag{2.122}$$

So, the resistance of a couple is a serial connection

$$R = R_p + R_n = \frac{\overline{\varrho_p} L_p}{A_{c,p}} + \frac{\overline{\varrho_n} L_n}{A_{c,n}} \tag{2.123}$$

as well as the thermal conductance

$$K = K_p + K_n = \frac{\overline{\kappa_p} A_{c,p}}{L_p} + \frac{\overline{\kappa_n} A_{c,n}}{L_n}. \tag{2.124}$$

In general, there is no simple relation between z_p, z_n, and Z; Z has to be regarded as a rather complicated average of z_p and z_n, but qualitatively, it is totally clear that the larger the figure of merit of the individual branches, the greater is Z for the couple. By adjusting the geometric parameters (e.g., length L_p, L_n, cross-sectional area $A_{c,p}$, $A_{c,n}$), the figure of merit can be optimized making the product of KR a minimum. The product can be evaluated by using Eqs. (2.123) and (2.124)

$$KR = \overline{\kappa_p \varrho_p} + \overline{\kappa_n \varrho_n} + \overline{\kappa_p \varrho_n} \frac{L_n A_{c,p}}{L_p A_{c,n}} + \overline{\kappa_n \varrho_p} \frac{L_p A_{c,n}}{L_n A_{c,p}}. \tag{2.125}$$

Differentiating the product with respect to $L_p A_{c,n} / L_n A_{c,p}$ and equating the result with zero result in

$$\left(\frac{L_n A_{c,p}}{L_p A_{c,n}}\right)^2 = \frac{\overline{\kappa_n \varrho_p}}{\overline{\kappa_p \varrho_n}}. \tag{2.126}$$

By using Eq. (2.126), which is an adjustment requirement for the dimensions of the p and n elements of the couple, in Eq. (2.125), the product is minimized

$$(KR)_{\min} = \left(\sqrt{\overline{\kappa_p \varrho_p}} + \sqrt{\overline{\kappa_n \varrho_n}}\right)^2 \tag{2.127}$$

which directly leads to the optimal figure of merit of the couple given by

$$Z_{\mathrm{opt}} = \frac{\overline{\alpha_{pn}}^2}{\left(\sqrt{\overline{\kappa_p \varrho_p}} + \sqrt{\overline{\kappa_n \varrho_n}}\right)^2}. \tag{2.128}$$

2.5.3.2 Influence of the Thomson Effect on the Performance of a TEG

Sunderland and Burak performed with their contribution a theoretical calculation as well as a comparison with the experimental data by assuming a constant Thomson coefficient [91]. Note that in their procedure, the definition of the Thomson coefficient contains an additional minus sign such that the differential equation for the Seebeck coefficient is

$$\frac{d\alpha}{dT} = -\frac{\tau}{T}, \tag{2.129}$$

which can be straightforwardly integrated because of the constant Thomson coefficient τ

$$\alpha = -\tau \ln\left(\frac{T}{T_b}\right), \tag{2.130}$$

where T_b denotes a constant of integration. There is another possibility to express the Seebeck coefficient, which is given by

$$\alpha = \alpha_m - \tau \ln\left(\frac{T}{T_{(m)}}\right), \tag{2.131}$$

where $\alpha_m = (\alpha_a + \alpha_s)/2$ is the mean value and $T_{(m)}$ is the temperature where $\alpha = \alpha_m$. Note that $T_{(m)}$ is in general not the mean temperature, that is, $T_m = (T_a + T_s)/2$. The second representation of the Seebeck coefficient, that is, Eq. (2.131), is better if materials with the same average Seebeck coefficient in the given temperature interval but different Thomson coefficients are compared. The solution of the heat balance and the resulting temperature profile with this kind of Seebeck coefficient (and constant Thomson coefficient) are of course not different from those of Eq. (2.114). The thermal efficiency for this particular case is expressed as

$$\eta = \frac{\alpha_m \Delta T + \tau[\Delta T - T_a \ln x + T_s \ln y]}{\frac{2\varrho\kappa}{\tau} + 2T_a \alpha_a + \frac{2\tau \Delta T - \alpha_m \Delta T - \tau[\Delta T - T_a \ln x + T_s \ln y]}{1 - \exp[-\tau/(2\varrho\kappa)\{\alpha_m \Delta T + \tau[\Delta T - T_a \ln x + T_s \ln y]\}]}}, \tag{2.132}$$

where $\Delta T = T_a - T_s (= T_h - T_c)$ and $x = T_a/T_{(m)}$ as well as $y = T_s/T_{(m)}$.

Based on Sunderland's and Burak's work [91], Shaw determined the influence of the Thomson effect on the optimized performance of a TEG in Ref. [92]. Again the basic configuration of a thermocouple is used, where two thermoelements, one n type and one p type material, are with the same length L and cross-sectional area A_c. The working conditions are taken the same as before, besides that the absorbing (hot) side with a fixed temperature $T_a = T_h$ is at $x = L$, whereas the sink (cold) side with $T_s = T_c$ is at $x = 0$. Between the junctions, there is a load resistance R_L, causing the thermoelements to carry a current I. As in Ref. [91], the heat balance equations for the thermoelements are used to calculate the temperature profiles and gradients in the thermocouple. To describe the thermal efficiency, some abbreviations are defined as follows:

$$E = \frac{\overline{\varrho}_n}{\overline{\varrho}_p}, \quad F = \sqrt{\frac{\overline{\varrho}_n \overline{\kappa}_n}{\overline{\varrho}_p \overline{\kappa}_p}}, \quad m = \frac{R_L}{R_i} = \frac{A_c R_L}{L(\overline{\varrho}_n + \overline{\varrho}_p)},$$

2.5 Temperature-Dependent Materials – Analytic Calculations

where the last term is the ratio between load and internal resistance. The temperature averages of the material properties are calculated as

$$\overline{\varrho}_i = \frac{1}{\Delta T} \int_{T_s}^{T_a} \varrho(T) dT, \quad \overline{\kappa}_j = \frac{1}{\Delta T} \int_{T_s}^{T_a} \kappa(T) dT \quad (j = n, p),$$

$$\text{and} \quad \overline{\alpha}_{pn} = \frac{1}{\Delta T} \int_{T_s}^{T_a} [\alpha_p(T) + \alpha_n(T)] dT.$$

Other constants considered are the dimensionless figure of merit (denoted here as ω, the Carnot factor η_C, and a combined value ψ

$$\omega = \left(\frac{\overline{\alpha}_{pn}}{\sqrt{\overline{\varrho}_p \overline{\kappa}_p} + \sqrt{\overline{\varrho}_n \overline{\kappa}_n}} \right)^2 T_a, \quad \eta_C = \frac{T_a - T_c}{T_a},$$

$$\text{and} \quad \psi_i = \tau_i \frac{IL}{\overline{\kappa}_j A_c} \quad (j = n, p).$$

The general definition of the thermal efficiency is

$$\eta = \frac{I \overline{\alpha}_{pn} - \frac{I^2}{A_c}(\overline{\varrho}_p + \overline{\varrho}_n)}{-\overline{\kappa}_p A_c \frac{dT}{dx}\big|_{x=L} - \overline{\kappa}_n A_c \frac{dT}{dx}\big|_{x=L} + I \alpha_{pn}(T_a) T_a}. \tag{2.133}$$

By setting all the given constants and averages, the efficiency reads as

$$\frac{\eta}{\eta_C} = \left(\frac{m}{m+1} \right) \omega \left\{ \frac{E+1}{(1+F)^2} \left(\frac{\psi_p}{1 - e^{\psi_p}} \right) \right.$$

$$\times \left[\frac{\psi_n F^2}{\psi_p E} \left(\frac{1 - e^{\psi_p}}{1 - e^{-\psi_n}} \right) - 1 \right] (m+1) + \frac{\omega \alpha_{pn}(T_a)}{\overline{\alpha}_{pn}}$$

$$\left. - \frac{\omega \eta_C}{(m+1)} \left(\frac{1}{1+E} \right) \left(\frac{1}{1 - e^{\psi_p}} + \frac{1}{\psi_p} \right) \left[1 + \frac{\psi_p E}{\psi_p} \left(\frac{1 - e^{\psi_p}}{1 - e^{-\psi_n}} \right) \right] \right\}^{-1}. \tag{2.134}$$

An often used simplification assumes that the (averaged) material properties are similar in both the branches (p, n) in the sense that

$$\overline{\varrho}_n = \overline{\varrho}_p = \overline{\varrho}, \overline{\kappa}_n = \overline{\kappa}_p = \overline{\kappa}, \tau_n = -\tau_p = \tau, \text{ and } \overline{\alpha}_n = \overline{\alpha}_p = \overline{\alpha}_{pn}/2 = \overline{\alpha},$$

which gives directly $E = F = 1$ and

$$\psi = \psi_n = -\psi_p = \frac{\tau \omega \eta_C}{\overline{\alpha}(1+M)} \tag{2.135}$$

These simplifications when used in Eq. (2.134) leads to the simplified (reduced) thermal efficiency

$$\frac{\eta}{\eta_C} = \frac{\omega^{m/(m+1)}}{\frac{\psi}{1-e^{-\psi}}(m+1) + \frac{\alpha(T_a)}{\overline{\alpha}}\omega - \left(\frac{1}{1-e^{-\psi}} - \frac{1}{\psi}\right)\frac{\omega \eta_C}{m+1}}. \tag{2.136}$$

Some limiting case is given where if the Seebeck coefficient is constant, that is, $\psi = 0$, then the limits

$$\lim_{\psi \to 0}\left(\frac{\psi}{1-e^{-\psi}}\right) = 1, \quad \lim_{\psi \to 0}\left(\frac{1}{1-e^{-\psi}} - \frac{1}{\psi}\right) = \frac{1}{2} \tag{2.137}$$

can be calculated with the help of l'Hôspital's rule and reduces Eq. (2.136) to the classical Ioffe-CPM result

$$\frac{\eta_0}{\eta_C} = \frac{\omega\, m/(m+1)^2}{1 + \omega/(m+1) - \omega \eta_C[2(m+1)^2]}. \tag{2.138}$$

This result can be taken to make a comparison and value the influence of the Thomson effect on the efficiency. The parameter ψ as the ratio of the Thomson heat that is generated in the material to the heat that would be conducted in the element under zero flux condition ($I = j = 0$) has an important physical significance. For a TEG, this quantity is much smaller than unity. This can be used after setting the condition Eq. (2.135) into Eq. (2.136) to substitute m by ψ. To find the condition for maximum efficiency, the zeroes of the derivative $\partial \eta / \partial \psi = 0$ have to be calculated. As mentioned earlier, $\psi \ll 1$, so the exponential function in Eq. (2.136) can be expressed in a series expansion and approximated by just expanding up to second-order terms. A tedious but straightforward calculation leads to an optimal value

$$m_{\text{opt}} = M = \left[\left(\frac{\eta_C b}{2} + \frac{\alpha(T_a)}{\overline{\alpha}} - \frac{\eta_C}{2} - \frac{\eta_C^2 b^2 \omega}{12}\right)\omega + 1\right]^{\frac{1}{2}} = \sqrt{1 + \Gamma \omega}$$

$$\Gamma = \frac{\eta_C b}{2} + \frac{\alpha(T_a)}{\overline{\alpha}} - \frac{\eta_C}{2} - \frac{\eta_C^2 b^2 \omega}{12}, \tag{2.139}$$

where $b = 2\tau/\overline{\alpha}$. The next step is to use this optimum value to determine the optimal efficiency

$$\frac{\eta_{\text{opt}}}{\eta_C} = \frac{M - 1}{[\Gamma - \eta_C/2 + b\eta_C^2/(2\Gamma)] + (\Gamma + \eta_C/2 - b\eta_C^2/4)M}. \tag{2.140}$$

A particular case is given if the Seebeck coefficient is independent of temperature, that is, $\alpha(T_a)/\overline{\alpha} = 1, b = 0$, and $\Gamma = 1 - \eta_C/2$, which reduces Eq. (2.140) to the classical Ioffe-CPM result [49]

$$M_0 = \sqrt{(1 - \eta_C/2)\omega + 1},$$

$$\frac{\eta_{\text{opt},0}}{\eta_C} = \frac{m_0 - 1}{1 - \eta_C + M_0}.$$

Another particular case is to assume a constant Thomson coefficient τ for which the second Kelvin relation can be integrated directly using the boundary values $\alpha(T = T_a) = \alpha_a$ and $\alpha(T = T_s) = \alpha_s$ resulting in

$$\tau = \frac{\alpha_s \zeta}{2\ln(1 - \eta_C)} \quad \text{with} \quad \zeta = \frac{\alpha_a - \alpha_c}{\alpha_c} \tag{2.141}$$

and for the average Seebeck coefficient, the relation

$$\bar{\alpha} = \alpha_s \left\{ 1 + \zeta \left[\frac{1}{\eta_C} + \frac{1}{\ln(1-\eta_C)} \right] \right\} \qquad (2.142)$$

can be deduced by integration of $\alpha(T)$ from T_s to T_a. Finally, the value of b can be determined by direct substitution

$$b = \left\{ \ln(1-\eta_C) \left[\frac{1}{\zeta} + \frac{1}{\eta_C} + \frac{1}{\ln(1-\eta_C)} \right] \right\}^{-1}. \qquad (2.143)$$

The effect of the Thomson heat can be studied. Depending on the sign of the Thomson coefficient, the efficiency of a TEG can be either increased or decreased by the Thomson effect. This result clearly indicates that the effect of Thomson heat has to be carefully considered for the optimum device design of a TEG.

2.5.4
Algebraic and General Temperature Dependence

A work similar to Burshtein's had been conducted by Moizhes [43]. Later on, almost the same calculation had been analyzed by Efremov and Pushkarsky [93]. For the calculation of the TEG efficiency, they provided approximations of the parts of Joule heat (n_J) and Thomson heat (n_T), which are returnable to the hot end of the TE element

$$n_J = 1 - \frac{\int_{T_s}^{T_a} \left[\kappa(T) \int_{T_s}^{T} \kappa(T') \varrho(T') dT' \right] dT}{\overline{\kappa}(\overline{\kappa \varrho})(\Delta T)^2} \qquad (2.144a)$$

and

$$n_T = 1 - \frac{\int_{T_s}^{T_a} \left[\kappa(T) \int_{T_s}^{T} \tau(T') dT' \right] dT}{\overline{\kappa \tau}(\Delta T)^2}. \qquad (2.144b)$$

The main assumption in that case was that the Joule heat and Thomson heat have only a slight effect on the temperature distribution in comparison to the contribution of the heat due to thermal conductivity (Fourier heat). Then the approximation is to consider the temperature distribution for the open-circuit case ($j = 0$) and determine the fractions of Joule and Thomson heat.

Note that Moizhes used another flow direction of the electrical current or boundary temperatures [$T_h = T_a$ at $x = L$, $T_c = T_s$ at $x = 0$, and I from $x = 0$ to $x = L$]. With the result given in Eq. (2.144), it can be shown that for $\varrho(T) =$ const., the Joule heat is divisible into two halves, that is, $n_J = 1/2$. It can be shown that this is exactly the case for all values of the current density if additionally $\tau = 0$. Furthermore, he gave an example for temperature-dependent material coefficients, where the temperature dependence has an algebraic character, that is,

$$\kappa(T) = \kappa_0 \left(\frac{T}{T_0}\right)^m, \quad \varrho(T) = \varrho_0 \left(\frac{T}{T_0}\right)^n, \quad \text{and} \quad \tau(T) = \tau_0 \left(\frac{T}{T_0}\right)^s$$

for which the contributions can be calculated. A decisive parameter in the result is the ratio between the heat reservoir temperatures $v = T_s/T_a$. For the value $v = 1/2$ or $T_s = 300K$ and $T_a = 600K$, respectively, two examples are calculated. On the one hand, $n = 2$ and $m = s = 0$ (solid solution), the value for $n_J = 0.642$ and $n_T = 0.500$; on the other hand, $m = -1$ (crystals of stoichiometric composition) are chosen, which result in $n_J = 0.608$ and $n_T = 0.550$. In summary, calculations show that the values for n_J and n_T do not greatly differ from $1/2$.

A further result of Moizhes' investigations is the generalized version of the maximum efficiency for a general temperature dependence of the material with the help of averages. First, he performed a geometric optimization. He assumed that the two elements of a *pn* couple have the same length L, then the cross-sectional areas have to be adjusted to achieve the maximum efficiency

$$\frac{A_c^{(p)}}{A_c^{(n)}} = \frac{\overline{\kappa_n}}{\overline{\kappa_p}} \sqrt{\frac{\overline{\varrho_p \kappa_p}}{\overline{\varrho_n \kappa_n}}}. \tag{2.145}$$

For a *pn* thermogenerator couple with a load-to-internal resistance ratio $\tilde{M} = R_L/R_i$, where $R_i = R_i^{(p)} + R_i^{(n)}$ and the geometric optimization, see Eq. (2.145), Moizhes then found

$$\eta = \frac{\eta_C}{\beta} \frac{\tilde{M}}{1+\tilde{M}} \frac{1}{1 - \frac{a}{1+\tilde{M}} + \gamma(1+\tilde{M}) - b} \tag{2.146a}$$

with the following notation

$$a = \frac{\eta_C}{\beta} \frac{n_J^{(p)} R_i^{(p)} + n_J^{(n)} R_i^{(n)}}{R_i}; \quad b = \eta_C \frac{n_T^{(p)} \overline{\tau_p} + n_T^{(n)} \overline{\tau_n}}{\alpha_a^{(p)} + \alpha_a^{(n)}}; \quad \beta = \frac{\alpha_a^{(p)} + \alpha_a^{(n)}}{\overline{\alpha_p} + \overline{\alpha_n}};$$

$$\gamma = \frac{1}{\beta T_a Z}, \tag{2.146b}$$

where the device-Z of the couple is

$$Z = \left(\frac{\overline{\alpha_p} + \overline{\alpha_n}}{\sqrt{\overline{\kappa_p \varrho_p}} + \sqrt{\overline{\kappa_n \varrho_n}}}\right)^2. \tag{2.146c}$$

The expression (2.146a) for the efficiency has an optimal value for

$$\tilde{M} = \sqrt{\frac{1+\gamma-b-a}{\gamma}} = \sqrt{1+(1-b-a)\beta T_a Z} \equiv M_0. \tag{2.147}$$

Substituting this resistance ratio M_0 (2.147) into Eq. (2.146), the maximum efficiency gets

$$\eta_{max} = \frac{\eta_C}{\beta} \frac{M_0 - 1}{M_0(1-b) + 1 - b - 2a}. \tag{2.148}$$

As a particular case of this maximum efficiency, we can verify the Ioffe-CPM approximation by using $b = 0$, $\beta = 1$, $n_J^{(p)} = n_J^{(n)} = 1/2$, and $a = (T_a - T_s)/(2T_a)$ in Eqs. (2.146) and (2.148).

2.5.5
Constant Thomson Coefficient Combined with Linear Temperature Dependence of Resistivity

Adamson and Sunderland's work [94] as a condensed part of Adamson's master thesis [95] deals with the influence of the Thomson coefficient connected with a variable resistivity on the performance of a thermoelectric cooler (TEC). For their investigation, they chose a standard single device configuration where the (hot) absorbing side with the temperature T_a is at $x = 0$, whereas the (cold) sink side with the temperature T_s is at $x = L$. The lateral surfaces of the TE element are perfectly insulated. The linear temperature dependence of the resistivity is expressed as

$$\varrho(T) = \varrho_{(m)} + B(T - T_m), \tag{2.149}$$

where $\varrho_{(m)} = \varrho(T = T_m)$. As for the use of semiconductors in heat pumps, the resistivity change with increasing temperature is supposed to be positive, the parameter B is considered to be positive in this respect. The linear dependence defined in Eq. (2.149) allows the comparison of materials with the same average resistivity represented by $\varrho_{(m)}$. Again the Thomson effect is considered due to the constant Thomson coefficient, which means a logarithmic behavior of the corresponding Seebeck coefficient, that is,

$$\tau = T\frac{d\alpha}{dT} \Rightarrow \alpha = \alpha_m + \tau \ln\left(\frac{T}{T_{(m)}}\right), \tag{2.150}$$

where $\alpha_m = {(\alpha_a + \alpha_s)}/{2}$ is the average Seebeck coefficient and $T_{(m)}$ is the temperature where this value of the Seebeck coefficient can be found. As before, it is helpful for the calculation to consider dimensionless quantities

$$\xi = \frac{x}{L}, \quad \Theta = \frac{T}{T_a - T_s}, \quad Y = I\frac{L}{A_c}\frac{\tau}{\kappa} \tag{2.151}$$

leading to

$$0 = \frac{d^2\Theta}{d\xi^2} - Y\frac{d\Theta}{d\xi} + \frac{\kappa B Y^2}{\tau^2}\Theta + \frac{\kappa Y^2(\varrho_{(m)} - BT_m)}{\tau^2(T_a - T_s)} \tag{2.152}$$

with the boundary conditions

$$\Theta(0) = \Theta_a \quad \text{and} \quad \Theta(1) = \Theta_s. \tag{2.153}$$

First of all, the temperature distribution has to be determined by the solution of Eq. (2.152), which directly depends on the relative magnitudes of τ^2, $4B\kappa$, and B such that a distinction of the cases has to be made. The simplest case is the CPM case for which $\tau = 0$ and $B = 0$ are fulfilled, resulting in

$$\Theta(\xi) = \Theta_a - \xi + \frac{I^2 \varrho_{(m)} L^2}{2\kappa A_c^2(T_a - T_s)}\xi(1 - \xi) \tag{2.154}$$

showing the typical parabolic behavior for the CPM case, see Section 2.3.2. If the case $B = 0$ is considered, then only the Thomson effect influences the shape of the temperature distribution, which turns out to exhibit an exponential character

$$\Theta(\xi) = [\exp(Y\xi) + \Theta_a] \left[\frac{\frac{\kappa Y}{(T_a - T_s)\tau^2} + 1}{1 - \exp(Y)} \right] + \frac{\kappa Y \varrho_{(m)} \xi}{(T_a - T_s)\tau^2}, \quad (2.155)$$

which shows the same behavior as described already in Section 2.5.3. Another particular case that is obvious is the case where the Thomson effect is neglected and only the linear temperature dependence of the resistivity is assumed, see Section 2.5.6. The temperature distribution for this case is given by

$$\Theta(\xi) = \Theta_m - \frac{\varrho_{(m)}}{B(T_a - T_s)} + \sin(F\xi) + E\cos(F\xi) + \left[\frac{E - 1 - E\cos(F)}{\sin(F)} \right] \quad (2.156)$$

with the following abbreviations

$$E = \frac{T_a - T_m + \frac{\varrho_{(m)}}{B}}{T_a - T_s} \quad \text{and} \quad F = \frac{IL\sqrt{B\kappa}}{\kappa A_c}. \quad (2.157)$$

If both effects (Thomson effect **and** temperature dependence of the resistivity) are present, then for $B > 0$, the cases $\tau^2 > 4B\kappa$, $\tau^2 = 4B\kappa$, and $\tau^2 < 4B\kappa$ have to be considered:

Case $\tau^2 > 4B\kappa$:

$$\Theta(\xi) = \Theta_m - \frac{\varrho_{(m)}}{B(T_a - T_s)} + \exp\left(\frac{1}{2}Y\xi\right) \left\{ E\cosh(G\xi) \right.$$

$$\left. + \frac{E - 1 - E\exp\left(\frac{1}{2}Y\right)\cosh(G)}{\exp\left(\frac{1}{2}Y\right)\sinh(G)} \sinh(G\xi) \right\} \quad (2.158)$$

with

$$G = \frac{Y\sqrt{\tau^2 - 4B\kappa}}{2\tau} \quad (2.159)$$

and E as defined before in Eq. (2.157).

Case $\tau^2 = 4B\kappa$:

$$\Theta(\xi) = \Theta_m - \frac{\varrho_{(m)}}{B(T_a - T_s)} + \exp\left(\frac{1}{2}Y\xi\right) \left[E + \frac{\left[E - 1 - E\exp\left(\frac{1}{2}Y\right)\right]\xi}{\exp\left(\frac{1}{2}Y\right)} \right]$$

$$(2.160)$$

Case $\tau^2 < 4B\kappa$:

$$\Theta(\xi) = \Theta_m - \frac{\varrho_{(m)}}{B(T_a - T_s)} + \exp\left(\frac{1}{2}Y\xi\right)\left\{E\cos(H\xi)\right.$$

$$\left. + \frac{E - 1 - E\exp\left(\frac{1}{2}Y\right)\cos(H)}{\exp\left(\frac{1}{2}Y\right)\sin(H)}\sin(H\xi)\right\} \quad (2.161)$$

with

$$H = \frac{Y\sqrt{4B\kappa - \tau^2}}{2\tau}. \quad (2.162)$$

To determine the performance equations, the corresponding temperature distribution has to be taken into account, for example, the heat removed from the cold junction (sink side)

$$\dot{Q}_s = \frac{\kappa A_c (T_a - T_s)}{L} \frac{d\Theta}{d\xi}\bigg|_{\xi=1} + \left(\alpha_m + \tau \ln \frac{\Theta_s}{\Theta_{(m)}}\right) \frac{Y\kappa A_c T_s}{L\tau} \quad (2.163)$$

The COP φ is defined as the ratio of the heat pumped to the total electrical power input, that is,

$$\varphi = \frac{\dot{Q}_s}{\frac{I^2}{A_c}\int_0^L \varrho\,dx + I\int_{T_s}^{T_a} \alpha\,dT}. \quad (2.164)$$

2.5.6
Linear Temperature Dependence of the Resistivity

In Reference [96], Kooi and coworkers consider, besides thermomagnetic effects, a thermoelectric engine where $\alpha = \alpha_0 = $ constant, $\varrho \propto T$, and $\kappa = \kappa_0 = $ constant. The calculation is performed for a TEC, but it could easily be extended to the TEG case. The heat balance equation results in

$$\frac{d^2T}{dx^2} + \frac{\varrho_0 j^2}{\kappa_0 T_0} T = \frac{d^2T}{dx^2} + \omega^2 T = 0$$

as $\varrho = \varrho_0 \, T/T_0$ is assumed as temperature dependence for the electrical resistivity. As abbreviation $\omega = \sqrt{\frac{\varrho_0}{\kappa_0 T_0}} j$ is defined. With constant boundary temperatures $T(x = 0) = T_a$ and $T(x = L) = T_s$, the integration ends up in

$$T(x) = T_a \cos(\omega x) + \left[\frac{T_s - T_a \cos(\omega L)}{\sin(\omega L)}\right]\sin(\omega x). \quad (2.165)$$

With the knowledge of the temperature profile, the efficiency is defined as

$$\eta = \frac{\text{electrical power}}{\text{heat absorbed (at } x=0)} = \frac{\int_0^L \left[\alpha_0 j \frac{dT}{dx} - \frac{\varrho_0 T}{T_0} j^2\right] dx}{\alpha_0 T_a j - \kappa_0 \left.\frac{dT}{dx}\right|_{x=0}}$$

$$\eta(\omega) = \frac{\Delta T \sin(\omega L) + 2\gamma T_m [1 - \cos(\omega L)]}{T_a \sin(\omega L) + \gamma T_a \cos(\omega L) - \gamma T_s}$$

$$= \frac{(1-v)\sin(\omega L) + \gamma(1+v)[1-\cos(\omega L)]}{\sin(\omega L) + \gamma \cos(\omega L) - \gamma v} \quad (2.166)$$

with

$$\gamma = (ZT)^{-\frac{1}{2}} \quad \text{and} \quad v = \frac{T_s}{T_a}.$$

To determine the optimum current density, $d\eta/d\omega = 0$ is calculated, leading to

$$2 + [(1-v)^2 - 2]\cos(\omega L) + \gamma(1-v^2)\sin(\omega L) = 0.$$

By using this result in Eq. (2.166), we obtain

$$\eta_{\max} = \eta_C \frac{M_\eta - 1}{M_\eta + v}, \quad (2.167)$$

which is the classical Ioffe-CPM result.

Ure and Heikes used in their remarkable book *Thermoelectricity: Science and Engineering*, in Chapter 15 "Theoretical Calculation of Device Performance" [97, Section 15.6, p. 498], Burshtein's example of a linear dependence of the resistivity, see [42], that is, $\varrho(T) = \varrho_0 + \varrho_1 T$ (κ and τ independent of temperature), to solve the thermal energy balance explicitly in terms of known analytic functions. Then the temperature gradient at $z = 0$ is given by

$$T'(0) = \left[T_a + \frac{\varrho_0}{\varrho_1}\right]\left[\frac{\tau j}{2\kappa} + Cj \coth(CjL)\right]$$
$$+ \left[T_s + \frac{\varrho_0}{\varrho_1}\right] Cj \exp\left(-\frac{\tau jL}{2\kappa}\right) \operatorname{csch}(CjL) \quad (2.168)$$

with $C = \frac{\sqrt{\tau^2 - 4\varrho_1 \kappa}}{2\kappa}$. A series expansion can be performed for Eq. (2.168), if $j \ll 2\kappa/\tau$ and $j \ll (CL)^{-1}$. It results in

$$T'(0) = -\frac{\Delta T}{L} + \frac{j^2 L}{2\kappa}\left[\varrho_0 + \varrho_1 \frac{(2T_a + T_s)}{3}\right]$$
$$+ \frac{\tau j}{2\kappa}\left[\Delta T - \frac{j^2 L^2 (\varrho_0 + \varrho_1 T_s)}{6\kappa} - \frac{\tau jL\Delta T}{6\kappa}\right] + \cdots \quad (2.169)$$

The first term is obviously due to heat conduction in the absence of an electrical current, whereas the second part represents the Joule contribution, and the remainder takes the Thomson heat into account. These three terms clearly have an influence on the temperature distribution, which is a superposition of (i) the linear profile observed in the absence of an electrical current, (ii) a term due to the Joule heat, which has a maximum near the center of the TE element, and (iii) a contribution, which originates in the Thomson heat. The Thomson term has different parts. The first term is caused by the normal linear distribution, the second has its origin in the gradient produced by the Joule heat, while the last one is due the Thomson heat. A general calculation or optimization of the performance of a TE device is not given in Ref. [97].

2.5.7
Linear Temperature Dependence of the Thermal Conductivity

In Reference [96] (Appendix B), another example is illustrated by taking into account $\alpha = \alpha_0$, $\varrho = \varrho_0$, and $\kappa = \kappa_0\, T/T_0$, that is, a linear temperature dependence of the thermal conductivity is considered for a thermoelectric device. The heat balance then becomes

$$\frac{d}{dx}\left[T\frac{dT}{dx}\right] + \frac{T_0 \varrho_0}{\kappa_0} j^2 = 0 \tag{2.170}$$

A straightforward integration gives the following temperature profile

$$T(x) = \left[T_a^2 + (T_s^2 - T_a^2)\frac{x}{L} + \frac{j^2 T_0 \varrho_0}{\kappa_0} x(L-x)\right]^{\frac{1}{2}}. \tag{2.171}$$

The efficiency is then

$$\eta(j) = \frac{\alpha_0 j(T_a - T_s) + \varrho_0 j^2 L}{\alpha_0 j T_a + \frac{\kappa_0}{2L T_0}(T_a^2 - T_s^2) - \frac{\varrho_0 j^2}{2} L} \tag{2.172}$$

With the temperature average of $\kappa(T)$, that is, $\overline{\kappa} = \frac{\kappa_0}{T_0} T_m$, Eq. (2.172) becomes

$$\eta(j) = \frac{\alpha_0 j(T_a - T_s) + \varrho_0 j^2 L}{\alpha_0 j T_a + \frac{\overline{\kappa}}{L}(T_a - T_s) - \frac{\varrho_0 j^2}{2} L}, \tag{2.173}$$

which results in a similar calculation as done following Eq. (2.53) by just substituting κ by its temperature average $\overline{\kappa} = \frac{\kappa_0}{T_0} T_m$.

2.6
The Influence of Contacts and Contact Resistances on the TE Performance

A key factor of the overall performance of a TE module is definitely the contact or the installation of the thermoelectric material in the modules and the corresponding contact resistances. A distinction is made between *thermal* and *electrical* contact resistances. The thermal contact resistance reduces the effective temperature difference, which really reaches the thermoelectric material. On the other hand, the electrical contact resistance represents an additional resistance to the internal resistance of a TE element, which contributes to an increase in the Joule heat loss. The smaller the contact resistances are, the better is the performance in comparison to the ideal case.

2.6.1
Thermoelectric Element with Contacting Bridge

To determine the influence of the contacts, we consider a TE element (p-type semiconductor without loss of generality) with a contacting bridge on each side. At each of the contact interfaces between the bridge and the TE material, an electrical and a thermal contact resistance are provided. For the sake of simplicity, the resistances are chosen symmetrically and homogeneously, that is, they are equal on both sides and homogeneously distributed over the contact interface.

As before, we suppose that there are no lateral heat losses due to radiation and convection. The heat flux and electrical current are in parallel in a specific direction (x-direction without loss of generality) and the problem is quasi-1D, because additionally, the cross-sectional area A_c is constant. To carry out the calculations analytically, the material properties are assumed to be constant, that is, there is no temperature or position dependence. For the determination of the performance, the temperature profile has to be calculated along the entire body, which has a length L. The body consists of three parts and is therefore separated into three intervals, for which the temperature profile has to be determined, that is,

$$T(x) = \begin{cases} T_1(x) & \text{with } 0 \leq x < L_1 = d \\ T_2(x) & \text{with } L_1 \leq x \leq L_2 \\ T_3(x) & \text{with } L_2 < x \leq L, \end{cases}$$

where the interfaces are situated at $x = L_1$ and $x = L_2$. Again a symmetric situation is supposed, that is, the contacting bridges have the same width d and are composed of the same material. Then the TE element has a length $L_E = L_2 - L_1 = L - 2d$. For the base set of equations, we use in all of the three intervals the Ioffe–CPM energy balance, see Eq. (2.13), where only the heat conduction (Fourier) part and the Joule heat are balanced due to the constant material properties.

$$\frac{d^2 T_1}{dx^2} = -\frac{j^2}{\kappa_1 \sigma_1} \quad \text{with} \quad 0 \leq x < L_1 = d \tag{2.174a}$$

$$\frac{d^2 T_2}{dx^2} = -\frac{j^2}{\kappa_2 \sigma_2} \quad \text{with} \quad L_1 \leq x \leq L_2 \tag{2.174b}$$

$$\frac{d^2 T_3}{dx^2} = -\frac{j^2}{\kappa_1 \sigma_1} \quad \text{with} \quad L_2 < x \leq L. \tag{2.174c}$$

The Peltier–Thomson part does not occur in this approximation. The integration of this set of equations leads to parabolic profiles in every interval, that is, we end up with six integration constants $C_1, \cdots, C_6$ in total throughout the solution, which can be determined from the boundary and matching conditions. Obviously, the hot and cold side temperatures, that is, $T_a = T_h$ at $x = 0$ and $T_s = T_c$ at $x = L$, determine two of the constants

$$T_1(x = 0) = T_a \quad \text{and} \quad T_3(x = L) = T_s. \tag{2.175}$$

2.6 The Influence of Contacts and Contact Resistances on the TE Performance

The missing conditions for the determination of the rest of the integration constants are fixed by the matching of the temperatures and heat flux at the two interfaces between the TE material and the conducting bridge. It is assumed that the heat fluxes at the interfaces are continuous, that is,

$$\left[-\kappa_1 \frac{dT_1}{dx} + j\alpha_1 T_1(x)\right]_{x=L_1-} - \frac{j^2}{2} R_{c,el} = \left[-\kappa_2 \frac{dT_2}{dx} + j\alpha_2 T_2(x)\right]_{x=L_1+}$$
$$+ \frac{j^2}{2} R_{c,el} \quad (2.176a)$$

$$\left[-\kappa_2 \frac{dT_2}{dx} + j\alpha_2 T_2(x)\right]_{x=L_2-} - \frac{j^2}{2} R_{c,el} = \left[-\kappa_1 \frac{dT_3}{dx} + j\alpha_1 T_3(x)\right]_{x=L_2+}$$
$$+ \frac{j^2}{2} R_{c,el}, \quad (2.176b)$$

where $R_{c,el}$ is the electric contact resistance of each of the two contacts, which is supposed to be the same for both the interfaces as said before. Similarly, a general assumption assumed for the distribution of Joule heat caused by the contact resistance, $j^2 R_{c,el}$, which is supposed to be symmetrically and equally distributed, is that one half enters into the contacting bridge while the other half into the TE material. At the contact between two solid materials, in general, a *thermal resistance* $R_{c,th}$ occurs. There are different factors that can have an influence on the thermal resistance, for example,

- the surface quality (surface roughness),
- the material combination,
- the intermediate material/gap medium (water, air, thermal grease, oxide),
- contact pressure and so on.

The values of the thermal contact resistance are determined empirically in most cases. Often, the heat transfer coefficient (reciprocal of the specific thermal contact resistance, contact coefficient) $\tilde{\alpha}_c$ is considered with

$$\tilde{\alpha}_c = (R_{c,th} A_c)^{-1}.$$

Due to the thermal contact resistance, there is a temperature step at the interface, and hence, the final two conditions for the determination of the integration constants are given

$$T_1(L_1-) = T_2(L_1+) + \frac{1}{\tilde{\alpha}_c} \left\{ \left[-\kappa_2 \frac{dT_2}{dx} + j\alpha_2 T_2(x)\right]_{x=L_1+} + \frac{j^2}{2} R_{c,el} \right\}, \quad (2.177a)$$

$$T_2(L_2-) = T_3(L_2+) + \frac{1}{\tilde{\alpha}_c} \left\{ \left[-\kappa_1 \frac{dT_3}{dx} + j\alpha_1 T_3(x)\right]_{x=L_2+} + \frac{j^2}{2} R_{c,el} \right\}. \quad (2.177b)$$

The solution of the system (2.174) related to the boundary temperature is given by the following relation:

$$T_1(x) = T_a - \frac{j^2}{2\kappa_1 \sigma_1} x^2 + C_1 x, \tag{2.178a}$$

$$T_2(x) = -\frac{j^2}{2\kappa_2 \sigma_2} x^2 + C_3 x + C_4, \tag{2.178b}$$

$$T_3(x) = T_s + \frac{j^2}{2\kappa_1 \sigma_1} (L^2 - x^2) + C_5 (x - L), \tag{2.178c}$$

see Eq. (2.175), too. The constants C_1, C_3, C_4, and C_5 to be determined have to fulfill the conditions at the interfaces, see Eqs. (2.176) and (2.177). The calculation is straightforward but leads to extended formulae for the constants and should not be listed here. The starting point of this investigation is the calculation under ideal conditions (no contacts, parallel fluxes and currents, no convective and radiative losses, etc.) for a TE element of length $L_E = 4$ mm and with material properties that are averaged over the temperature range, see Table 2.5.

By using the given parameters, the optimal performance under ideal conditions in the framework of the Ioffe-CPM approximation can be determined. In Table 2.6, these results are compared with the numerical calculation of the temperature-dependent material parameters. The temperature averages of the material properties can be found in Table 2.5. The numerical calculations are executed with

Table 2.5 Temperature averages of the material properties for the p-type semiconductor material in the temperature interval $T = 50°\,C, \cdots, 350°\,C$ [323.15 K, $\cdots$, 623.15 K] with a middle temperature of $T_m = 473.15$ K.

$\overline{\alpha}$ (μV/K)	$\overline{\kappa}$ (W/m K)	$\overline{\sigma}$ (S/m)	f (W/K² m)	$z = \alpha^2 \sigma / \kappa$ (K^{-1})	$z T_m$ (1)
110.46	3.125	128028	$1.562 \cdot 10^{-3}$	$5 \cdot 10^{-4}$	0.24

Table 2.6 Optimal current and current density, respectively, for the maximum power output and maximum efficiency and their corresponding values for constant (temperature averaged, case A–Ioffe-CPM approximation) and temperature-dependent material properties (case B–numerical).

	Case A		Case B	
	Power output	Efficiency	Power output	Efficiency
j_{opt} (A/m²)	530333	502213	523465	500351
I_{opt} (A)	13.26	12.56	13.09	12.51
Max. (abs.)	0.2197 W	3.31%	0.2174 W	3.26%
Max. (per area A_c)	8787.25 W/m²	3.31%	8694.26 W/m²	3.26%

FEM; for a description of the method and an example calculation, see for example, Chapter 6.

For the material chosen here and the given geometric parameters as well as the boundary conditions, the Ioffe-CPM gives a good approximation.

To calculate the influence of the contacts, we assume additionally to the pure semiconductor TE element metallic contacting bridges. The metallic material of the contacts has a thermal conductivity $\kappa = 23$ W/(m K) and an electrical resistivity of $\varrho = 70\ \mu\Omega \cdot$ cm. Further, we assume that the Seebeck coefficient of the bridge material can be neglected in comparison to the TE material, because usually, the values for the Seebeck coefficient of the TE material are much larger than for metals. The width of the bridges is assumed as $d = 2.5$ mm. A single bridge has a bulk value for the (electrical) resistance of $R_B = 7$ mΩ. For the electrical contact resistance, we assumed a value of $[10^{-8} \Omega m^2]$. An absolute value for the contact resistance of $R_{c,el} = 40$ mΩ is found for one interface. The internal resistance of the TE element is given by $R_i = L/(A_c \bar{\sigma}) = 1.25$ mΩ ($L = 4$ mm and $A_c = 1$ mm^2). In contrast to this internal resistance, the overall resistance of the contacts is $R_{c,ges} = 94$ mΩ, a value that is almost two orders of magnitude greater than the internal resistance. This indicates that the maximal performance of the TE device is significantly decreased due to the contacts. Furthermore, there is an additional temperature drop caused by the thermal contact resistance, and this leads to a decrease in the effective temperature difference, which can be found at the boundaries of the TE element. This is an additional factor that reduces the performance.

We found a good agreement between the simulation results and the analytical solution. The temperature difference directly at the TE element is decreased from $\Delta T = 300$ K to an effective value of $(\Delta T)_{eff} = 265.38$ K.

The degradation of the performance is overstated as already explained, which can be seen in the tremendous decrease of the power output and the efficiency due to the bridges and the corresponding contact resistance. The maximum performance parameters without and with contacts are given and compared in Table 2.7 where the difference becomes obvious.

Table 2.7 Optimal current and current density, respectively, for the power output and the efficiency and their corresponding values for temperature-dependent material properties without contacts (ideal) and with contacts.

	Ideal		With contacts	
	Power output	Efficiency	Power output	Efficiency
j_{opt} (A/m^2)	523465	500351	6640	6920
I_{opt} (A)	13.09	12.51	0.166	0.173
Maximum (absolute)	0.2174 W	3.26%	0.00224 W	0.052%
Maximum (per area A_c)	8694.26 W/m^2	3.26%	89.6 W/m^2	0.052%

2.6.2
Numerical Example for the Influence of the Electrical Contact Resistance on the Performance

In this section, we want to highlight the influence of the electrical contact resistance on the performance parameters, especially the maximum power output density and efficiency. We use a quasi-1D arrangement with an element of length $L = 5$ mm and a cross-sectional area of $A_c = 1$ mm². On both sides of the element, we assumed a conducting layer made of copper. The resistivity of copper is taken as $\varrho = 1.7 \cdot 10^{-8}$ Ωm, while for the thermal conductivity, we take a value of around $\kappa = 400$ W/(m K). As the Seebeck coefficient of copper is much smaller than the Seebeck coefficient of the TE material, we neglected its influence. As a TE material, we choose Zn_4Sb_3 for which temperature-averaged material properties (temperature interval between $T_s = 325$ K to $T_a = 525$ K $\Rightarrow \Delta T = 200$ K) are given in Table 2.8. From the average value of the electrical conductivity, the internal resistance can be calculated approximately as $R_i = L/(A_c \bar{\sigma}) = 125$ mΩ. In Figure 2.16, the arrangement is displayed. Here we want to vary the electrical contact resistivity which can be found at the interfaces between the copper and the TE material. Besides the bulk resistance of the copper contacting bridge ($R_{c,bulk} = 3.4$ μΩ), there is then another contribution that leads to an additional Joule loss at both the interfaces. We assume an electrical contact resistivity of $\varrho_c = 100$ μΩcm² according to $R_{c,el} = 10$ mΩ per interface. With these parameters, the ratio of the total contact resistances

Table 2.8 Temperature averages of the material properties for a Zn_4Sb_3 sample in the temperature interval $T = 325$ K, $\cdots$, 525 K with a middle temperature $T_m = 425$ K.

$\bar{\alpha}$ (μV/K)	$\bar{\kappa}$ (W/m K)	$\bar{\sigma}$ (S/cm)	f (W/K² m)	$z = \alpha^2 \sigma / \kappa$ (K⁻¹)	zT_m (1)
175.25	0.7682	401.97	$1.235 \cdot 10^{-3}$	$1.607 \cdot 10^{-3}$	0.683

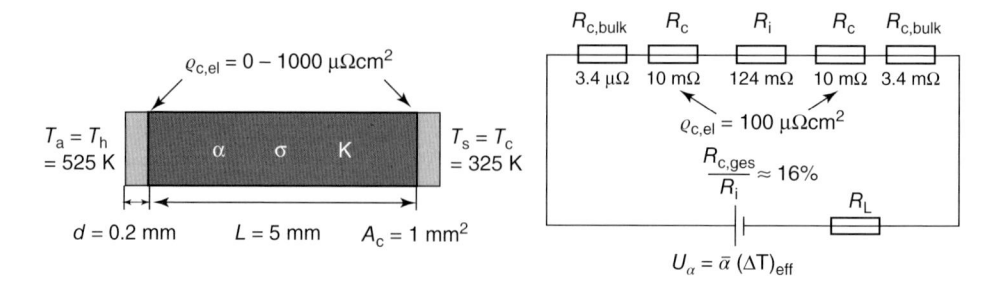

Figure 2.16 Variation of the electrical contact resistivity.

(interfaces and bridges) to the approximate internal resistance R_i is about 16%. Here we want to investigate how the increase in the contact resistivity has an influence on the performance. To show that, we varied ϱ_c in the range of $\varrho_c = 0, \cdots, 1000\,\mu\Omega\mathrm{cm}^2$ (0; 10; 50; 100; 250; 500; 750; 1000). In Figure 2.17(a) and (c), the power output densities and efficiency in dependence on the electrical current are shown for the eight values of ϱ_c. The ideal case, where the contact resistivity $\varrho_c = 0$ vanishes, yields the curves with the best performance. In the direction of the arrows, the contact resistivity is increased due to the eight values given, which results in a decrease of performance, that is, the maximum power factor and efficiency as well as the corresponding current values are decreased. The results for the maximum performance parameters are calculated for every single value of the contact resistivity and represented as red points/lines in Figure 2.17(a) and (c). The results are then summarized in Figure 2.17(b) and (d), where the maximum performance values are shown in dependence on the contact

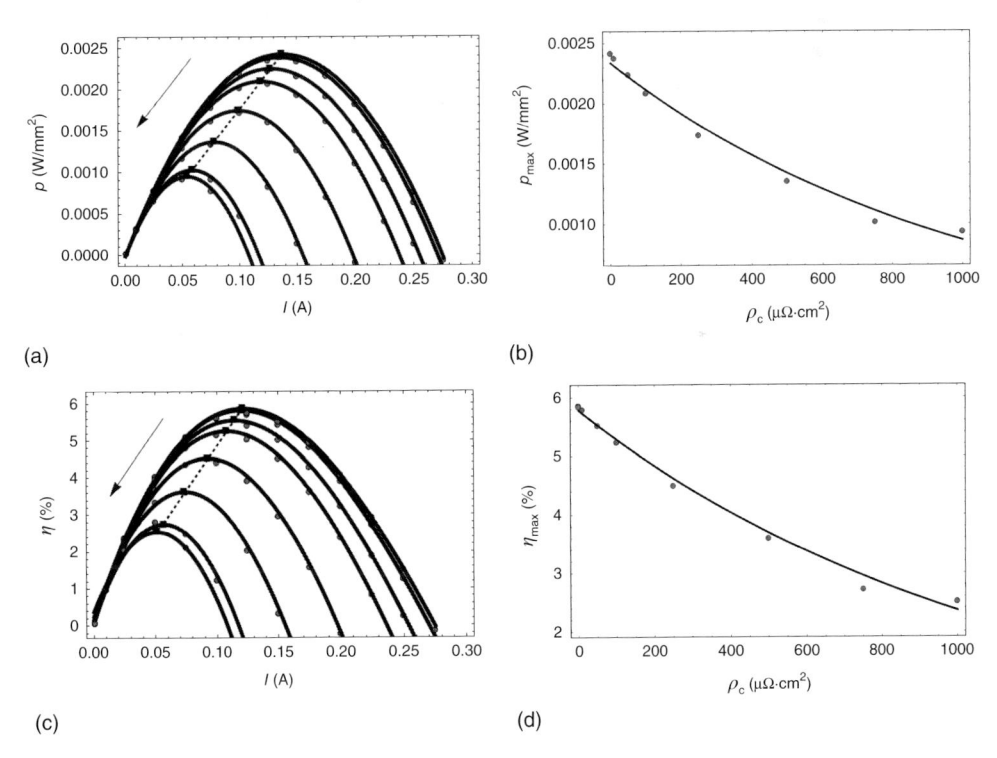

Figure 2.17 Performance parameters for different $R_{c,el}$. The contact resistivity is varied between $\varrho_c = 0, \cdots, 1000\,\mu\Omega\mathrm{cm}^2$ (0; 10; 50; 100; 250; 500; 750; 1000). The maximum values are calculated for these specific points, and the solid lines represent polynomial fits, which serve as a guidance for the general trend. (a) Power output density; (b) Maximum power output densities; (c) Efficiency; (d) Maximum efficiency.

2 Continuum Theory of TE Elements

Table 2.9 Optimal performance values for the power density and efficiency in dependence on the contact resistivity.

ϱ_c ($\mu\Omega\text{cm}^2$)	$R_{c,el}$ (mΩ)	$\frac{R_{c,tot}}{R_i}$ (%)	p_{max} (W/mm^2)	$\frac{p_{max}}{p_{max}^0}$ (%)	η_{max} (%)	$\frac{\eta_{max}}{\eta_{max}^0}$ (%)
0	0	0.0055	0.002428	100	5.86	100
10	1	1.62	0.002387	98.34	5.81	99.14
50	5	8.07	0.002252	92.77	5.54	94.53
100	10	16.13	0.002099	86.47	5.26	89.66
250	25	40.33	0.001747	71.97	4.52	77.08
500	50	80.65	0.001366	56.25	3.62	61.79
750	75	120.97	0.001027	42.30	2.74	46.81
1000	100	161.30	0.0009477	39.03	2.54	43.24

resistivity. The figures can be found in Table 2.9, from which it can be seen that the total electrical contact resistance should be less than 10% of the internal resistance of the TE element such that the decrease in the optimal performance is below 10%. As the bulk value for the resistance of the copper bridge is about 0.006% of the internal resistance, it can be neglected, and the value of the contact resistivity should be less than $\varrho_c = 100\,\mu\Omega\text{cm}^2$. Of course, this is only valid for the chosen materials and temperature range. In Figure 2.18, the normalized performance parameters for different ratios of total contact resistance $R_{c,tot} = 2R_{c,bulk} + 2R_{c,el}$ to the internal resistance are shown.

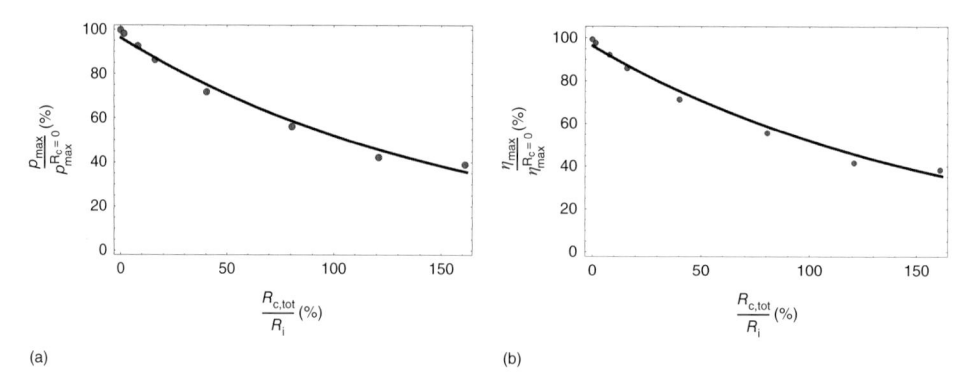

Figure 2.18 Normalized performance parameters for different ratios of total contact resistance $R_{c,tot}$ to the internal resistance R_i. The solid lines represent polynomial fits, which serve as a guidance for the general trend. (a) Maximum power output densities; (b) Maximum efficiency.

2.7
Dissipative Coupling between the TEG and the Heat Baths

2.7.1
Finite-Time Thermodynamics Optimization

Evaluation of the upper bounds of the performance of energy conversion processes is central in thermodynamics since it is the starting point of optimization studies. The archetypal quantity to be optimized has long been the efficiency of heat-to-work conversion processes, with the Carnot efficiency as an unrealistic target in mind. The main drawback of works based on a Carnot-like analysis is that they neglect two factors that are closely connected: (i) for practical applications, heat engines should produce *power*, so the question of rates and time arises and becomes crucial; (ii) real-life systems are always *dissipatively* connected to the temperature reservoirs, and it is through such connections that conversion processes occur in *finite times*. It is acknowledged by many that the very beginnings of finite-time thermodynamics, a field of research that is a extension of irreversible thermodynamics, date back to the 1950s when physicists and engineers had to consider the problem of improving the performance of the then newly built atomic power plants, the operation of which depends very much on the constraints imposed by the working conditions [98–100]. The *efficiency at maximum power* became the target of interest. Finite-time thermodynamics really took off in the 1970s with the pedagogical work published by Curzon and Ahlborn [101], followed by those of Andresen [102–104]. In a nutshell, one seeks to optimize the power–efficiency trade-off or, as Bjarne Andresen put it in his recent review [105]: *The cornerstone of finite-time thermodynamics is all about the price of haste and how to minimize it.*

The mathematical expression for the efficiency at maximum power derived by Curzon and Ahlborn for a heat engine dissipatively connected to two heat sources at temperatures T_h and T_c reads

$$\eta_{CA} = 1 - \sqrt{\frac{T_c}{T_h}}. \tag{2.179}$$

This formula is interesting for a variety of reasons: (i) though the heat engine is connected to the reservoirs through finite-conductance exchangers, η_{CA} depends only on the temperatures T_h and T_c, but not on the exchangers' conductance; (ii) it is very simple and general since the specifics of the heat engine are not described by this expression; and it is very similar to the Carnot efficiency, η_C.

2.7.2
Thermoelectric Generator Model

We consider a single-leg module placed between two heat reservoirs with fixed temperatures T_h and T_h, respectively. The thermal contacts between the thermogenerator and the heat baths are ensured by a heat exchanger on each side; these are completely characterized by constant finite thermal conductance, K_h and K_c, respectively. The temperatures at both ends of the module, which we denote as T_{hM} and T_{cM}, are different from those of the heat baths, and they may vary during the operation of the module. With the thermal conductances being associated in series, the equivalent contact conductance, $K_{contact}$, is $K_{contact}^{-1} = K_c^{-1} + K_h^{-1}$. In the case of nonlinear heat transfer or radiative processes, an extension to varying thermal conductances may be considered, but for illustrative purposes, we restrict our analysis to constant thermal conductances. The setup is depicted in Figure 2.19.

The internal electrical resistance is given by $R = L/\sigma_T A_c$, with A_c and L being the section and the length, respectively, of the leg; we assume that it is isothermal. The open voltage is $V_{oc} = \alpha \Delta T_{TEG}$, where α is the Seebeck coefficient and $\Delta T_{TEG} = T_{hM} - T_{cM}$ is the temperature difference across the module, which is smaller than the total temperature difference $\Delta T = T_h - T_c$. The thermal conductance K_{TEG}, to which electrons and phonons contribute, depends on the working conditions and varies during operation. It reduces to the conductance $K_{V=0}$, under zero voltage (electrical short circuit), and to the conductance $K_{I=0}$ at zero electrical current (open circuit). An electrical-current-dependent form of the thermal conductance of the module K_{TEG} is derived next. The figure of merit ZT is given by $ZT = \alpha^2 T/(RK_{I=0})$.

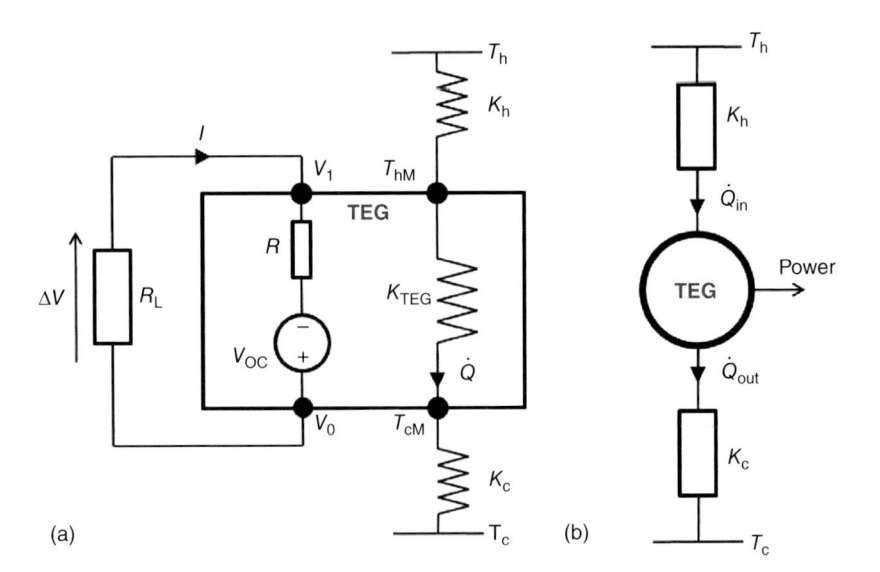

Figure 2.19 Thermoelectric (a) and thermodynamic (b) pictures of the thermoelectric generator and the load.

2.7.3
Thermal Flux and Electrical Current

By assuming that the temperature difference across the TEG, ΔT_{TEG}, is not very large, the incoming and outgoing heat fluxes are linear in the temperature difference ΔT_{TEG}. The linear force–flux formalism thus yields a simple description of the TEG characteristics:

$$\begin{pmatrix} I \\ \dot{Q} \end{pmatrix} = \frac{1}{R} \begin{pmatrix} 1 & \alpha \\ \alpha T_m & \alpha T_m^2 + RK_{I=0} \end{pmatrix} \begin{pmatrix} \Delta V \\ \Delta T_{TEG} \end{pmatrix}, \tag{2.180}$$

where I is the electrical current through the load, which, for simplicity, we consider as a resistance, $\dot{Q}$ is the thermal flux through the TEG, and ΔV is the voltage across the TEG. The average temperature in the module is simply considered as $T_m = (T_{cM} + T_{hM})/2$.

The thermal current is due to convective heat transfer, that is, heat transported within the electrical current, and steady-state conduction, usually associated with Fourier's law:

$$\dot{Q} = \alpha T_m I + K_{I=0} \Delta T_{TEG} \tag{2.181}$$

Using Ohm's law: $\Delta V = -R_L I$, we find the electrical current I:

$$I = \frac{\Delta V + \alpha \Delta T_{TEG}}{R} = \frac{\alpha \Delta T_{TEG}}{R_L + R}. \tag{2.182}$$

Substitution of the electrical current I into Eq. (2.181) yields the TEG thermal conductance K_{TEG}:

$$\dot{Q} = \left(\frac{\alpha^2 T_m}{R_L + R} + K_{I=0} \right) \Delta T_{TEG} = K_{TEG} \Delta T_{TEG}. \tag{2.183}$$

The thermal conductivity K_{TEG} thus depends on the electrical operating point imposed by the external load. Note that an expression of K_{TEG} may also be obtained, starting from the relationship between the two thermal conductances $K_{V=0}$ and $K_{I=0}$ of the TEG [40]:

$$K_{V=0} = K_{I=0}(1 + ZT_m) \tag{2.184}$$

which may be generalized to:

$$K_{TEG}(I) = K_{I=0}\left(1 + \frac{I}{I_{sc}} Z T_m \right), \tag{2.185}$$

where $I_{sc} = \alpha \Delta T_{TEG}/R$ is the short-circuit current such that $K_{TEG}(I_{sc}) = K_{V=0}$. Note that Eqs. (2.185) and (2.183) provide the same expression for the thermal conductance K_{TEG}. It is also important to understand that there is no closed-form solution for the global distributions of electrical and heat currents and potentials in the device because of the dependence of the short-circuit current I_{sc} on the effective temperature difference across the TEG, ΔT_{TEG}.

As depicted in Figure 2.19, the electrical component of the TEG may be viewed as the association of a perfect generator and a resistance, which is the physical internal resistance of the generator. The Seebeck voltage depends on the temperature difference across the TEG, ΔT_{TEG}, which in turn depends on the electrical load because of the presence of finite thermal contact conductances. Therefore, V_{oc}, which explicitly reads:

$$V_{\text{oc}} = \alpha \Delta T \frac{K_{\text{contact}}}{K_{I=0} + K_{\text{contact}}} - IR \frac{ZT_m}{1 + K_{\text{contact}}/K_{I=0}} \equiv V'_{\text{oc}} - IR' \tag{2.186}$$

and which is not the output voltage of a perfect Thévenin generator. It is given by the sum of two terms, the first of which is independent of the electrical load, while the latter depends on the electrical current. The open-circuit voltage V'_{oc} and the internal resistance $R_{\text{TEG}} = R + R'$ as defined in Eq. (2.186) now permit the characterization of a perfect Thévenin generator.

2.7.4
Calculation of the Temperature Difference across the TEG

For practical and modeling purposes, it is useful to obtain expressions of power and efficiency as functions of the *fixed* temperature difference between the two reservoirs, $\Delta T = T_h - T_c$, rather than ΔT_{TEG}. In Ioffe's approach [49], the incoming heat flux $\dot{Q}_{\text{in}}$ and outgoing heat flux $\dot{Q}_{\text{out}}$ may be expressed as:

$$\dot{Q}_{\text{in}} = \alpha T_{\text{hM}} I - \frac{1}{2} R I^2 + K_{I=0}(T_{\text{hM}} - T_{\text{cM}}) \tag{2.187}$$

$$\dot{Q}_{\text{out}} = \alpha T_{\text{cM}} I + \frac{1}{2} R I^2 + K_{I=0}(T_{\text{hM}} - T_{\text{cM}}) \tag{2.188}$$

Now, since $\dot{Q}_{\text{in}} = K_h(T_h - T_{\text{hM}})$ and $\dot{Q}_{\text{out}} = K_c(T_{\text{cM}} - T_c)$, we end up with a 2×2 system, which establishes a relationship between T_{cM} and T_{hM} to T_c and T_h:

$$\begin{pmatrix} T_h + \frac{1}{2}\frac{RI^2}{K_h} \\ -T_c - \frac{1}{2}\frac{RI^2}{K_c} \end{pmatrix} = \begin{pmatrix} \mathcal{M}_{11} & \mathcal{M}_{12} \\ \mathcal{M}_{21} & \mathcal{M}_{22} \end{pmatrix} \begin{pmatrix} T_{\text{hM}} \\ T_{\text{cM}} \end{pmatrix}, \tag{2.189}$$

where $\mathcal{M}_{11} = K_{I=0}/K_h + \alpha I/K_h + 1$, $\mathcal{M}_{12} = -K_{I=0}/K_h$, $\mathcal{M}_{21} = K_{I=0}/K_c$, and $\mathcal{M}_{22} = \alpha I/K_c - K_{I=0}/K_c - 1$ are the expressions of the four dimensionless matrix elements. The temperatures T_{hM} and T_{cM} are obtained by matrix inversion, but the exact expression of ΔT_{TEG} as a function of T_h and T_c is cumbersome and truly not necessary for the analyses and discussions that follow. An approximate and straightforward relationship between ΔT_{TEG} and ΔT may be obtained using an analogue of the voltage divider formula, by assuming that the thermal flux is constant in the whole system:

$$\Delta T_{\text{TEG}} = T_{\text{hM}} - T_{\text{cM}} \approx \frac{K_{\text{contact}}}{K_{\text{TEG}} + K_{\text{contact}}} \Delta T. \tag{2.190}$$

2.7.5
Maximization of Power and Efficiency with Fixed ZT_m

2.7.5.1 Maximization of Power by Electrical Impedance Matching

The TEG electrical power output may be expressed as

$$P = \frac{{V'_{oc}}^2 R_L}{(R_{TEG} + R_L)^2}, \qquad (2.191)$$

which shows that, for a given thermal configuration, power maximization is achieved if

$$R_L = R_{TEG}. \qquad (2.192)$$

By using the ratio $m = R_L/R$ introduced by Ioffe [49], the condition for maximization also reads:

$$m_{P=P_{max}} = 1 + \frac{ZT_m}{K_{contact}/K_{I=0} + 1}, \qquad (2.193)$$

Note that in the ideal case, when the thermal conductance of the heat exchangers is infinite, the electrical impedance matching condition, Eq. (2.193), corresponds to the condition $m = 1$, or, equivalently, to $R_L = R$; but, as we consider finite thermal conductance of the contacts to the heat baths, the ideal condition cannot be fulfilled, and it becomes Eq. (2.193), as previously reported by Freunek and coworkers [106]. The maximum output power then reads:

$$P_{max} = \frac{(K_{contact}\Delta T)^2}{4(K_{I=0} + K_{contact})T_m} \frac{ZT_m}{1 + ZT_m + K_{contact}/K_{I=0}}, \qquad (2.194)$$

2.7.5.2 Maximization of Power by Thermal Impedance Matching

By assuming a situation when the total conductance $K_{contact}$ is fixed by an external constraint, optimization of power is then performed with respect to $K_{I=0}$ under the condition:

$$\frac{K_{contact}}{K_{I=0}} = 1 + \frac{ZT_m}{1 + m}, \qquad (2.195)$$

which amounts to satisfying:

$$K_{contact} = K_{TEG}. \qquad (2.196)$$

This equality is similar to that derived by Stevens in Ref. [107], who concluded that the condition for thermal impedance matching is the equality between the thermal contact resistance and that of the TEG. The difference between Stevens' result and ours, Eq. (2.196), is that the thermal resistance used in Ref. [107] for the thermoelectric system is derived by assuming an open-circuit condition, which does not account for the convective part of the thermal current, while K_{TEG} defined in Eq. (2.183) does. The similarity between electrical and thermal impedance matching given by Eqs. (2.193) and (2.195), respectively, is striking.

2.7.5.3 Simultaneous Thermal and Electrical Impedance Matching

Now, assume that the TEG's environment imposes a particular configuration. The *joint* optimization of the electrical and thermal conditions yields the optimal operating point of the TEG. This is performed by solving Eqs. (2.193) and (2.195) *simultaneously*:

$$\frac{K_{contact}}{K_{I=0}} = \sqrt{ZT_m + 1}, \tag{2.197}$$

$$m_{P=P_{max}} = \sqrt{ZT_m + 1}. \tag{2.198}$$

Equation (2.197) was previously published by Freunek and coworkers [106], while Yazawa and Shakouri obtained both conditions [108]. Under these two impedance matching conditions, the maximum power produced by the TEG is given by:

$$P_{max} = \frac{K_{contact} Z T_m}{(1+\sqrt{1+ZT_m})^2} \frac{(\Delta T)^2}{4 T_m}. \tag{2.199}$$

2.7.5.4 On the Importance of Thermal Impedance Matching

Three curves of the maximum power P_{max} as a function of the ratio $K_{I=0}/K_{contact}$ [see Eq. (2.194)] are shown on Figure 2.20; these were obtained for three values of the figure of merit ZT_m. The higher the value of ZT_m is, the greater is the maximum of P_{max} and the larger is the width at half maximum; this is what one may obviously expect, but the curves displayed in Figure 2.20 also highlight the importance of thermal impedance matching: a high value of ZT_m does not guarantee a greater P_{max} for *any* value of the thermal conductance at zero electrical current $K_{I=0}$; for

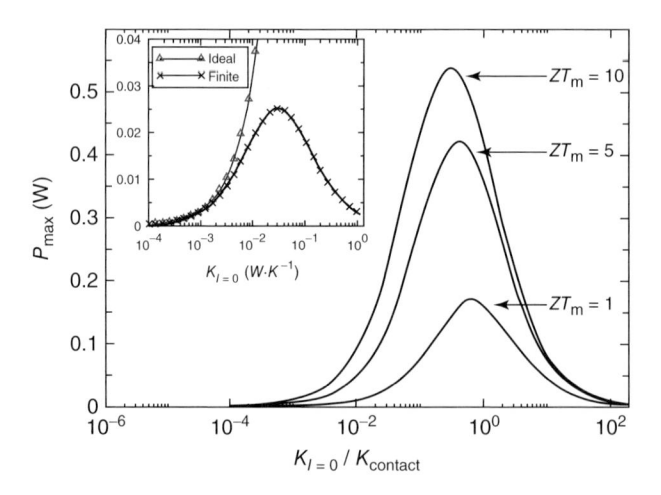

Figure 2.20 Maximum power as function of the ratio $K_{I=0}/K_{contact}$ for various T_m values at fixed $K_{contact}$ (notice the use of a logarithmic scale for the abscissa axis). In the inset, the curves are computed with the data of Ref. [109]: $K_{I=0} = 3 \times 10^{-3}, 6 \times 10^{-3}$, and 1.2×10^{-2} W·K^{-1}.

instance, P_{max} at $K_{I=0} = K_{contact}$ for $ZT = 1$ is greater than P_{max} at $K_{I=0} = 5K_{contact}$ for $ZT = 10$.

In addition, two curves represent the maximum power as a function of $K_{I=0}$ for both finite and perfect thermal contacts, respectively, in the inset of Figure 2.20; we thus see why the TEG with the highest $K_{I=0}$ shows the largest degradation of performance. This was previously observed by Nemir and Beck [109], who, in their analysis of the impact of thermal contacts on device performance, considered different configurations leading to the same value for the figure of merit ZT_m. For a given value of contact thermal conductance, they found that the device performance is strongly influenced by the way the figure of merit of the thermoelectric module is fixed.

2.7.5.5 Maximum Efficiency

The conversion of the heat current $\dot{Q}$ into the electric power P is performed with efficiency $\eta = P/\dot{Q}$, which also reads:

$$\eta = \frac{K_{contact} + K_{TEG}}{K_{contact} K_{TEG}} \frac{P}{\Delta T}, \qquad (2.200)$$

considering Eqs. (2.183) and (2.190). The given expression assumes the following form:

$$\eta = \eta_C \times \frac{m}{1 + m + (ZT_h)^{-1}(1+m)^2 - \eta_C/2} \qquad (2.201)$$

where $\eta_C = 1 - T_c/T_h$ is the Carnot efficiency, in the case of ideal thermal contacts.
We find that the value of m, which maximizes the efficiency Eq. (2.200), is:

$$m\bigg|_{\eta=\eta_{max}} = \sqrt{(1+ZT_m)\left(1+ZT_m \frac{K_{I=0}}{K_{contact} + K_{I=0}}\right)} \qquad (2.202)$$

and that it explicitly depends on the thermal conductances $K_{contact}$ and $K_{I=0}$.

2.7.5.6 Analysis of Optimization and Power–Efficiency Trade-Off

If the working conditions modify $K_{contact}$, as for liquid–gas heat exchangers, the operating point of the thermoelectric device changes. Whether $m|_{\eta=\eta_{max}}$ is bounded or not as the ratio $K_{I=0}/K_{contact}$ varies is a question that must be studied. Without loss of generality for the discussion that follows, it is convenient to assume that the figure of merit is fixed, which is the case if the mean temperature T_m varies slightly with $K_{contact}$.

When finite thermal contacts are accounted for in the model system, the optimal values of the electrical load to achieve maximum power or efficiency differ from those of the ideal case. The optimal parameters m_{opt} [$m|_{\eta=\eta_{max}}$ for maximum efficiency in Eq. (2.202), and $m|_{P=P_{max}}$ for maximum power, in Eq. (2.193)] are plotted against $K_{contact}$ (scaled to $K_{I=0}$) in Figure 2.21, for two particular values for the figure of merit: $ZT_m = 1$ and $ZT_m = 3$. For a given value of ZT_m and conditions close to perfect thermal contacts, that is, $K_{contact} \gg K_{I=0}$, the maximum power and

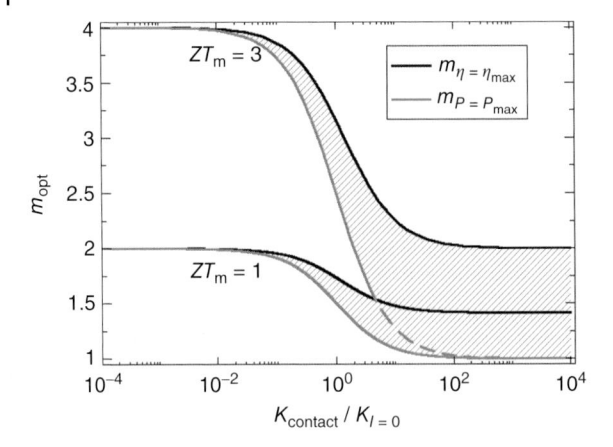

Figure 2.21 Variations of the optimal parameters $m_{\eta=\eta_{max}}$ (dashed line) and $m_{P=P_{max}}$ (dashed–dotted line) as functions of $K_{contact}$ scaled to $K_{I=0}$, for $ZT_m = 1$ and $ZT_m = 3$. The shaded areas correspond to the region of best optimization.

maximum efficiency are well separated: $m_{\eta=\eta_{max}} \to \sqrt{1+ZT_m}$, and $m_{P=P_{max}} \to 1$. Furthermore, the separation between both quantities can only increase with ZT_m.

For $K_{contact} \ll K_{I=0}$, we obtain $m_{\eta=\eta_{max}} \to 1+ZT_m$, which is the upper bound to $m_{P=P_{max}}$ as well; hence, both optimal parameters coincide. The convergence of these two optimal parameters toward the same value might appear profitable in the sense that this implies that there is no need for a trade-off between efficiency and power; but this situation also implies a significant performance decrease. As the optimal regions for the power–efficiency trade-off are those lying between each pair of curves, the narrowing of these zones, which also occurs along with the lower values of ZT_m, is undesirable since this situation offers less flexibility in terms of the working conditions for the thermoelectric generator.

Two power–efficiency curves are shown in Figure 2.22; they correspond to $K_{contact} = 10 K_{I=0}$ and $K_{contact} = 50 K_{I=0}$, respectively. As the contact thermal conductance decreases, we observe a narrowing of the optimal zone consistent with our analysis of Figure 2.21. The maximal values that P_{max} and η_{max} may take are indicated with arrows. For very poor thermal contacts, the power–efficiency curve shrinks and reduces to a point located at the origin of the limit of zero thermal conductance.

2.8
Shaped Thermoelectric Elements

In this section, an overview of the influence of the leg geometry on the performance of a thermoelectric element is provided. For this purpose, the representative literature references are shortly discussed. In the 1950s, Gehloff and coworkers investigated TE elements with arbitrary shapes [110]. Their work shows

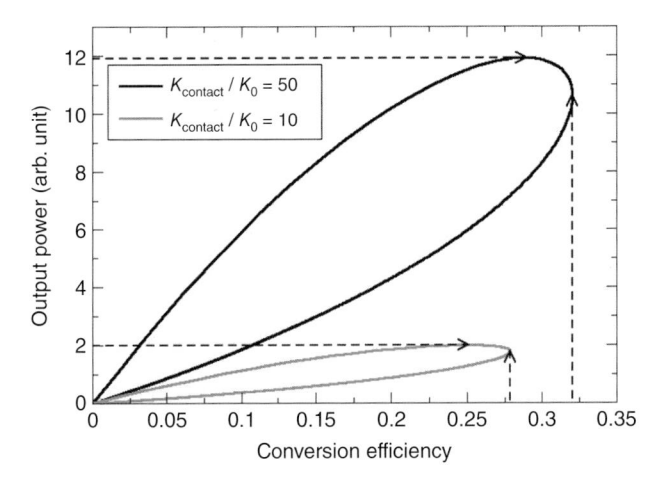

Figure 2.22 Power versus efficiency curves for two cases with a fixed figure of merit $ZT_m = 1$. Only the value of $K_{contact}$ varies.

a generalization of Altenkirch's investigations [66, 81, 111] for noncylindrical thermoelements, that is, elements with arbitrary shape under the condition of $\kappa/\sigma T$ being a constant. The statement of the authors regarding the influence of the shape on the performance (of a TEC) is that there are no greater amounts of critical cooling and cooling power, upon comparing cylindrical and noncylindrical TE elements. In Reference [112], Boerdijk pointed out that Gehlhoff's work is not general enough for modern semiconductors because of the possibility of a considerable variation of the material properties with temperature. Another critical point was that the main equations are based on the work of Kohlrausch and Diesselhorst [113, 114]. For those equations, Domenicali showed that they are not completely consistent with the concepts of nonequilibrium thermodynamics [12]. Domenicali summarizes in a comprehensive review [3] the basics of the so-called Onsager–de Groot–Callen theory applied to thermoelectricity [115–124].

As a part of his PhD thesis, Peschke published in 1957 a condensed work [51], which was the first to our knowledge, considering (correctly) thermoelements of different shapes. Besides the classical prismatic thermoelements, he investigated conical and cylindrical elements. Especially, the efficiencies for these differently shaped elements or thermocouples are calculated and combined in one formula by the use of different electrical current path lengths for the different shapes.

A rather general approach based on the concepts of the field of thermodynamics had been given by Clingman [125], where it was stated the determination of the efficiency of a TE device is closely related to the concept of entropy production. In a TE device, entropy is produced due to an irreversible process. Because of this process, the Carnot efficiency $\eta_C = \frac{T_a - T_c}{T_a}$ is reduced. The irreversible processes in a TE device cannot be eliminated such that the entropy production becomes zero and the Carnot efficiency is reached. To optimize the efficiency means to minimize the entropy production, which can be performed by designing the TE appropriately, for example, through the shape, although this is not performed explicitly in Ref. [125].

An explicit investigation of the influence of the shape on the performance of a TE element with a temperature-dependent material had been performed by Brandt [126], who performed a numerical calculation by solving a generalized heat balance equation by varying different influential parameters. One focused parameter is the "element shape factor," which leads to a deviation from a cylindrical element as the investigated shape of the elements was a truncated cone. In Brandt's calculation, the hot side temperature had been fixed. At the cold junction, all of the energy crossing and generated at the junction is supposed to be dissipated by a convective fin, which leads to the van Neumann boundary conditions.

In the work of Boerdijk [112], nonequilibrium thermodynamics is applied for thermocouples that consist of (i) elements with arbitrary shape, (ii) have temperature-dependent material properties, and (iii) are inhomogeneous and anisotropic under certain restrictions. Again it could be shown that the maximal values of the performance parameters are independent of the shape of the elements. Besides the stationary behavior, the transient behavior is investigated in this work.

Based on [112], Boerdijk introduced in Ref. [127] a theory where the material properties are expressed in terms of a MacLaurin series in T and u (a position coordinate). The performance parameters are derived in these approximations and the accuracy is discussed. The calculation had been performed for arbitrarily shaped elements.

Semenyuk showed for a TEC in Ref. [128] that the heat flux equations for arbitrarily shaped elements have the same form as the well-known ones calculated for prismatic elements as long as the contact surfaces are equipotential and isothermal while the remaining surfaces are thermally and electrically insulated. By assuming constant properties material [49], that is, no temperature and position dependence of the material properties, and uniform isotropic conditions, Semenyuk stated that only the zT value is uniquely defining the limiting efficiency even for an arbitrarily shaped TE element. He was critically discussing the work of Boerdijk [112], who did investigate the effect of the element's shape on the performance if the material has an arbitrary dependence on the spatial coordinates as well as the temperature. Boerdijk introduced a certain scalar function as a kind of potential u. The gradient of this function gives the vector of the electrical current density meaning that the electric field in a nonuniform TE element is a potential field. With regard to this, Semenyuk's critics is that this is not true, in general, in reality because thermal and physical inhomogeneities in thermoelectric element can lead to additional voltage contributions and thus the electric field is not a potential field any more.

In Reference [60], disk-shaped thermocouples in the cooling mode are investigated and the maximum temperature difference is determined if the current flow direction is radially inward or outward. The ratio of the inner and outer radii is varied and the impact on the maximum temperature difference discussed. Another shape, that is, a rotationally symmetric body with a lateral surface in form of a hyperbola, is pointed out to be able to produce lower temperatures. This work had been followed up by the same author in Refs [61, 62].

In 1957, Shiliday [129] performed a calculation for circular TE elements and made experimental measurements with this design. He found out that in the framework of his approximations, the linear and annular geometries have the same power consumed and provide the same amount of cooling. The same had been pointed out by Lund [130], who additionally stated that the advantage of a radial current flow in these kind of TE elements is to possibly reach greater heat pumping rates per unit capital costs, that is, less material has to be used for this design to produce the same heat pumping rate.

Directly pointing to the problem is Thacher's work [131], which is based on his PhD thesis [132], showing the major results and some subsequent results. The calculations are performed in a one-dimensional approach using variational calculus, which is more prominently used in thermoelectricity nowadays, see, for example, [133, 134]. Euler–Lagrange equations are formulated in an integral form of the TEG efficiency. The extremum of the efficiency is found by varying the temperature distribution and the cross-sectional area distribution (shape) of a TE element. Again it had been pointed out that minimizing the total entropy production rate (per unit power output) means maximizing the thermal efficiency of a TEG. From the calculated "shape relation," it can easily be deduced that if the element is laterally adiabatic (no heat transfer over the lateral surface of the TE element), the cross-sectional area must be constant regardless of the temperature dependence of the material properties. The derived relation is independent of the choice of the independent variable, r or z. If r is chosen, then the Landecker configuration can be confirmed in the adiabatic case.

Zhang and coworkers found in a 3D thermoelectric finite element analysis that an increased maximum cooling or a higher peak cooling temperature can be observed by varying the contact interface size in a TEC [135].

In Lienhard's textbook, there is a rather comprehensive and general overview of the influence of the geometry and shape on the heat transfer in the multidimensional situation [136].

In the more recent past, there are other contributions considering ring-shaped elements as this kind of shape is interesting for applications [63–65]. It is pointed out that the electrical resistance R and the thermal conductance K are influenced by the shape, so do the power output and the thermal load. As the Seebeck coefficient is in general independent of the geometry and shape, the device's figure of merit is independent of it as well, as in the calculation we find only the product of R and K, which is independent of the shape. With that, the efficiency is supposed to be independent of the shape.

In Reference [137], Palma and coworkers treated TE elements with variable cross-sectional area in their hyperbolic heat conduction model (Cattaneo model), which they solved with help of a nonlinear finite element method by using Hoyos' arguments [138, 139]. In his works, Hoyos investigated the transient behavior of a TEC. He proposed a tapered semiconductor leg and a reduced copper tab at the cold side of the TEC to reduce the recovery time and improve the transient response. The tapered (conical) shape of the elements leads to an asymmetric distribution of the Joule heat, which preferentially is conducted toward the hot junction. This means that less than half of the Joule heat reaches the cold junction. Under the given conditions, that is, Ioffe-CPM approximation, there is an equal distribution of Joule heat in the cylindrical legs, meaning that one half of the heat goes to the cold junction whereas the other half goes to the hot junction. As the cold junction has a smaller size, it is possible to reduce the size of the copper tab at this junction. This reduction leads to a decrease in the thermal capacitance and by that an "increase" in the transient response can be observed.

A detailed study on the performance of thermoelectric elements with nonconstant cross-sectional areas had been conducted by Arenas and coworkers [140]. In a one-dimensional approach, the electrical current flow is supposed to be in one spatial direction, that is, without loss of generality in z direction, a variable cross-sectional area $A_c = A_c(z)$ is assumed to investigate the influence on the performance of a TE element. Further assumptions are that the TE material is physically and chemically homogeneous (Ioffe-CPM approach) and that the lateral surfaces are considered adiabatic. The boundary temperatures are supposed to be fixed (Dirichlet boundary conditions) such that a constant thermal gradient ΔT is applied between the ends.

Recently, Sahin and Yilbas studied the effect of leg geometry on the performance of a TEG (thermocouple). Their procedure is based on a 1D approach with constant material. The shape parameter is varied and the influence on the performance of a TE thermocouple is discussed [141].

The performance parameters of shaped thermoelectric elements have been considered and mentioned in passing in Ref. [142]. This work is more concentrated on the thermomechanical behavior of modules built with shaped TE elements.

2.9
Other Influences on the Performance of TE Devices

In the previous sections, influential factors as the temperature dependence of the material properties or the influence of the contacts on the performance of a thermoelectric device are discussed in great detail. For these factors, it was possible to find analytical solutions in special cases. There are also other factors that are beyond the ideal TE device, which can have an effect on the performance but usually are neglected. Some of them are briefly discussed here without claiming completeness.

2.9.1
Lateral Heat Losses, Convective and Radiative Heat Transfer

2.9.1.1 Convection Losses and Benefits
Telkes already recognized that the lateral (or radial) heat losses should be as small as possible [143]. These losses can be minimized by surrounding the

thermoelectric elements with an adequate thermal insulation material. Bjørk *et al.* [144] as well as Ziolkowski *et al.* [145] compared different heat loss mechanisms and their effect on the degradation of the performance of a TEG by numerical calculations. It was found that the effects can be significant. In both the articles, the previous works are summarized and discussed. It is stated that most works only consider a single heat loss mechanism, and in most cases, these heat loss mechanisms are oversimplified.

Other works show that for example, convective heat transfer can have a beneficial effect on the performance. For example, convective heat transfer over the lateral surface of a TE element was investigated by Rollinger, Ybarrondo, and Sunderland [20, 146–148]. In their approaches, differential equations are used for the description of the temperature distribution under steady-state conditions. For the sake of simplicity, constant material properties are supposed in a thermoelement. Further, a convective heat transfer from its longitudinal surface is assumed, and under these conditions, the results are presented. The differential equations are solved by analytical methods. By doing this, they found that the analytical solutions given in closed-form built the basis for the investigation of a single-element TEG. The effect of surface heat transfer represented by adequate heat-transfer parameters on the performance of the TEG, that is, power output and efficiency, was determined by a variation of the relevant parameters [146].

It was found in all cases that the surface heat transfer increased the power output under matched load conditions. Furthermore, it was shown that an increase in both the power output and the thermal efficiency can be achieved in many cases if surface heat transfer is used.

One conclusion is that surface heat transfer can be beneficial in many applications. Further, it was found, on the one hand, that especially in the cases where maximum power output is of primary importance, a benefit can be reached and, on the other hand, an improvement in the performance can be of help in the cases where the cold-junction fin area is greatly restricted by weight, space, or cost considerations.

Similar results could be found for application of convective heat transfer on a single-element TEC where the focus was laid on the performance parameters of a TEC, that is, the maximum temperature difference, the maximum heat pumping, and the maximum COP, and their sensitivity to changes in surface heat transfer parameters [148].

Again it could be found that an increase of each of these three performance criteria is possible due to the use of surface heat transfer. For the maximum heat pumping, gains in the order of 200% were observed by the use of the convective heat transfer mechanism. It was found that the greater the convective heat transfer coefficient, the greater the gains in performance are. From these results, it was deduced that a part of the element near the cold junction always has to be insulated in order to use the surface heat transfer for an improvement in the performance such that the fraction of the element to be insulated is an additional parameter in the calculation. For any given set of conditions, an optimum value of this fraction can be calculated from the given expressions.

Division of the longitudinal surface of a single-element TEG into three regions was introduced by Ybarrondo and Sunderland [147]. The middle region is supposed to be thermally insulated, whereas one region is exposed to convection with the environment at the cold junction and, on the opposite side, the remaining region is exposed to convection with the environment at the hot junction. The relative sizes of these three regions are then additional variation parameters.

The solutions found in Ref. [147] show again improvements in performance that can be gained through the use of surface heat transfer. Once more substantial gain can be reached especially for the power output in the TEG element.

The use of surface heat transfer makes it possible to greatly reduce the amount of fin surface, which is needed at the hot and cold junctions. It is shown, with a numerical example in which no fins are used at the hot and cold junctions that 97% of the maximum theoretical output can be achieved by using surface heat transfer.

Kafafy and Hameed investigated the thermoelectric devices that are subject to lateral heat convection in a quasi-1D model [149, 150]. By their investigations, two important nondimensional physical parameters to shape the temperature distribution were identified. These two parameters are called the *heat resistance ratio* (HRR) and the *energy growth ratio* (EGR). The first is the ratio between the conductive thermal resistance and the convective thermal resistance. The conductive thermal resistance $R_{th,cond}$ is an important factor controlling the temperature gradient along the TE element, whereas $R_{th,conv}$ is the factor that controls the heat exchange between the solid TE element and the flowing gas, which passes along the surface of the TE element. The second factor, the energy growing rate, is the ratio between electrical Joule heating and Fourier heat conduction. The authors' investigations were concentrated on the effect of the HRR and EGR on the temperature gradient. Furthermore, nonuniform TE legs were observed and the effect of the TE aspect ratio on HRR and EGR has also been investigated. The results lead to the conclusion that larger aspect ratios are more favorable for fast switching from effective heat exchange between the gas and the TE walls.

2.9.1.2 Radiation Losses and Benefits

If the hot side temperatures of a TEG are more elevated, then radiation losses are considered to be present and can lead to degradation of the performance. In principle, two types of radiations are present. On the one hand, it is the surface-to-ambient radiation, and on the other hand, it is the surface-to-surface radiation. In Reference [144], the difference between the case where both the radiation mechanisms are considered and the case where only the surface-to-ambient radiation is considered is found to be small (less than $\approx 0.35\%$). Therefore, it is justified that only the surface-to-ambient contribution is taken into account to estimate the heat losses due to radiation. This approach is carried out in most cases because it is much faster than the other method. In Reference [145], the effect of radiation on the performance was investigated for different pellet lengths. One result of this investigation was that, for longer pellets with broader surface, a reduced heat exchange via radiation was observed, because the open surfaces in the gap "saw" a more balanced thermal environment. The main influential factor for radiation losses is the aspect ratio of the gaps between the pellets.

Recently, the interest in hybrid systems where solar radiation is used as heat source and a TEG is used for direct energy conversion has increased rapidly [151–165]. Photovoltaic and solar thermal conversion are established technologies. The concept of direct conversion of solar energy to electricity is more than a century old. In the historical review in Ref. [166], several patents from the late nineteenth century to the beginning of the twentieth century are mentioned. In her work [166], Telkes had performed both experimental work and theoretical calculations and estimations. The key challenge for a solar thermoelectric generator (STEG) is to achieve a significant temperature difference while maintaining a low solar radiation flux. In Reference [151], a temperature difference of less than $\Delta T \leq 5K$ under general conditions and geometrical design of the TEG component was estimated. There are different possibilities to increase the temperature difference, for example, an optical concentration of the solar radiation on the TE device with the help of lenses and mirrors. Telkes found for 50 times optical concentration by a lens, that is, $C_{opt} = 50$, a temperature difference of 247K, see Table VII in Ref. [166]. In Reference [151], another approach, a flat-panel absorber configuration with a highly solar-absorbing surface, is introduced where the heat originated in the solar radiation is thermally concentrated. For such a device, a peak efficiency of the whole flat-panel device is measured in the range of 4.6–5.2%. In general, the whole efficiency η is the product between the optothermal efficiency η_{ot} and the efficiency of the TEG η_{TE}, that is, $\eta = \eta_{ot}\eta_{TE}$, such that there are two "adjusting screws" in the system. The flat-panel device is situated in an evacuated enclosure, which significantly reduces the losses due to convection.

Other loss mechanisms are reflection–transmission losses, reradiation losses, and losses of the optical devices if they are used as concentrators due to inaccurate focusing. The TEG part can be improved by the traditional way, that is, by using a better material, for example, nanostructured TE material [151] or using segmented/cascaded TE devices [152, 167].

2.9.2
Anisotropic Thermoelectric Elements

One main assumption, which is made in the performance calculation of thermoelectric devices, is that the elements the device is built of are isotropic. It is a physical principle that both the thermal and the electrical current follow the path of least resistance. For an isotropic material used in an element, these paths lie parallel to the applied thermal gradient and/or the electric field, as in an isotropic material all directions are equivalent. In an anisotropic material, this is not the case, because there is a directional dependence of the transport properties. Therefore, it is clear that the path of least resistance will not lie parallel to the applied thermal gradient or the electric field in general. This implies that the thermal and electrical currents are not found to be necessarily in parallel (Figure 2.23).

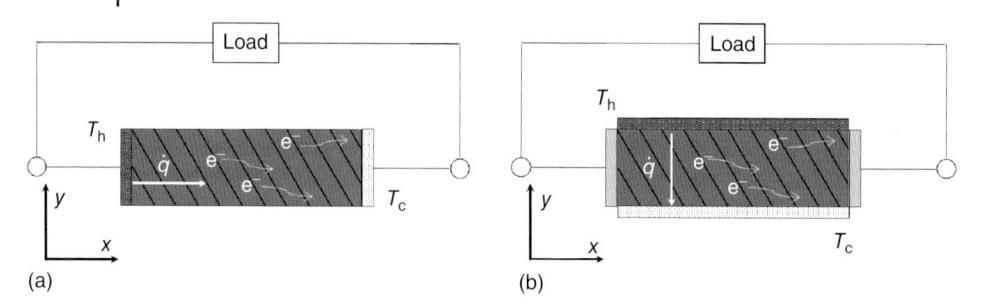

Figure 2.23 Two arrangements of a thermoelectric element made of an anisotropic material. The direction of the electrical current is supposed to be always along the x-axis, whereas the heat flux is either *parallel* (a) or *perpendicular* (b). (Figures adapted from Ref. [168, p. 80]).

Because of the anisotropic character of the material, it is possible to build a working thermoelectric module based on a single material. Two arrangements are distinguished in this case.

The calculation of effective (scalar) transport parameters for an anisotropic cell in the perpendicular configuration is shown in great detail in Chapter 6 in Ref. [168, p. 78ff]. This is done to guess an anisotropic equivalent figure of merit.

The definition of a figure of merit for an anisotropic material is given in Ref. [4] from a thermodynamic point of view. The performance of an anisotropic TEC in a magnetic field is investigated in Ref. [5].

Artificial anisotropy can be created by multilayer structures. Nonvanishing off-diagonal elements of the transport properties tensor can lead to transverse Seebeck and Peltier effects, which can be used in devices. Recently, there has been many works conducted in this field, see, for example, [7, 169–180].

References

1. Miller, D.G. (1960) Thermodynamics of irreversible processes. The experimental verification of the Onsager reciprocal relations. *Chem. Rev.*, **60** (1), 15–37.
2. Newnham, R.E. (2005) *Properties of Materials: Anisotropy, Symmetry, Structure: Thermoelectricity*, Chapter 21, Oxford University Press, pp. 234–242.
3. Domenicali, C.A. (1954) Irreversible thermodynamics of thermoelectricity. *Rev. Mod. Phys.*, **26** (2), 237–275.
4. Buda, I.S., Lutsyak, V.S., Khamets, U.M., and Shcherbina, L.A. (1991) Thermodynamic definition of the thermoelectric figure of merit of an anisotropic medium. *Phys. Status Solidi A*, **123** (2), K139–K143
5. Buda, I.S. and Lutsyak, V.S. (1995) The theory of an anisotropic thermoelectric cooler in a magnetic field. *Phys. Status Solidi A*, **147** (2), 491–496.
6. Snarskii, A.A., Palti, A.M., and Ascheulov, A.A. (1997) Anisotropic thermoelements - review. *Fizika i Tekhnika Poliprovodnikov*, **31**, 1281–1298. (in Russian).
7. Bulat, L.P. and Snarskii, A.A. (2006) Anisotropic thermoelements, in *CRC Handbook of Thermoelectrics: Macro to Nano*, Chapter 45 (ed. D.M. Rowe), Taylor and Francis, Boca Raton, FL.
8. Silk, T.W. and Shofield, A.J. (2008) Thermoelectric effects in anisotropic systems: measurement and applications, arXiv:0808.3526v1.

9. Prybyla, A.V. (2010) Spiral anisotropic and anisotropic thermoelements. *J. Thermoelectric.*, **4**, 77–86.
10. Anatychuk, L.I. (2012) Thermoelectric induction in power generation: prospects and proposals, in *CRC Handbook of Thermoelectrics: Thermoelectrics and Its Energy Harvesting*, Chapter 2 (ed. D.M. Rowe), CRC Press, Boca Raton, FL.
11. Domenicali, C.A. (1953) Irreversible thermodynamics of thermoelectric effects in inhomogeneous, anisotropic media. *Phys. Rev.*, **92** (4), 877–881.
12. Domenicali, C.A. (1954) Stationary temperature distribution in an electrically heated conductor. *J. Appl. Phys.*, **25** (10), 1310–1311.
13. Buist, R.J. (1995) The extrinsic Thomson effect, In *Proceedings of the 14th International Conference on Thermoelectrics*, A.F. Ioffe Physical-Technical Institute, St. Petersburg, pp. 301–304.
14. Bridgman, P.W. (1926) Thermal conductivity and thermal E.M.F. of single crystals of several non-cubic metals. *Proc. Am. Acad. Arts Sci.*, **61** (4), 101–134.
15. Snyder, G.J. (2006) Thermoelectric power generation: efficiency and compatibility, in *CRC Handbook of Thermoelectrics: Macro to Nano*, Chapter 9 (ed. D.M. Rowe), Taylor and Francis, Boca Raton, FL.
16. Kaliazin, A.E., Kuznetsov, V.L., and Rowe, D.M. (2001) Rigorous calculations related to functionally graded and segmented thermoelements. Proceedings of the 20th International Conference on Thermoelectrics (ICT 2001), pp. 286–292.
17. Rowe, D.M. (ed.) (2006) *CRC Handbook of Thermoelectrics: Macro to Nano*, CRC Press, Boca Raton, FL.
18. Seifert, W., Ueltzen, M., and Müller, E. (2002) One-dimensional modelling of thermoelectric cooling. *Phys. Status Solidi A*, **1** (194), 277–290.
19. Seifert, W., Müller, E., and Walczak, S. (2008) Local optimization strategy based on first principles of thermoelectrics. *Phys. Status Solidi A*, **205** (12), 2908–2918.
20. Ybarrondo, L.J. (1964) Effects of surface heat transfer and spatial property dependence on the optimum performance of a thermoelectric heat pump. PhD thesis. Georgia Institute of Technology.
21. Ybarrondo, L.J. and Sunderland, J.E. (1965) Influence of spatially dependent properties on the performance of a thermoelectric heat pump. *Adv. Energy Convers.*, **5** (4), 383–405.
22. Zabrocki, K., Müller, E., Seifert, W., and Trimper, S. (2011) Performance optimization of a thermoelectric generator element with linear material profiles in a 1d setup. *J. Mater. Res.*, **26** – Focus Issue – Adv. Thermoelectric. Mater. (15), 1963–1974.
23. Sherman, B., Heikes, R.R., and Ure, R.W. Jr. (1960) Calculation of efficiency of thermoelectric devices. *J. Appl. Phys.*, **31** (1), 1–16.
24. Sherman, B., Heikes, R.R., and Ure, R.W. Jr. (1960) Calculation of efficiency of thermoelectric devices, in Thermoelectric Materials and Devices, *Materials Technology Series*, Chapter 15 (eds I.B. Cadoff and E. Miller), Reinhold Publishing Cooperation, New York, pp. 199–226.
25. Mahan, G.D. (1991) Inhomogeneous thermoelectrics. *J. Appl. Phys.*, **70** (8), 4551–4554.
26. Anatychuk, L.I., Luste, O.J., and Vikhor, L.N. (1996) Optimal functions as an effective method for thermoelectric devices design. 15th International Conference on Thermoelectrics (ICT 1996), pp. 223–226.
27. Anatychuk, L.I. and Vikhor, L.N. (1997) Functionally graded materials and new prospects for thermoelectricity use. Proceedings of the 16th International Conference on Thermoelectrics (ICT 1997), pp. 588–591.
28. Helmers, L., Müller, E., Schilz, J., and Kaysser, W.A. (1998) Graded and stacked thermoelectric generators - numerical description and maximisation of output power. *Mater. Sci. Eng., B*, **56** (1), 60–68.
29. Mahan, G.D. (2000) Density variations in thermoelectrics. *J. Appl. Phys.*, **87** (10), 7326–7332.

30. Müller, E., Walczak, S., Seifert, W., Stiewe, C., and Karpinski, G. (2005) Numerical performance estimation of segmented thermoelectric elements, in *Proceedings of the 24th International Conference on Thermoelectrics (ICT 2005)* (ed. T.M. Tritt), Institute of Electrical and Electronics Engineers, Inc., pp. 352–357.
31. Seifert, W., Müller, E., and Walczak, S. (2006) Generalized analytic description of one-dimensional non-homogeneous TE cooler and generator elements based on the compatibility approach, in *25th International Conference on Thermoelectrics, Vienna, Austria, 06–10 August* (ed. P. Rogl), IEEE, Piscataway, NJ, pp. 714–719.
32. Müller, E., Walczak, S., and Seifert, W. (2006) Optimization strategies for segmented Peltier coolers. *Phys. Status Solidi A*, **203** (8), 2128–2141.
33. Bian, Z. and Shakouri, A. (2006) Beating the maximum cooling limit with graded thermoelectric materials. *Appl. Phys. Lett.*, **89** (21), 212101.
34. Bian, Z., Wang, H., Zhou, Q., and Shakouri, A. (2007) Maximum cooling temperature and uniform efficiency criterion for inhomogeneous thermoelectric materials. *Phys. Rev. B: Condens. Matter Mater. Phys.*, **75** (24), 245208.
35. Hogan, T.P. and Shih, T. (2006) Modeling and characterization of power generation modules based on bulk materials, in *CRC Handbook of Thermoelectrics: Macro to Nano*, Chapter 12 (ed. D.M. Rowe), Taylor and Francis, Boca Raton, FL.
36. Müller, E., Zabrocki, K., Goupil, C., Snyder, G.J., and Seifert, W. (2012) Functionally graded thermoelectric generator and cooler elements, in *CRC Handbook of Thermoelectrics: Thermoelectrics and Its Energy Harvesting*, Chapter 4 (ed. D.M. Rowe), CRC Press, Boca Raton, FL.
37. Anatychuk, L.I. and Vikhor, L.N. (2012) *Functionally Graded Thermoelectric Materials*, Thermoelectricity, vol. **4**, Institute of Thermoelectricity Bukrek Publishers, Chernivtsi.
38. Zabrocki, K., Müller, E., and Seifert, W. (2010) One-dimensional modeling of thermogenerator elements with linear material profiles. *J. Electron. Mater.*, **39**, 1724–1729.
39. Clingman, W.H. (1961) New concepts in thermoelectric device design. *Proc. IRE*, **49** (7), 1155–1160.
40. Goupil, C., Seifert, W., Zabrocki, K., Müller, E., and Snyder, G.J. (2011) Thermodynamics of thermoelectric phenomena and applications. *Entropy*, **13** (8), 1481–1517.
41. Kedem, O. and Caplan, S.R. (1965) Degree of coupling and its relation to efficiency of energy conversion. *Trans. Faraday Soc.*, **61**, 1897–1911.
42. Burshtein, A.I. (1957) An investigation of the steady-state heat flow through a current-carrying conductor. *Sov. Phys.-Tech. Phys.*, **2**, 1397–1406. Translated from Zhurnal Tekhnicheskoi Fiziki.
43. Moizhes, B.Ya. (1960) The influence of the temperature dependence of physical parameters on the efficiency of thermoelectric generators and refrigerators. *Sov. Phys. Solid State*, **2**, 671–680. Translated from Fizika Tverdogo Tela, **2** (4), 728–737, 1960.
44. Ursell, T.S. and Snyder, G.J. (2002) Compatibility of segmented thermoelectric generators. 21st International Conference on Thermoelectrics, Piscataway, NI USA, 25–29 September 2002, IEEE (Institute of Electrical and Electronics Engineers), IEEE, pp. 412–417.
45. Snyder, G.J. and Ursell, T.S. (2003) Thermoelectric efficiency and compatibility. *Phys. Rev. Lett.*, **91** (14), 148301.
46. Drabkin, L.A. and Ershova, L.B. (2006) Comparison of approaches to thermoelectric modules mathematical optimization, in *25th International Conference on Thermoelectrics, Vienna, Austria, 06-10 August* (ed. P. Rogl), IEEE, Piscataway, pp. 476–479.
47. Pontryagin, L.S., Boltyanski, V.G., Gamkrelidze, R.S., and Mischenko, E.F. (1976) *The Mathematical Theory of Optimal Processes*, Nauka, Moscow. (in Russian).

48. Ioffe, A.F. (1950) *Energeticheskie osnovy termoelektricheskikh bataryei iz poluprovodnikov*, USSR Academy of Sciences.
49. Ioffe, A.F. (1957) *Semiconductor Thermoelements and Thermoelectric Cooling*, Infosearch, Ltd., London.
50. Goldsmid, H.J. (1960) *Applications of Thermoelectricity*, Butler & Tanner Ltd.
51. Peschke, K. (1957) Thermoelemente und Gleichstromthermogeneratoren. *Electr. Eng. (Arch. Elektrotech.)*, **43**, 328–354.
52. Harman, T.C. and Honig, J.M. (1967) *Thermoelectric and Thermomagnetic Effects and Applications*, McGraw-Hill Book Company, New York.
53. Cobble, M.H. (1995) Calculations of generator performance, in *CRC Handbook of Thermoelectrics*, Chapter 39 (ed. D.M. Rowe), CRC Press, Boca Raton, FL, pp. 489–501.
54. Gryaznov, O.S., Moizhes, B.Ya., and Nemchinskii, V.A. (1978) Generalized thermoelectric efficiency. *Sov. Phys.- Tech. Phys.*, **23** (8), 975–980. Translated from Zhurnal Tekhnicheskoi Fiziki, **48**, 1720–1728.
55. Korzhuev, M.A., Ivanova, L.D., Petrova, L.I., Granatkina, Yu.V., and Svechnikova, T.E. (2007) The multistage thermoelectric devices with inhomogeneous legs. Proceedings of the 5th European Conference on Thermoelectrics, Odessa, Ukraine, September 2007.
56. Korzhuev, M. (2009) Increase in the efficiency of thermoelectric heaters by using inhomogeneous branches. *Tech. Phys.*, **54**, 1241–1243.
57. Korzhuev, M. (2010) Symmetry analysis of thermoelectric energy converters with inhomogeneous legs. *J. Electron. Mater.*, **39**, 1381–1385.
58. Korzhuev, M. and Avilov, E. (2010) Use of the Harman technique for figure of merit measurements of cascade thermoelectric converters. *J. Electron. Mater.*, **39**, 1499–1503.
59. Korzhuev, M.A. and Katin, I.V. (2010) On the placement of thermoelectric generators in automobiles. *J. Electron. Mater.*, **39**, 1390–1394.
60. Landecker, K. (1976) Some aspects of the performance of refrigerating thermojunctions with radial flow of current. *J. Appl. Phys.*, **47** (5), 1846–1851.
61. Landecker, K. (1977) On power-generating thermojunctions with radial flow of current. *Sol. Energy*, **19** (5), 439–443.
62. Landecker, K. (1978) Heat transport in coaxial thermoelectric disks with radial flow of current. *J. Appl. Phys.*, **49** (9), 4939–4941.
63. Riffat, S.B. and Qiu, G.Q. (2006) Design and characterization of a cylindrical, water-cooled heat sink for thermoelectric air-conditioners. *Int. J. Energy Res.*, **30**, 67–80.
64. Min, G. (2006) Thermoelectric module design theories, in *CRC Handbook of Thermoelectrics: Macro to Nano*, Chapter 11 (ed. D.M. Rowe), Taylor and Francis, Boca Raton, FL.
65. Min, G. and Rowe, D.M. (2007) Ring-structured thermoelectric module. *Semicond. Sci. Technol.*, **22** (8), 880–883.
66. Altenkirch, E. (1911) Elektrothermische Kälteerzeugung und reversible elektrische Heizung. *Phys. Z.*, **12**, 920–924.
67. Goldsmid, H.J. (1986) *Electronic Refrigeration*, Pion, London.
68. Freedman, S.I. (1966) Thermoelectric power generation, in *Direct Energy Conversion*, Inter-University Electronic Series, vol. **3**, Chapter 3 (ed. G.W. Sutton), McGraw-Hill Book Company, pp. 105–180.
69. Heikes, R.R. and Ure, R.W. Jr. (1961) *Thermoelectricity: Science and Engineering*, Interscience Publishers, Inc., New York.
70. Egli, P.H. (1960) *Thermoelectricity*, John Wiley & Sons, Inc., New York.
71. Seifert, W., Pluschke, V., Goupil, C., Zabrocki, K., Müller, E., and Snyder, G.J. (2011) Maximum performance in self-compatible thermoelectric elements. *J. Mater. Res.*, **26** - Focus Issue - Adv. Thermoelectric. Mater. (15), 1933–1939.
72. Ybarrondo, L.J. (1967) Improved expressions for the efficiency of an infinite stage thermoelectric heat pump

and generator. *Solid State Electron.*, **10**, 620–622.
73. Norwood, M.H. (1961) A comparison of theory and experiment for a thermoelectric cooler. *J. Appl. Phys.*, **32** (12), 2559–2563.
74. Bergman, D.J. and Levy, O. (1991) Composite thermoelectrics - exact results and calculational methods, in *Modern Perspectives on Thermoelectrics and Related Materials*, vol. **234** (eds G.A. Slack, D.D. Allred, and C.B. Vining), MRS Proceedings, Materials Research Society (MRS) Pittsburgh, PA, pp. 39–45.
75. Bergman, D.J. and Levi, O. (1991) Thermoelectric properties of a composite medium. *J. Appl. Phys.*, **70** (11), 6821–6833.
76. Bergman, D.J. and Fel, L.G. (1999) Enhancement of thermoelectric power factor in composite thermoelectrics. *J. Appl. Phys.*, **85** (12), 8205–8216.
77. Bergman, D.J. and Fel, L.G. (1999) Enhancement of thermoelectric power factor in composite thermoelectrics. 18th International Conference on Thermoelectrics, pp. 76–79.
78. Yang, Y., Ma, F.Y., Lei, C.H., Liu, Y.Y., and Li, J.Y. (2013) Is thermoelectric conversion efficiency of a composite bounded by its constituents? *Appl. Phys. Lett.*, **102** (5), 053905.
79. Chen, M. and Liao, B. (2004) Comment on 'optimal design of a multi-couple thermoelectric generator'. *Semicond. Sci. Technol.*, **19** (5), 659–660.
80. Chen, M., Lu, S.-S., and Liao, B. (2005) On the figure of merit of thermoelectric generators. *J. Energy Resour. Technol.*, **127** (1), 37–41.
81. Altenkirch, E. (1909) Über den Nutzeffekt der Thermosäulen. *Phys. Z.*, **10**, 560–580.
82. Cadoff, I.B. and Miller, E. (eds) (1960) *Thermoelectric Materials and Devices*, Materials Technology Series, Reinhold Publishing Cooperation, New York. Lectures presented during the course on Thermoelectric Materials and Devices sponsored by the Department of Metallurgical Engineering in cooperation with the Office of Special Services to Business and Industry, New York, June 1959 and 1960.
83. Gordon, J.M. (1991) Generalized power versus efficiency characteristics of heat engines: the thermoelectric generator as an instructive illustration. *Am. J. Phys.*, **59** (6), 551–555.
84. Yan, Z. and Chen, J. (1993) Comment on "Generalized power versus efficiency characteristics of heat engines: the thermoelectric generator as an instructive illustration," by J.M. Gordon [Am. J. Phys. **59**, 551–555 (1991)]. *Am. J. Phys.*, **61** (4), 380.
85. Gordon, J.M. (1993) A response to Yan and Chen's "Comment on 'Generalized power versus efficiency characteristics of heat engines: the thermoelectric generator as an instructive illustration'". *Am. J. Phys.*, **61** (4), 381.
86. Castro, P.S. and Happ, W.W. (1960) Performance of a thermoelectric converter under constant heat flux operation. *J. Appl. Phys.*, **31** (8), 1314–1317.
87. Thacher, E.F. (1985) Heat loss and thermoelectric generator design. *Energy Convers. Manage.*, **25**, 519–525.
88. Lee, J.S. (1969) The influence of variable thermal conductivity and variable electrical resistivity on thermoelectric generator performance. *Energy Convers.*, **9** (3), 91–97.
89. Cohen, R.W. and Abeles, B. (1963) Efficiency calculations of thermoelectric generators with temperature varying parameters. *J. Appl. Phys.*, **34** (6), 1687–1688.
90. Ioffe, A.F., Stil'bans, L.S., Iordanishvili, E.K., and Stavitskaya, T.S. (1956) *Termoelektricheskoe Okhlazhdemie*, U.S.S.R. Academy of Sciences. (in Russian). Translated in English as "Semiconductor Thermoelements and Thermoelectric cooling" [49].
91. Sunderland, J.E. and Burak, N.T. (1964) The influence of the Thomson effect on the performance of a thermoelectric power generator. *Solid-State Electron.*, **7** (6), 465–471.
92. Shaw, D.T. (1966) The optimized performance of a thermoelectric generator with the Thomson effect. *Solid-State Electron.*, **9** (3), 282–285.

93. Efremov, A.A. and Pushkarsky, A.S. (1971) Energy calculation of thermoelements with arbitrary temperature dependence of thermoelectric properties of materials by heat balance technique. *Energy Convers.*, **11** (3), 101–104.
94. Adamson, W.L. and Sunderland, J.E. (1966) The influence of the Thomson coefficient and variable resistivity on thermoelectric heat pump performance. *Solid-State Electron.*, **9** (2), 105–112.
95. Adamson, W.L. (1965) The effects of the Thomson coefficient and variable resistivity on thermoelectric heat pump performance. Master's thesis. Georgia Institute of Technology.
96. Kooi, C.F., Horst, R.B., and Cuff, K.F. (1968) Thermoelectric-thermomagnetic energy converter staging. *J. Appl. Phys.*, **39** (9), 4257–4263.
97. Ure, R.W. Jr. and Heikes, R.R. (1961) *Theoretical Calculation of Device Performance*, Chapter 15, Interscience Publishers, Inc., New York, pp. 458–517.
98. Yvon, J. (1955) The Saclay reactor: two years of experience in the use of a compressed gas as a heat transfer agent. Proceeding of the International Conference of Peaceful Uses of Atomic Energy.
99. Chambadal, P. (1957) *Les centrales nucléaires*, Armand Colin.
100. Novikov, I.I. (1958) The efficiency of atomic power stations. *J. Nucl. Energy II*, **7**, 125. Translated from Atomnaya Energiya, **3**, 409, 1957.
101. Curzon, F.L. and Ahlborn, B. (1975) Efficiency of a Carnot engine at maximum power output. *Am. J. Phys.*, **43**, 22–24.
102. Andresen, B., Salamon, P., and Berry, R.S. (1977) Thermodynamics in finite time: extremals for imperfect heat engines. *J. Chem. Phys.*, **66** (4), 1571–1577.
103. Andresen, B., Berry, R.S., Nitzan, A., and Salamon, P. (1977) Thermodynamics in finite time. I. The step-Carnot cycle. *Phys. Rev. A*, **15** (5), 2086–2093.
104. Salamon, P., Andresen, B., and Berry, R.S. (1977) Thermodynamics in finite time. II. Potentials for finite-time processes. *Phys. Rev. A*, **15** (5), 2094–2102.
105. Andresen, B. (2011) Current trends in finite-time thermodynamics. *Angew. Chem. Int. Ed.*, **50** (12), 2690–2704.
106. Freunek, M., Müller, M., Ungan, T., Walker, W., and Reindl, L.M. (2009) New physical model for thermoelectric generators. *J. Electron. Mater.*, **38** (7), 1214–1220.
107. Stevens, J.W. (2001) Optimal design of small ΔT thermoelectric generation systems. *Energy Convers. Manage.*, **42**, 709–720.
108. Yazawa, K. and Shakouri, A. (2011) Cost-efficiency trade-off and the design of thermoelectric power generators. *Environ. Sci. Technol.*, **45**, 7548–7553.
109. Nemir, D. and Beck, J. (2010) On the significance of the thermoelectric figure of merit z. *J. Electron. Mater.*, **39** (9), 1897–1901.
110. Gehlhoff, P.O., Justi, E., and Kohler, M. (1950) Verfeinerte Theorie der elektrothermischen Kälteerzeugung. *Abh. Braunschweig. Wiss. Ges.*, **2**, 149–164.
111. Unger, S. and Schwarz, J. (2010) *Edmund Altenkirch - Pionier der Kältetechnik*, Statusbericht des Deutschen Kälte- und Klimatechnischen Vereins, vol. 23, DKV, Hannover. (in German).
112. Boerdijk, A.H. (1959) Contribution to a general theory of thermocouples. *J. Appl. Phys.*, **30** (7), 1080–1083.
113. Kohlrausch, F. (1900) Ueber den stationären Temperaturzustand eines elektrisch geheizten Leiters. *Ann. Phys.*, **306** (1), 132–158.
114. Diesselhorst, H. (1900) Ueber das Problem eines elektrisch erwärmten Leiters. *Ann. Phys.*, **306** (2), 312–325.
115. Callen, H.B. (1947) On the theory of irreversible processes. PhD thesis. Massachusetts Institute of Technology - (M.I.T.). Cambridge, MA.
116. Callen, H.B. (1948) The application of Onsager's reciprocal relations to thermoelectric, thermomagnetic, and galvanomagnetic effects. *Phys. Rev.*, **73** (11), 1349–1358.
117. Callen, H.B. and Greene, R.F. (1952) On a theorem of irreversible thermodynamics. *Phys. Rev.*, **86** (5), 702–710.

118. Callen, H.B. (1960) *Thermodynamics*, John Wiley & Sons, Inc.
119. Bernard, W. and Callen, H.B. (1959) Irreversible thermodynamics of non-linear processes and noise in driven systems. *Rev. Mod. Phys.*, **31** (4), 1017–1044.
120. de Groot, S.R. and van Kampen, N.G. (1954) On the derivation of reciprocal relations between irreversible processes. *Physica*, **21** (1-5), 39–47.
121. de Groot, S.R. (1963) *Thermodynamics of Irreversible Processes*, North-Holland Publishing Company, Amsterdam.
122. Onsager, L. (1931) Reciprocal relations in irreversible processes. I.. *Phys. Rev.*, **37** (4), 405–426.
123. Onsager, L. (1931) Reciprocal relations in irreversible processes. II.. *Phys. Rev.*, **38** (12), 2265–2279.
124. Onsager, L. (1945) Theories and problems of liquid diffusion. *Ann. N. Y. Acad. Sci.*, **46**, 241–265.
125. Clingman, W.H. (1961) Entropy production and optimum device design. *Adv. Energy Convers.*, **1**, 61–79.
126. Brandt, J.A. (1962) Solutions to the differential equations describing the temperature distribution, thermal efficiency, and power output of a thermoelectric element with variable properties and cross sectional area. *Adv. Energy Convers.*, **2**, 219–230.
127. Boerdijk, A.H. (1961) Zero-, first-, and second-order theories of a general thermocouple. *J. Appl. Phys.*, **32** (8), 1584–1589.
128. Semenyuk, V.A. (1977) Efficiency of cooling thermoelectric elements of arbitrary shape. *J. Eng. Phys. Thermophys.*, **32** (2), 196–200.
129. Shiliday, T.S. (1957) Performance of composite Peltier junctions of Bi_2Te_3. *J. Appl. Phys.*, **28**, 1035–1042.
130. Lund, T. (1978) Refrigerating thermojunctions with radial flow of current. *J. Appl. Phys.*, **49** (9), 4942.
131. Thacher, E.F. (1982) Shapes which maximize thermoelectric generator efficiency. Proceedings of the 4th International Conference on Thermoelectric Energy Conversion, Arlington, TX, pp. 67–74.
132. Thacher, E.F. (1980) Optimal thermoelectric generator shapes. PhD thesis. New Mexico State University, Las Cruces, NM.
133. Seifert, W., Zabrocki, K., Snyder, G.J., and Müller, E. (2010) The compatibility approach in the classical theory of thermoelectricity seen from the perspective of variational calculus. *Phys. Status Solidi A*, **207** (3), 760–765.
134. Gerstenmaier, Y.C. and Wachutka, G. (2012) Unified theory for inhomogeneous thermoelectric generators and coolers including multistage devices. *Phys. Rev. E*, **86**, 056703.
135. Zhang, Y., Bian, Z., and Shakouri, A. (2005) Improved maximum cooling by optimizing the geometry of thermoelectric leg elements. 24th International Conference on Thermoelectrics, 2005 (ICT 2005), pp. 248–251.
136. Lienhard, J.H. IV and Lienhard, J.H. V (2008) *A Heat Transfer Textbook*, 3rd edn, Phlogiston Press, Cambridge, MA.
137. Palma, R., Pérez-Aparicio, J.L., and Taylor, R.L. (2012) Non-linear finite element formulation applied to thermoelectric materials under hyperbolic heat conduction model. *Comput. Methods Appl. Mech. Eng.*, **213–216**, 93–103.
138. Hoyos, G.E., Rao, K.R., and Jaeger, D. (1977) Fast transient response of novel Peltier junctions. *Energy Convers.*, **17**, 45–54.
139. Hoyos, G.E. (1977) Fast transient response of novel Peltier junctions. PhD thesis. The University of Texas at Arlington.
140. Arenas, A., Vázquez, J., and Palacios, R. (2002) Performance analysis of thermoelectric pellets with non-constant cross sections. Proceedings of the 7th European Workshop on Thermoelectrics.
141. Sahin, A.Z. and Yilbas, B.S. (2013) The thermoelement as thermoelectric power generator: effect of leg geometry on the efficiency and power generation. *Energy Convers. Manage.*, **65**, 26–32.
142. Erturun, U., Erermis, K., and Mossi, K. (2014) Effect of various leg geometries on thermo-mechanical and power generation performance of thermoelectric devices. *Appl. Therm. Eng.*, **73** (1), 126–139.

143. Telkes, M. (1947) The efficiency of thermoelectric generators. I. *J. Appl. Phys.*, **18** (12), 1116–1127.
144. Bjørk, R., Christensen, D.V., Eriksen, D., and Pryds, N. (2014) Analysis of the internal heat losses in a thermoelectric generator. *Int. J. Therm. Sci.*, **85**, 12–20.
145. Ziolkowski, P., Pionas, P., Leszczynski, J., Karpinski, G., and Müller, E. (2010) Estimation of thermoelectric generator performance by finite element modeling. *J. Electron. Mater.*, **39** (9), 1934–1943.
146. Rollinger, C.N. and Sunderland, J.E. (1961) The performance of a convectively cooled thermoelement used for power generation. *Solid-State Electron.*, **3** (3–4), 268–277.
147. Ybarrondo, L.J. and Sunderland, J.E. (1962) Effects of surface heat transfer on the performance of a thermoelectric generator. *Solid-State Electron.*, **5** (3), 143–154.
148. Rollinger, C.N. and Sunderland, J.E. (1963) The performance of a thermoelectric heat pump with surface heat transfer. *Solid-State Electron.*, **6** (1), 47–57.
149. Kafafy, R. and Hameed, A. (2011) Modeling of non-uniform thermoelement devices subject to lateral heat convection, in *Defect and Diffusion Forum*, vol. **312–315**, Trans Tech Publications, Switzerland, pp. 782–787.
150. Hameed, A.H. and Kafafy, R. (2012) Thermoelectrically controlled micronozzle - a novel application for thermoelements. *J. Mech. Sci. Technol.*, **26** (11), 3631–3641.
151. Kraemer, D., Poudel, B., Feng, H.-P., Caylor, J.C., Ju, B., Yan, X., Ma, Y., Wang, X., Wang, D., Muto, A., Chiesa, M., Ren, Z., and Chen, G. (2011) High-performance flat-panel solar thermoelectric generators with high thermal concentration. *Nat. Mater.*, **10**, 532–538.
152. McEnaney, K., Kraemer, D., Ren, Z., and Chen, G. (2011) Modeling of concentrating solar thermoelectric generators. *J. Appl. Phys.*, **110**, 074502.
153. Baranowski, L.L., Warren, E.L., and Toberer, E.S. (2014) High-temperature high-efficiency solar thermoelectric generators. *J. Electron. Mater.*, **43**, 2348–2355.
154. Deng, Y., Zhu, W., Wang, Y., and Shi, Y. (2013) Enhanced performance of solar-driven photovoltaic-thermoelectric hybrid system in an integrated design. *Sol. Energy*, **88**, 182–191.
155. Chen, W.-H., Wang, C.-C., Hung, C.-I., Yang, C.-C., and Juang, R.-C. (2014) Modeling and simulation for the design of thermal-concentrated solar thermoelectric generator. *Energy*, **64**, 287–297.
156. Toberer, E.S., Baranowski, L.L., and Warren, E.L. (2014) Solar thermoelectric generator, US Patent Appl. 14/190,064.
157. Date, A., Date, A., Dixon, C., and Akbarzadeh, A. (2014) Theoretical and experimental study on heat pipe cooled thermoelectric generators with water heating using concentrated solar thermal energy. *Sol. Energy*, **105** (0), 656–668.
158. de Leon, M.T., Chong, H., and Kraft, M. (2014) Solar thermoelectric generators fabricated on a silicon-on-insulator substrate. *J. Micromech. Microeng.*, **24** (8), 085011.
159. Fisac, M., Villasevil, F.X., and López, A.M. (2014) High-efficiency photovoltaic technology including thermoelectric generation. *J. Power Sources*, **252**, 264–269.
160. Jeyashree, Y., Juliet, A.V., and Joseph, A.A. (2014) Experimental analysis of thermoelectric generator using solar energy. Smart Structures and Systems (ICSSS), 2014 International Conference on, pp. 67–71.
161. Li, C., Zhang, M., Miao, L., Zhou, J., Kang, Y.P., Fisher, C.A.J., Ohno, K., Shen, Y., and Lin, H. (2014) Effects of environmental factors on the conversion efficiency of solar thermoelectric co-generators comprising parabola trough collectors and thermoelectric modules without evacuated tubular collector. *Energy Convers. Manage.*, **86** (0), 944–951.
162. Nia, M.H., Nejad, A.A., Goudarzi, A.M., Valizadeh, M., and Samadian, P. (2014) Cogeneration solar system using thermoelectric module and Fresnel

lens. *Energy Convers. Manage.*, **84**, 305–310.
163. Trinh, A.-K., González, I., Fournier, L., Pelletier, R., Sandoval, J.C., and Lesage, F.J. (2014) Solar thermal energy conversion to electrical power. *Appl. Therm. Eng.*, **70** (1), 675–686.
164. Date, A., Date, A., Dixon, C., Singh, R., and Akbarzadeh, A. (2015) Theoretical and experimental estimation of limiting input heat flux for thermoelectric power generators with passive cooling. *Sol. Energy*, **111**, 201–217.
165. Da, Y. and Xuan, Y.M. (2015) Perfect solar absorber based on nanocone structured surface for high-efficiency solar thermoelectric generators. *Sci. China Technol. Sci.*, **58** (1), 19–28.
166. Telkes, M. (1954) Solar thermoelectric generators. *J. Appl. Phys.*, **25** (6), 765–777.
167. Baranowski, L.L., Snyder, G.J., and Toberer, E.S. (2013) Effective thermal conductivity in thermoelectric materials. *J. Appl. Phys.*, **113**, 204904.
168. Matthews, J.E. (2011) Thermoelectric and heat flow phenomena in mesoscopic systems. PhD thesis. University of Oregon.
169. Mann, B.S. (2006) Transverse thermoelectric effects for cooling and heat flux sensing. Master's thesis. Virginia Polytechnic Institute and State University.
170. Derryberry, R.A. (2007) Artificial anisotropy for transverse thermoelectric heat flux sensing. Master's thesis. Virginia Polytechnic Institute and State University, Blacksburg, VA.
171. Goldsmid, H.J. (2010) *Introduction to Thermoelectricity: Transverse Devices*, Chapter 11, Springer-Verlag, pp. 191–201.
172. Reitmaier, C., Walther, F., and Lengfellner, H. (2010) Transverse thermoelectric devices. *Appl. Phys. A*, **99** (4), 717–722.
173. Goldsmid, H.J. (2011) Application of the transverse thermoelectric effects. *J. Electron. Mater.*, **40** (5), 1254–1259.
174. Goldsmid, H.J. (2012) Transverse thermoelectric effects and their application, in *CRC Handbook of Thermoelectrics: Thermoelectrics and Its Energy Harvesting*, Chapter 1 (ed. D.M. Rowe), CRC Press, Boca Raton, FL.
175. Grayson, M., Zhou, C., and Tang, Y. (2012) Transverse Thermoelectrics as Monolithic Peltier Coolers for Infrared Detectors. Technical report, Electrical Engineering & Computer Science, Northwestern University.
176. Zhou, C., Birner, S., Tang, Y., Heinselman, K., and Grayson, M. (2013) Driving perpendicular heat flow: $(p \times n)$-type transverse thermoelectrics for microscale and cryogenic Peltier cooling. *Phys. Rev. Lett.*, **110**, 227701 (5).
177. Ali, S.A. (2013) Computational study of transverse Peltier coolers. Master's thesis. Ohio State University.
178. Ali, S.A. and Mazumder, S. (2013) Computational study of transverse Peltier coolers for low temperature applications. *Int. J. Heat Mass Transfer*, **62**, 373–381.
179. Monroe, D. (2013) A new direction for thermoelectric cooling. *Physics*, **6**, 63 (1 page).
180. Crawford, C. (2014) Transverse thermoelectric effect. PhD thesis. University of New Orleans.

3
Segmented Devices and Networking of TE Elements

Knud Zabrocki, Christophe Goupil, Henni Ouerdane, Eckhard Müller, and Wolfgang Seifert[†]

3.1
Segmented Devices

In the previous chapter, we have seen that the behavior of a thermoelectric element can be comprehensively described by analytical formulae if constant TE properties (constant properties model, CPM) are assumed. Optimum operation conditions are known, which yield maximum output power/cooling power or maximum efficiency/maximum coefficient of performance (COP), and the rules on how to tune load or current to reach these conditions have been deduced. On the other hand, it has been observed that the performance of a thermoelectric leg is sensitive to inhomogeneity in the local distribution of the TE properties. There are some types of inhomogeneity that will clearly reduce the performance because of involved loss mechanisms such as poor-conducting regions connected in series to the element acting similarly to nonideal contacts, or losses such as electrical or thermal bypasses, or any kind of lateral (directed across the temperature gradient) variation of the Seebeck coefficient, which gives rise to closed current loops ("eddy currents") in the material. The latter convert part of the thermoelectrically generated electrical power into Joule heat inside the TE element instead of exporting it to the load. Apart from these obvious detrimental effects of material inhomogeneity, the question of whether suitably arranged local variation of the thermoelectric material properties can help to improve the achievable performance on the background of available materials arises. It can be shown that such improvement, if achievable anyway, can be related to suitable material profiles (i.e., spatial variation of material properties) along the temperature gradient only but not across.

This chapter presents a numerical tool that allows for accurate calculation of the device behavior of a TE element containing an arbitrary spatial variation of the material properties but retains, at least segmentwise, the advantage of analytical description of the temperature profile. The intention behind is to study the element performance in typical gradient situations on an empirical basis by means of numerical calculations. Given examples are based on monotonic spatial profiles of the material properties $\alpha(x)$, $\sigma(x)$, $\kappa(x)$, which are formulated analytically from

Continuum Theory and Modelling of Thermoelectric Elements, First Edition. Edited by Christophe Goupil.
© 2016 Wiley-VCH Verlag GmbH & Co. KGaA. Published 2016 by Wiley-VCH Verlag GmbH & Co. KGaA.

practicable parameters: spatial average value, relative change of the property all over the element, as well as the profile's curvature. Serial numerical experiments give an overview of how certain gradients of the TE properties may interact. The numerical method is based on subdivision of the TE element into several segments with each of them individually treated as a CPM element. The main principle of the approach is demonstrated for a twofold segmentation and then extended to an arbitrary number of segments, thus allowing the treatment of any shape of gradient scheme strictly from the physics point of view and numerically accurately. From a practical point of view, it turns out that the effects achievable by continuous gradients can be almost entirely accomplished even by segmentations involving merely a low number of different segments. Going beyond this one-dimensional approach, the Millman theorem introduced in the following offers a flexible way to treat 1D, 2D, or 3D inhomogeneous geometries represented by a discrete network of nodes and coupling elements in an iterative computation directly referring to the Onsager approach.

In a later chapter, a strict treatment on optimum profile combinations based on the self-compatibility approach is discussed. In fact, it can be concluded that there are no further physical principles behind the optimization of TE elements by gradients rather than local compatibility and high material performance (zT, power factor). The tool for numerical parameter studies presented in this chapter, as can be easily extended to further classes of profile functions beyond the ones implemented here, may be a helpful tool in conjunction with strict derivations to cross-check their results numerically by an independent route and to give an idea, for instance, about how shallow or steep the resulting maxima are—showing the practical accessibility.

To judge on the usefulness of specific gradient configurations requires beforehand to separate gradient effects from those related to the material performance. As a key question, not yet clearly answered, the following problem remains: which constraints have to be chosen for the performance of the materials representing the material property profiles in a device optimization that consists in selecting the "best" profiles. It is obvious that higher average material performance will improve the device performance and thus could eventually overlay gradient-borne beneficial effects if it is implicitly varied when the profiles are varied in the optimization procedure. However, the effect of material performance should be eliminated from the procedure to learn about the actual effect of the gradients. One approach to this problem is the choice of suitable averages of the TE properties, which are fixed during the optimization to describe "equally good" materials. Fixing the spatial average of zT was used as an intuitive limit to make the effect of gradients visible since, otherwise, the achieved performance is dominated by implicit zT variation related to the variation of the profiles.

In this manner, the free numerical variation of gradients and gradient combinations will provide a suitable intuitive guide on tendencies for favorable and detrimental gradient configurations but will finally not be suitable to give a strict answer on optimum configurations, also due to the limitation to a certain predefined profile function class but mainly due to the lack of a clear physically based constraint to quantify "equal material performance" behind different grading configurations. What can be shown by these numerical studies anyway is that optimal grading strategies cannot be formulated generally but depend on the operation mode and boundary conditions and that CPM is not the optimum configuration but is close to it in most of the practically relevant situations. One certain exception to this general experience has been found, namely ΔT_{max} and COP optimization of a cooling device at large temperature difference, where CPM performance can be substantially exceeded. Intuitively, it seems obvious that a performance improvement above the CPM element should be possible since it is not completely self-compatible and thus not optimal. It has to be mentioned that material gradients may have a strong effect in reducing performance if unsuitably combined, whereas optimum grading schemes may be achieved including strong gradients.

An alternative option of selecting limits of the material properties, which was not accomplished here, could be to fix instead of just the local averages of α, σ, κ (or zT instead of κ as done here) the equivalent device parameters of a TE element, that is, α_{eff}, R, and K, which, as a physically correct equivalence, have to be formed by the temperature average value of the Seebeck coefficient and from the spatial averages values of $1/\sigma$ and $1/\kappa$, respectively. This chapter illustrates that quantitative differences between the cases of, that is, fixing the spatial averages of σ or of ϱ will amount to substantial magnitude if the profiles get steeper. With fixing α_{eff}, R, and K, a direct comparison to CPM is given, relating the maximum performance of a graded element to the well-known respective CPM values. This approach would be an instructive example to gain a qualitative imagination of the physical action of thermoelectric-functional gradients and how they will affect the overall performance. At zero current, no difference in the behavior of any configuration of this kind can be observed. This will change under current flow since, then, additional sources and drains of heat related to the current will appear, whose local distribution over the element will depend on the shape of the profiles of α, σ, κ. Compared to CPM, the amount of Peltier heat absorbed/released at the junctions will change with the difference in the amount being shifted from the junctions into the volume of the element appearing there as Thomson and distributed Peltier heat. The larger the difference over the element is, the larger will be the Peltier heat at one side. Varying the shape of the Seebeck profile will shift the local distribution of Thomson and distributed Peltier heat along the element. This will change the portions of this heat conducted to the one and the other terminal of the element, resulting in an overall change of the transport of Peltier heat along the element. The closer the distributed Peltier heat appears to the terminal of opposite sign of Peltier heat in the element, the lower is the thermal resistance between both terminals and the larger will be the fraction of Peltier heat flowing back by Fourier conduction. A gradient of the electrical conductivity will result in an asymmetry of Joule heat release in the element and thus a preferential drain to the closer terminal, whereas a gradient in thermal conductivity will overlay the distribution of any internal heat source (Thomson, distributed Peltier, Joule) to the terminals according to the related asymmetry of thermal conductance between any source and both terminals, respectively. Furthermore, there will be an interplay between

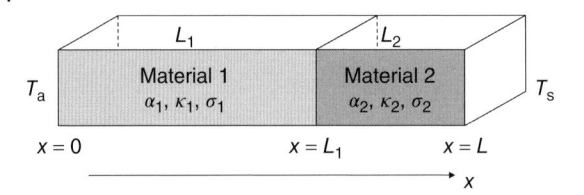

Figure 3.1 Element built up from two segments with different material properties.

all of these effects since each of them will modify the temperature profile and with that also the distribution of Thomson heat and the profile of each TE property if dependent on temperature. Another, different, approach to formulate the limits to the material quality as a constraint for the variation of material gradients might be proposed from the point of view of a material developer based on practical availability of materials. A temperature-dependent upper envelope function to any experimental zT value can be set as a limit, complemented by the lower envelope to thermal conductivity measurements.

3.1.1
Double-Segmented Element

In order to understand the formalism of segmented elements, we start with a discussion on an element consisting of two different segments or sections. In both segments, CPM conditions are assumed; see Figure 3.1.

If the temperature profile is known, all performance parameters can be calculated. The fundamental equation for determining the temperature profile is the energy balance equation, see Eq. (1.89). A major ideal assumption is that there is a perfect contact at the interface between both materials such that there is no temperature drop at all caused by the contact. Another assumption is that the envelope faces are perfectly insulated, both electrically and thermally. By that assumption, the general 3D problem is reduced to a quasi-1D problem. The cross section of the element A_c is kept constant and the element length L is the sum of the lengths of segment one L_1 and segment two L_2. At one end $x = 0$ (the absorbing side), we apply a temperature T_a, and at the other end $x = L$ (sink side), it is T_s, which is fixed. These Dirichlet boundary conditions are used in order to describe both the generator case ($T_a \equiv T_{hot}$ and $T_s \equiv T_{cold}$) and the Peltier cooling case ($T_a \equiv T_{cold}$ and $T_s \equiv T_{hot}$) in an combined manner. As a direct consequence of the chosen boundary conditions, the flow of the electrical charge and that of heat are directed in parallel along the x-direction (which is chosen without loss of generality as the flow direction). Furthermore, the electrical current density $j = I/A_c$ is constant along the x-axis (element).

The ansatz for the temperature profile in each section is a parabola

$$T_1(x) = -\frac{j^2}{2\kappa_1 \sigma_1} x^2 + C_1 x + C_2 \quad \text{and} \quad T_2(x) = -\frac{j^2}{2\kappa_2 \sigma_2} x^2 + C_3 x + C_4, \quad (3.1)$$

where $T(x)$ is subdivided into $T_1(x)$, which is valid in the first segment ($0 \leq x < L_1$) and $T_2(x)$ in the second element ($L_1 < x \leq L = L_1 + L_2$). The four free constants C_i can be determined from both boundary conditions for temperature and matching conditions for temperature and total heat flux at the junction between both elements. The heat fluxes $q_1 = -\kappa_1 T_1'(x) + j\alpha_1 T_1(x)$ and $q_2 = -\kappa_2 T_2'(x) + j\alpha_2 T_2(x)$ are defined in the corresponding segments. Besides the usual boundary conditions $T_1(x = 0) = T_a$ and $T_2(x = L) = T_s$, additional matching conditions at the interface are given by a temperature match $T_1(x = L_1) = T_2(x = L_1)$ and a heat flux match $q_1(x = L_1) = q_2(x = L_1)$. The calculation of the free constants of the temperature profile is straightforward but tedious. As the constants are not needed for further steps, the calculation is omitted.

To calculate the power output density from the energy balance all over the element by taking into account the Peltier heat at the terminals and at the inner interface reduced by the Joule heat, we consider Eqs. (2.48) and (2.49), see Section 2.4,

$$p_{\text{out}} = -j^2 L \sigma_{\text{eff}}^{-1} - j T_{\text{int}} \alpha_{12} + j(T_a \alpha_1 - T_s \alpha_2). \tag{3.2}$$

In Eq. (3.2), an effective electrical conductivity is used, which is defined as

$$\sigma_{\text{eff}}^{-1} = \frac{1}{L}\left(\frac{L_1}{\sigma_1} + \frac{L_2}{\sigma_2}\right). \tag{3.3}$$

The parameter α_{12} is just the difference between the Seebeck coefficients, that is, $\alpha_1 - \alpha_2$. Note that the interface temperature is a function of the electrical current density $T_{\text{int}} \equiv T_{\text{int}}(j)$ which can be easily seen from the continuity of heat flow at the inner interface,

$$T_{\text{int}}(j) = \frac{\frac{j^2}{2}\frac{L}{\sigma_{\text{eff}}} + T_a \frac{\kappa_1}{L_1} + T_s \frac{\kappa_2}{L_2}}{\frac{\kappa_1}{L_1} + \frac{\kappa_2}{L_2} - j\alpha_{12}}. \tag{3.4}$$

Because of the nonlinear (nontrivial) dependence of the interface temperature on j, finding the maximum power output density from an extreme value problem analytically is much more complicated in general than for the CPM case, see again Eqs. (2.52), (2.54), and (2.55).

3.1.1.1 Effective Electrical Conductivity
A particular case occurs if we assume $\alpha_1 = \alpha_2$ or $\alpha_{12} = 0$.[1] Then the power output density is

$$p_{\text{out}}^{\alpha_{12}=0} = -j^2\left(\frac{L_1}{\sigma_1} + \frac{L_2}{\sigma_2}\right) + j\alpha(T_a - T_s) = -\frac{j^2 L}{\sigma_{\text{eff}}} + j\alpha \Delta T,$$

which shows a similar functional form as Eq. (2.52) with the **effective** electrical conductivity σ_{eff}, see Eq. (3.3). Analogously, the maximum power output in this particular case is

$$p_{\text{max}}^{\alpha_{12}=0} = \alpha^2 \sigma_{\text{eff}} \frac{(\Delta T)^2}{4L}. \tag{3.5}$$

---
[1] Although this means that the key parameter giving the thermoelectric coupling, the Seebeck coefficient, is constant.

It is obvious that in the case of constant Seebeck coefficient, we find that the total resistance of our element is built up as a serial connection of the internal resistances of the two segments. In our setup, we find a constant current I and a density j over the length of the whole element. We find a voltage drop over the element denoted as V, whereas over the first segment, it is V_1, and over the second segment, V_2. As we have a serial connection

$$V = V_1 + V_2 \quad \text{and} \quad RI = R_1 I + R_2 I$$

dividing by I and using the classical formula for a *planar* geometry, for example, a prismatic or cylindrical element

$$\frac{L}{A_c \sigma_{\text{eff}}} = \frac{L_1}{A_c \sigma_1} + \frac{L_2}{A_c \sigma_2},$$

the formula found in Eq. (3.3) for the effective electrical conductivity is verified.

There is a relationship between the internal resistance of the element R_i and the effective electrical conductivity σ_{eff}, which is an averaged quantity

$$R_i = \frac{L}{A_c \sigma_{\text{eff}}}.$$

Note that the reciprocal relation $\varrho(x) = 1/\sigma(x)$ holds locally, but there is no generally valid relation between the averages of $\sigma(x)$ and $\varrho(x)$, $\sigma_{\text{av}} L = \int_0^L \sigma(x)\,dx$ and $\varrho_{\text{av}} L = \int_0^L \varrho(x)\,dx$. That means $\sigma_{\text{av}} \neq \varrho_{\text{av}}^{-1}$ in general. On the contrary there is a direct correlation between the effective values $\sigma_{\text{eff}} = \varrho_{\text{eff}}^{-1}$, where $\varrho_{\text{eff}} = (L_1 \varrho_1 + L_2 \varrho_2)/L$, see example [1]. In the limit of continuously graded material (consisting of infinitesimal segments), the correlation between the global effective value and the local value is given by

$$\sigma_{\text{eff}}^{-1} = \frac{1}{L} \int_0^L \frac{1}{\sigma(x)}\,dx = \left(\frac{1}{\sigma}\right)_{\text{av}} = \varrho_{\text{av}} = \varrho_{\text{eff}}. \tag{3.6}$$

Let us discuss a parameterized functional expression of a linear spatial dependence of the electrical conductivity as an example. Fixing the average, see, for example, [2, 3], we have

$$\sigma(x; L, \sigma_{\text{av}}, \xi_\sigma) = \frac{2\sigma_{\text{av}}}{1+\xi_\sigma}\left[\xi_\sigma + (1-\xi_\sigma)\frac{x}{L}\right]$$

with $\xi_\sigma = \sigma_a/\sigma_s$ being the relative change over the element. Setting this profile into Eq. (3.6), we get for the relation between σ_{eff} and the average σ_{av}

$$\sigma_{\text{eff}} = -\frac{2(1-\xi_\sigma)}{(1+\xi_\sigma)\ln \xi_\sigma} \sigma_{\text{av}} = \zeta \sigma_{\text{av}} \tag{3.7}$$

From Eq. (3.7), we get that $\zeta(\xi_\sigma) \leq 1$, which means that the effective value is always smaller (or equal at $\xi_\sigma = 1$) than the spatial average $\sigma_{\text{eff}} \leq \sigma_{\text{av}}$, see Figure 3.2. This relation is pointed out here since it is of essential importance for any attempt to study the influence of a material gradient of the electrical conductivity on the performance of a TE element. This kind of study needs to be based on a valid

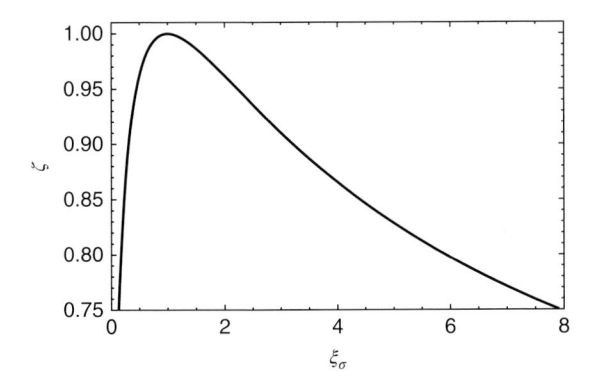

Figure 3.2 The correlation factor ζ between σ_{eff} and σ_{av} in dependence of ξ_σ.

comparison to a reference system, which should be, most suitably, a CPM element of the same σ_{eff}, that is, of the same R, but hence generally not the same σ_{av}. By selecting σ_{av} instead of σ_{eff} as a reference, erroneous results will be obtained, which overestimate the performance of systems with a strong gradient of the electrical conductivity simply since they have a resistance lower than that of the reference element. The same holds accordingly for the thermal conductivity, where, similarly, the effective value $\kappa_{\text{eff}} = 1/L \int_0^L \kappa^{-1} dx$ is the correct reference but not κ_{av}, according to an equally assumed thermal conductance K for compared segmented/graded systems.

3.1.1.2 Effective Thermal Conductivity

An analogous consideration as in the previous section can be made for heat conduction at zero current where the thermal conductivity κ behaves formally the same as σ. For a CPM element and Dirichlet boundary conditions, this ends up in a linear temperature profile, which means that $dT/dx \equiv \Delta T/L$ holds exactly. If heat conduction through an element consisting of two segments (with constant κ_1 and κ_2) with ideal coupling is considered, it is the temperature drop that is important. For the Fourier parts of the heat fluxes, the following relations can be found:

$$q_{F,1} = -\kappa_1 \frac{dT_1}{dx} = -\kappa_1 \frac{\Delta T_1}{L_1},$$

$$q_{F,2} = -\kappa_2 \frac{dT_2}{dx} = -\kappa_2 \frac{\Delta T_2}{L_2}, \quad \text{and} \quad q_F = -\kappa_{\text{eff}} \frac{dT}{dx} = -\kappa_{\text{eff}} \frac{\Delta T}{L}. \quad (3.8)$$

Here a comparison between the two segmented elements and an equivalent CPM element with the same complete length and effective material properties is performed.

A straightforward relation between the temperature drops is

$$\Delta T = \Delta T_1 + \Delta T_2. \quad (3.9)$$

By combining these relations with the previous one observed from Eq. (3.8), it follows

$$-\frac{L q_F}{\kappa_{\text{eff}}} = -\frac{L_1 q_{F,1}}{\kappa_1} - \frac{L_2 q_{F,2}}{\kappa_2}. \tag{3.10}$$

The heat flux through the element is constant $q_F = q_{F,1} = q_{F,2}$, so an analogous expression for the effective thermal conductivity can be found

$$\frac{L}{\kappa_{\text{eff}}} = \frac{L_1}{\kappa_1} + \frac{L_2}{\kappa_2}. \tag{3.11}$$

If the element is subdivided into N segments with the width Δx_n ($n = 1, \ldots, N$), then for the effective thermal conductivity, the following relation

$$\kappa_{\text{eff}}^{-1} = \frac{1}{L} \sum_{n=1}^{N} \frac{\Delta x_n}{\kappa_n} \tag{3.12}$$

can be found. After applying the (thermodynamic) limit[2] $\Delta x_n \to 0$ and $N \to \infty$, there is a transition from summation to integration

$$\kappa_{\text{eff}}^{-1} = \frac{1}{L} \int_0^L \frac{dx}{\kappa(x)}. \tag{3.13}$$

Now it is possible to construct the thermal conductance K (as a global variable of the element) from the corresponding effective thermal conductivity and the geometric parameters of a geometry with constant A_c, for example, a prismatic or cylindrical,

$$K = \frac{A_c \kappa_{\text{eff}}}{L} \tag{3.14}$$

and the reciprocal of K is the thermal resistance R_{th}.

A straightforward calculation based on Eqs. (3.2) and (3.4) together with an expression for q_h would give a complex expression for z_{eff} for a two-segment stack as a function of the current density.

3.1.2
Algorithm of Multisegmented Elements

What has been performed analytically for twofold segmentation can be generalized for multisegmented elements. The algorithm is suitable for investigating segmented (ideal) devices, but it can also be extended into an approximation of excellent accuracy for continuously graded elements within the framework of a one-dimensional approach. Heat transfer is assumed along the x-axis only, and heat losses by radiation and contact resistance at any junction are neglected.

To begin with, we consider a TE element with N segments and constant material properties $\alpha_n, \kappa_n, \sigma_n$ in each segment ($n = 1, \ldots, N$) representing the effective values. Let n be the number of segments counted from the absorbing side (first

---

2) Switching to infinitesimal segments and continuously graded material.

3.1 Segmented Devices

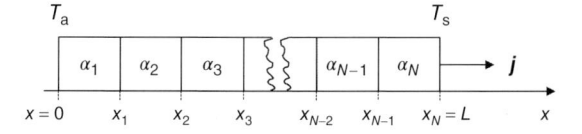

Figure 3.3 One-dimensional model of a multisegmented TE element (total length L, N segments, p-type if $\alpha_n > 0 \forall n$, constant electric current density $\mathbf{j} = j\mathbf{e}_x$ through a constant cross-sectional area).

segment in contact with temperature $T_a = T(x = 0)$) to the heat sink side (last segment in contact with temperature $T_s = T(x = L)$, see Figure 3.3).

The total length of all N segments together is set to L. The length of each segment is $L_n = L/N$, and x_n ($n = 1, \ldots, N-1$) denotes the position of any of the $N-1$ internal junctions between the segments. All material parameters $\alpha_n, \kappa_n, \sigma_n$ of the segments are assumed to be independent of temperature. Then, the temperature profile in each segment n is a parabolic one,

$$T^{(n)}(x) = -\frac{c_o^{(n)}}{2}x^2 + c_1^{(n)} x + c_2^{(n)}, \quad c_o^{(n)} = \frac{j^2}{\sigma^{(n)}\kappa^{(n)}} \quad \forall n = 1, \ldots, N. \tag{3.15}$$

To determine the free constants, conservation of heat at any intermediate junction between segment n and segment $n+1$ requires that both temperature and overall heat transport are continuous functions

$$T^{(n)}(x_n) = T^{(n+1)}(x_n) \quad \forall n = 1, \ldots, N-1,$$
$$q^{(n)}(x_n) = q^{(n+1)}(x_n) \quad \forall n = 1, \ldots, N-1. \tag{3.16}$$

Since $q^{(n)}$ denotes the overall (convective plus conductive) heat transport, the Peltier heat occurring at each interface does not appear explicitly in this balance but causes a step in the Fourier heat q_F at each interface, in general.

These $2(N-1)$ equations, combined with the two boundary conditions for temperature,

$$T^{(0)}(x = 0) = T_a, \quad T^{(N)}(x = L) = T_s, \tag{3.17}$$

generate a linear equation system for the $2N$ unknown coefficients $c_1^{(n)}, c_2^{(n)}$. In matrix form, the system of $2N$ equations can be written as

$$\mathbf{Ac} = \mathbf{Y} \tag{3.18}$$

with the matrix

$$\mathbf{A} = \begin{pmatrix} 0 & 1 & 0 & 0 & 0 & 0 & \cdots & 0 & 0 \\ x_1 & 1 & -x_1 & -1 & 0 & 0 & \cdots & 0 & 0 \\ K_1^1 & K_2^1 & -K_1^2 & -K_2^2 & 0 & 0 & \cdots & 0 & 0 \\ 0 & 0 & x_2 & 1 & -x_2 & -1 & \cdots & 0 & 0 \\ 0 & 0 & K_1^2 & K_2^2 & -K_1^3 & -K_2^3 & \cdots & 0 & 0 \\ \cdots & \cdots & \cdots & \cdots & \cdots & \cdots & \cdots & \cdots & \cdots \\ 0 & 0 & \cdots & 0 & 0 & x_{N-1} & 1 & -x_{N-1} & -1 \\ 0 & 0 & \cdots & 0 & 0 & K_1^{N-1} & K_2^{N-1} & -K_1^N & -K_2^N \\ 0 & 0 & \cdots & 0 & 0 & 0 & 0 & x_N & 1 \end{pmatrix} \tag{3.19}$$

and the vectors

$$\mathbf{c}^T = \left(c_1^{(1)} \ c_2^{(1)} \ c_1^{(2)} \ c_2^{(2)} \ \ldots \ c_2^{(N-1)} \ c_1^{(N)} \ c_2^{(N)} \right) \tag{3.20}$$

and

$$\mathbf{Y}^T = (T_a \ Y_{1,1} \ Y_{1,2} \ Y_{2,1} \ Y_{2,2} \ \ldots \ Y_{N-1,1} \ Y_{N-1,2} \ Y_N) \tag{3.21}$$

with the following coefficients

$$K_1^n = \alpha_n j x_n - \kappa_n, \quad K_2^n = \alpha_n j, \quad Y_N = T_s + c_o^{(n)} L^2,$$
$$Y_{n,1} = \left[c_o^{(n)} - c_o^{(n+1)} \right] x_n^2/2,$$
$$Y_{n,2} = \left[\kappa_{n+1} c_o^{(n+1)} - \kappa_n c_o^{(n)} \right] x_n + [\alpha_n c_o^{(n)} - \alpha_{n+1} c_o^{(n+1)}] j \frac{x_n^2}{2}. \tag{3.22}$$

This linear equation system can be solved by means of standard methods. As a result, we have now an exact numerical tool in our hands, which allows for a straightforward calculation of the system behavior and performance of any arbitrarily graded TE element, including the elements whose gradients are based on their temperature-dependent properties. In the latter case, the calculation has to be repeated iteratively to achieve self-consistency between the temperature profile and the material profiles depending on it.

3.1.2.1 Numerical Parameter Studies on Graded Elements

Fundamental parameter studies [4, 5] have been carried out to demonstrate a quantitative improvement of performance achievable by material grading. Numerical calculations were based on a "model-free" setup where an independent and free variability of the spatial material profiles is assumed as a starting point although the solid-state-based interrelations between the individual properties are well known. To analytically describe continuously graded elements, arbitrarily but suitably chosen continuous monotonic spatial gradient functions for all material profiles have been applied (see e. g., [6])

$$y(x) = y_a + \frac{1 - \exp(k_y x/L)}{1 - \exp(k_y)} (y_s - y_a), \quad y = \alpha, \sigma, \kappa, \tag{3.23}$$

which fix the values y_a and y_s at the absorbing and sink side, respectively, but allow variation of the curvature by the numerical parameter k_y; for example, see Figure 3.4.

Having calculated the temperature profile

$$T(x) = \sum_{n=1}^{N} T^{(n)}(x), \tag{3.24}$$

all application-relevant device properties and performance parameters can be derived. Executing the calculation in a loop while varying the current allows the selection of optimum operation parameters (optimum load resistance and current) and deduction of the maximum performance values.

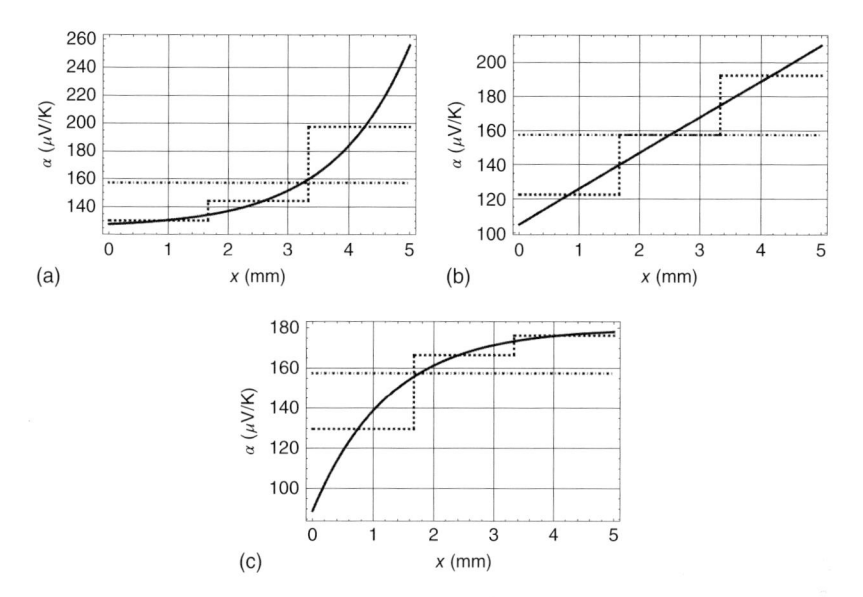

Figure 3.4 Continuously graded spatial model functions of the Seebeck coefficient (solid lines) based on Eq. (3.23). All the three example profiles have in common the value of the spatial average $\bar{\alpha} = 157.5$ µV/K (dashed–dotted horizontal lines) and the ratio $\alpha_a/\alpha_s = 0.5$ but vary in curvature. To approximate the continuous profiles, spatial averages have been calculated segmentwise to enter the multisegment numerical calculation scheme. Examples for $N = 3$ elements are shown (dashed lines); practically larger values for N, for example, $N = 20$–100 are used to well approximate arbitrary continuous profiles (a) $k_\alpha = 4$ (b) $k_\alpha = 0$ (c) $k_\alpha = -4$.

The difference between the temperature profile $T(x)$ obtained from the segmented algorithm (parabolas in each segment given by Eq. (3.15)) and the profile $T(x)$ derived from Domenicali's 1D thermal energy balance [7] under isotropic conditions (see Eq. (2.7)), which can be rewritten as

$$\kappa(x)\,T''(x) + \kappa'(x)\,T'(x) - j\alpha'(x)\,T(x) = -\varrho(x)j^2 \qquad (3.25)$$

becomes smaller with increasing number of segments, whereas convergence is ensured for $N \to \infty$ if the constant material properties in each segment are defined as spatial averages of $\alpha(x), \kappa(x), \sigma(x)$ over each segment's length.

Generally, the algorithm of multisegmented elements has proved to be useful for separating and quantifying the influence of material gradients and for determining optimal segmentation schemes based on the spatial profiles of the TE properties. Evidence has been provided that a stack made of $N = 3$–5 segments yields significant improvement of performance [5] whereas only little improvement has been achieved beyond $N \approx 5$ (quasi-continuity).

It was found that best performance is achieved close to linear vanishing gradient profiles (CPM) in certain cases (see Figure 3.5), whereas unfavorable combination of slopes such as strong gradients of α and σ oriented in parallel leads to a

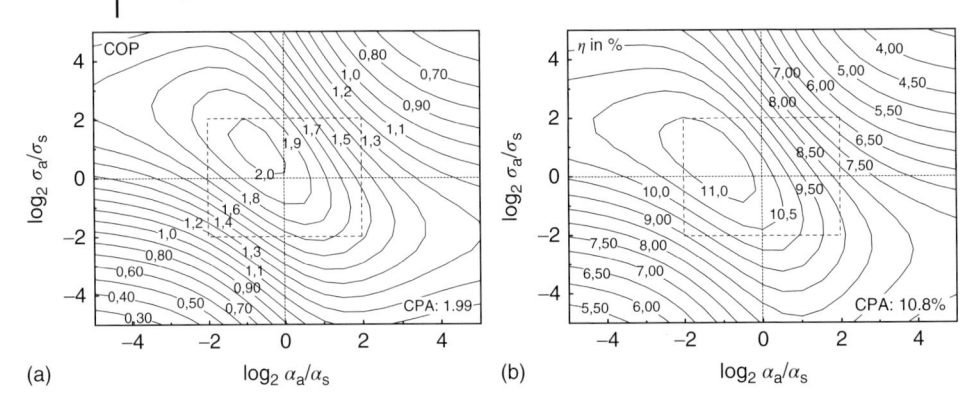

Figure 3.5 Variation of the maximum COP φ_{max} of a TEC ($T_a = 280\,\text{K}$, $\kappa_a/\kappa_s = 1$) high (a) and maximum efficiency η_{max} of a TEG ($T_a = 600\,\text{K}$, $\kappa_a/\kappa_s = 1$) (b). The sink side temperature is fixed to $T_s = 300\,\text{K}$. Both are plotted versus the chosen constant gradients (i.e., linear spatial profiles) of the Seebeck coefficient and electrical conductivity (see also [5]); under the constraint of $(zT)_{av} = \text{const.}$; double-logarithmic plot: $\log_2(\alpha_a/\alpha_s)$ (abscissa), $\log_2(\sigma_a/\sigma_s)$ (ordinate) in both figures; $\kappa = \text{const.}$ The inner frame marks the region of small-to-moderate gradients. The background colors indicate the maximum zT, which is found locally within the respective gradient configuration, which gives a hint on the practical feasibility of certain combinations of gradients. The central point $(0,0)$ marks the CPM material; note that the performance can be slightly improved above the CPM case for moderate gradients in a similar configuration for both cooling COP and TEG efficiency. Further, note that the performance is limited by the constraints of linear profiles; curved profiles may alter for somewhat higher maximum performance.

significant drop in the performance due to mistuning of local compatibility along the element.

Further, it is worth mentioning that, in relation to the higher optimum current density, achievable effects are greater for Peltier coolers than for a thermoelectric generator (TEG) although the operating temperature difference was assumed to be much larger for generators. The greatest improvement for Peltier elements can be achieved at the maximum temperature difference if steep material gradients are selected. This is understandable based on the compatibility approach, see Chapter 5: at low current density, the related current density $u(x)$ will vary only slightly along the element as will the compatibility factor $s(x)$, whereas this changes for a thermoelectric cooler (TEC) with optimal current. With $u(x)$ varying greatly along a CPM element than $s(x)$, material gradients may reduce local incompatibility while affecting both $u(x)$ and $s(x)$ in a different manner. For a detailed discussion, we refer to [4, 5]. The aim of such investigations is to identify optimum combinations of material gradients along a graded generator or cooler element. However, the practical accessibility of optimum gradient schemes is limited by the constraint of maximum zT of available materials, by the magnitude of the compositional gradient (extremely steep profiles are difficult to prepare in a controlled manner and might not be stable at elevated temperature due to diffusion processes), and by interrelations between the thermoelectric properties due to the solid-state nature of the TE materials.

For an overview on material grading, we also refer to [8].

3.2 Networks

3.2.1 Presentation

An alternative way of describing a given thermoelectric material is to consider the local equilibrium assumption, which is the basis of the Onsager approach. In other words, a thermoelectric medium may be divided into unit cells where the intensive variables μ and T may be considered as constant. More precisely, this means that, in each cell and for every intensive variable, the ratio between the average value and its standard deviation is large enough so we can assume an equilibrium value of both μ and T. The question is whether the determination of the correct size of a cell is a physical or a numerical problem. From the physical point of view, the cell should be large enough to be considered as a thermodynamic system. This is not critical in the present approach since we do not consider the statistics at the particle scale. From the numerical point of view, the size of the cell determines the grain size of the particle and can be adjusted freely (see Figure 3.6).

The thermoelement is considered to be electrically and thermally polarized at a given temperature and electrochemical potential differences, ΔT and $\Delta \mu$. Then, each cell undergoes the potential differences $dT_{ij} = T_i - T_j$ and $d\mu_{ij} = \mu_i - \mu_j$.

By defining the size of the unit cell as $dV = dx\,dy\,dz$, we can derive the corrects forces at the boundaries of the cell as the gradient of the potentials (see Figure 3.7). As we know, many expressions of the gradient lead to forces but only specific ones fulfill the symmetry of the off-diagonal terms of the Onsager matrix. When considering the energy and particles fluxes, the correct forces are given by the gradients of the potentials for energy and particles, see Eq. (1.45)

$$\mathbf{F}_N = \nabla \left(-\frac{\mu}{T}\right), \quad \mathbf{F}_E = \nabla \left(\frac{1}{T}\right).$$

The choice of energy and particle fluxes, instead of heat and particle fluxes as usually considered, is due to the conservation property of these two fluxes, which is a

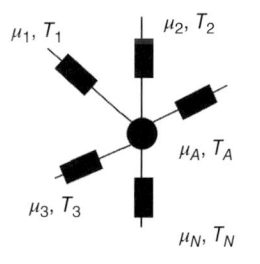

Figure 3.6 Sketch of the general Millman theorem with the boxes representing unit cells surrounding a central node A.

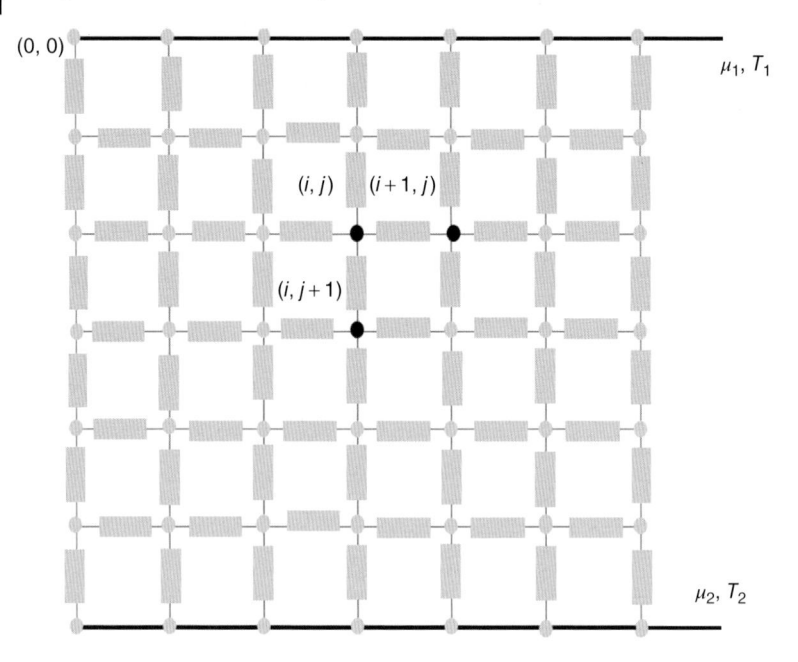

Figure 3.7 Application of the Millman theorem to an inhomogeneous material.

central point for the validity of the model. Nevertheless, a mapping of the energy and particles fluxes may easily be transformed into a heat and particle fluxes mapping since we have, see Eq. (1.44),

$$\mathbf{j}_E = \mathbf{j}_Q + \mu \mathbf{j}_N.$$

Finally, the unit cell is defined by the following Onsager matrix [see also Section 1.7.2 just below Eq. (1.45)]

$$\begin{bmatrix} \mathbf{j}_N \\ \mathbf{j}_E \end{bmatrix} = \begin{bmatrix} L_{NN} & L_{NE} \\ L_{EN} & L_{EE} \end{bmatrix} \begin{bmatrix} \nabla\left(-\frac{\mu}{T}\right) \\ \nabla\left(\frac{1}{T}\right) \end{bmatrix},$$

where

$$L_{NE} = L_{EN}$$

and

$$L_{11} = L_{NN} = \frac{\sigma_T}{e^2} T,$$

$$L_{NE} = \frac{\sigma_T S_J T^2}{e^2} + \mu \frac{\sigma_T}{e^2} T, \tag{3.26}$$

$$L_{EE} = T^2 \left[\kappa_E + 2\mu \frac{\sigma_T S_J}{e^2} \right] + \mu^2 \frac{\sigma_T}{e^2} T. \tag{3.27}$$

We have now all the ingredients of the model. With a correct algorithm, we may now obtain a mapping of the different potentials and currents inside a given thermoelectric element connected to temperature and electrochemical reservoirs. In addition to the potentials and the currents, the local heat and entropy production can also be obtained.

3.2.2
Useful Expressions

For the sake of simplicity of the notations, the electrochemical potential is simply noted μ without indices for the following treatments. After a complete simulation, we obtain a mapping of the variables T, μ, $I = jA_c$, and $I_E = j_E A_c$. I and I_E are obtained by integration of the j and j_E fluxes over the sample. From these, we can extract the following information:

- Local average electrical potential: $\Delta V = RI + \alpha \Delta T$.
- Entropy production: $v_S = I_E \Delta \left(\frac{1}{T}\right) + I \Delta \left(-\frac{\mu}{eT}\right)$.

The corresponding local expressions are in matrix form:

$$\begin{bmatrix} j \\ j_E \end{bmatrix} = \begin{bmatrix} \sigma_T & \alpha \sigma_T \\ \sigma_T \alpha T + \mu \frac{\sigma_T}{e} & \kappa_E + \mu \frac{\sigma_T \alpha}{e} \end{bmatrix} \begin{bmatrix} \mathbf{E} \\ -\nabla T \end{bmatrix}$$

with $\mathbf{E} = -\frac{\nabla \mu}{e}$ and $\mathbf{j} = e\mathbf{j}_N$, and $\nabla \left(\frac{1}{T}\right) = -\frac{1}{T^2} \nabla T$.

Then, we obtain the general conductivity matrix,

$$[\sigma] = \begin{bmatrix} \sigma_T & \alpha \sigma_T \\ \sigma_T \alpha T + \mu \frac{\sigma_T}{e} & \kappa_E + \mu \frac{\sigma_T \alpha}{e} \end{bmatrix}.$$

One can notice that the Seebeck coefficient is not the only coupling term between the transport of energy and particles. In a material with $\alpha = 0$ and $\mu = 0$, the conductivity matrix reduces to a diagonal form $\begin{bmatrix} \sigma_T & 0 \\ 0 & \kappa_E \end{bmatrix}$ where Ohm's law and Fourier's law are decoupled and no energy is transferred by the particles, nor by the energy flux. It should be noticed that for numerical reasons, the coupling between heat and electrical current leads to convergence problems. As a possible solution, the user may solve the complete mapping for small Seebeck coefficients first and then reiterate with increasing values of α. It may also be very useful to first test the computer code with $\alpha = 0$ in order to check its validity under pure electrical and pure thermal conditions.

3.2.3
Discretization

Let us consider one cell of length L and transverse section S. The incoming electrical and energetic currents are:

$$\begin{bmatrix} I \\ I_E \end{bmatrix} = \frac{S}{L} \begin{bmatrix} \sigma_T & \alpha \sigma_T \\ \sigma_T \alpha \overline{T} + \overline{\mu} \frac{\sigma_T}{e} & \kappa_E + \overline{\mu} \frac{\sigma_T \alpha}{e} \end{bmatrix} \begin{bmatrix} -\Delta V \\ -\Delta T \end{bmatrix},$$

$$\begin{bmatrix} I \\ I_E \end{bmatrix} = \frac{S}{L} \begin{bmatrix} \sigma_T & \alpha\sigma_T \\ \sigma_T \alpha \overline{T} + \overline{\mu}\frac{\sigma_T}{e} & \kappa_E + \overline{\mu}\frac{\sigma_T \alpha}{e} \end{bmatrix} \begin{bmatrix} -\Delta\frac{\mu}{e} \\ -\Delta T \end{bmatrix},$$

$$\begin{bmatrix} I \\ I_E \end{bmatrix} = \frac{S}{L} \begin{bmatrix} \frac{\sigma_T}{e} & \alpha\sigma_T \\ \frac{\sigma_T \alpha}{e}\overline{T} + \overline{\mu}\frac{\sigma_T}{e^2} & \kappa_E + \overline{\mu}\frac{\sigma_T \alpha}{e} \end{bmatrix} \begin{bmatrix} -(\mu_{\text{out}} - \mu_{\text{in}}) \\ -(T_{\text{out}} - T_{\text{in}}) \end{bmatrix},$$

$$\begin{bmatrix} I \\ I_E \end{bmatrix} = \begin{bmatrix} Y_{11} & Y_{12} \\ Y_{21} & Y_{22} \end{bmatrix} \begin{bmatrix} -(\mu_{\text{out}} - \mu_{\text{in}}) \\ -(T_{\text{out}} - T_{\text{in}}) \end{bmatrix}.$$

We note the importance of the central node where the average values are considered. These average values are consistent with the definition of the local equilibrium of the cell.

3.2.4
Solution: General Millman Theorem

The classical Millman theorem, frequently used in circuit theory, can easily be extended to multiple and coupled potentials as we show now. Let us consider a node A where we want to determine μ_A and T_A.

The Millman theorem extended to coupled currents gives

$$\mu_A \sum_i^N Y_{11i} + T_A \sum_i^N Y_{12i} = \sum_i^N \mu_i Y_{11i} + \sum_i^N T_i Y_{12i},$$

$$\mu_A \sum_i^N Y_{21i} + T_A \sum_i^N Y_{22i} = \sum_i^N \mu_i Y_{21i} + \sum_i^N T_i Y_{22i}.$$

Writing

$$A = \sum_i^N \mu_i Y_{11i} + \sum_i^N T_i Y_{12i},$$

$$B = \sum_i^N \mu_i Y_{21i} + \sum_i^N T_i Y_{22i},$$

we get,

$$\mu_A \sum_i^N Y_{11i} + T_A \sum_i^N Y_{12i} = A,$$

$$\mu_A \sum_i^N Y_{21i} + T_A \sum_i^N Y_{22i} = B,$$

and finally,

$$\begin{bmatrix} A \\ B \end{bmatrix} = \begin{bmatrix} \sum_i^N Y_{11i} & \sum_i^N Y_{12i} \\ \sum_i^N Y_{21i} & \sum_i^N Y_{22i} \end{bmatrix} \begin{bmatrix} \mu_A \\ T_A \end{bmatrix}.$$

The chemical potential μ_A and the temperature T_A on a specific node A are then simply given by the matrix inversion, so,

$$\begin{pmatrix} \mu_A \\ T_A \end{pmatrix} = \begin{pmatrix} \sum_i^N Y_{11i} & \sum_i^N Y_{12i} \\ \sum_i^N Y_{21i} & \sum_i^N Y_{22i} \end{pmatrix}^{-1} \begin{pmatrix} A \\ B \end{pmatrix}.$$

One should bear in mind that the matrix elements Y_{21i} and Y_{22i} also contain the average values $\overline{T}$ and $\overline{\mu}$. As a consequence, the resolution of the complete mapping cannot be obtained by a global matrix inversion of the overall network. The solution is then obtained by iterations.

It should be noticed that the Millman theorem does not need any definition of the dimensionality of the network. With a correct definition of the indices, 1D, 2D, and 3D networks can be solved using this theorem.

3.2.5
Implementation

We consider a rectangular 2D array so that each node has four nearest neighbors. Each node is defined by its proper temperature and electrochemical potential. The location of a unit cell is then in between two adjacent nodes. As one can see, an additional set of cells or nodes for calculation is also required around the array for boundary conditions. These additional cells will be filled with the same values of potential as their neighbors leading to zero currents, in order to ensure satisfaction of conservation laws. The calculation may start at $(1, 1)$ and end at (n, m).

Then, for each node (i, j), we calculate the temperature and:

$$\begin{pmatrix} \mu_{A_{ij}} \\ T_{A_{ij}} \end{pmatrix} = \begin{pmatrix} \Sigma_{11} & \Sigma_{12} \\ \Sigma_{21} & \Sigma_{22} \end{pmatrix}^{-1} \begin{pmatrix} A_{ij} \\ B_{ij} \end{pmatrix}$$

with

$$A_{ij} = \mu_{i-1,j} Y_{11i-1,j} + \mu_{i+1,j} Y_{11i+1,j} + \mu_{i,j-1} Y_{11i,j-1} + \mu_{i,+1} Y_{11i,j+1}$$
$$+ T_{i-1,j} Y_{12i-1,j} + T_{i+1,j} Y_{12i+1,j} + T_{i,j-1} Y_{12i,j-1} + T_{i,j+1} Y_{12i,j+1},$$
$$B_{ij} = \mu_{i-1,j} Y_{21i-1,j} + \mu_{i+1,j} Y_{21i+1,j} + \mu_{i,j-1} Y_{21i,j-1} + \mu_{i,+1} Y_{21i,j+1}$$
$$+ T_{i-1,j} Y_{22i-1,j} + T_{i+1,j} Y_{22i+1,j} + T_{i,j-1} Y_{22i,j-1} + T_{i,j+1} Y_{22i,j+1}$$

and,

$$\Sigma_{11} = Y_{11i-1,j} + Y_{11i+1,j} + Y_{11i,j-1} + Y_{11i,j+1},$$
$$\Sigma_{12} = Y_{12i-1,j} + Y_{12i+1,j} + Y_{12i,j-1} + Y_{12i,j+1},$$
$$\Sigma_{21} = Y_{21i-1,j} + Y_{21i+1,j} + Y_{21i,j-1} + Y_{21i,j+1},$$
$$\Sigma_{22} = Y_{22i-1,j} + Y_{22i+1,j} + Y_{22i,j-1} + Y_{22i,j+1}.$$

The initial conditions are given by fixed values of the temperature and electrochemical potential of some nodes. Then for each node $A(i, j)$, except the fixed ones, $T_{A_{ij}}$ and $\mu_{A_{ij}}$ can be estimated.

3 Segmented Devices and Networking of TE Elements

The $T_{A_{ij}}$ and $\mu_{A_{ij}}$ calculation is then repeated until the convergence to stable values is obtained.

3.2.6
Numerical Illustration

In the following example, we consider the case of a system, which is divided into two parts as depicted in Figure 3.8. The system is composed of a thermoelectric network of 25×17 cells. The network consists of two uniform domains with two different Seebeck coefficients α_1 and α_2 with

$$\alpha_1 = \frac{\alpha_2}{2} = 10^{-9} \cdot (22224.0 + 930.6\,T - 0.9905\,T^2) \cdot 0.5. \tag{3.28}$$

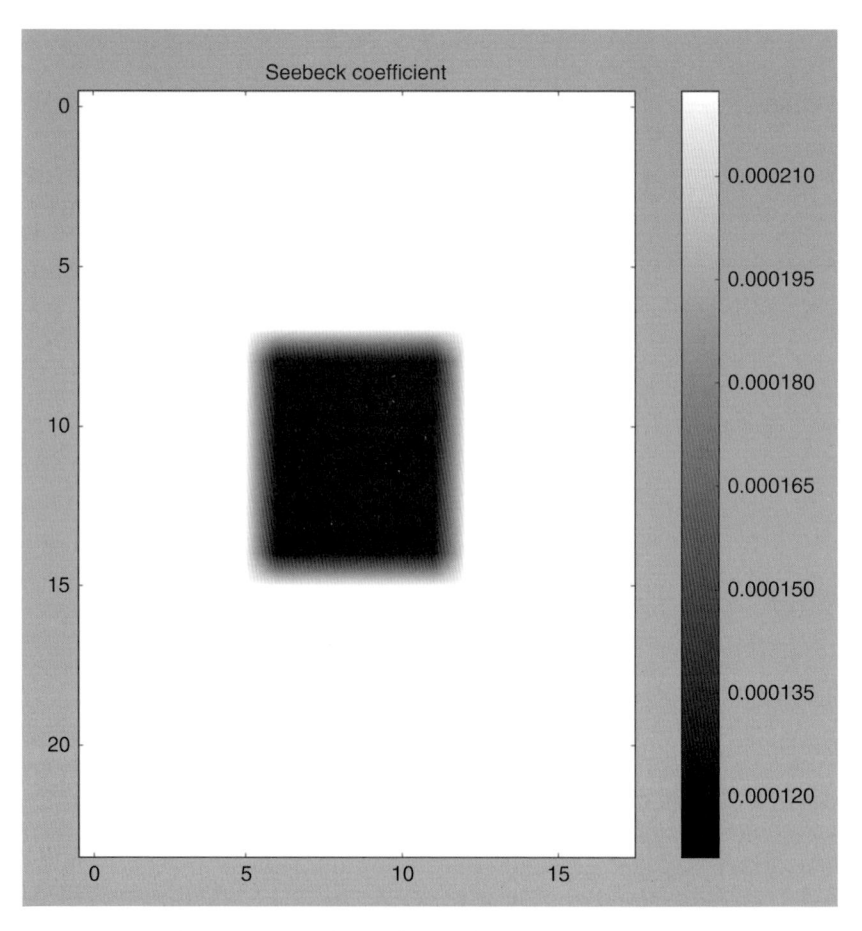

Figure 3.8 Distribution of the Seebeck values inside the network. The lower values are in the center zone.

The conductivities are similar for both cells,

$$\varrho(T) = 10^{-10} \cdot (5112.0 + 163.4\,T + 0.6279\,T^2) \tag{3.29}$$

and

$$\kappa(T) = 10^{-4} \cdot (62605.0 - 277.7\,T + 0.4131\,T^2). \tag{3.30}$$

The temperature dependence of the material properties in latter formulae is given in SI units. The temperature difference along the system is $\Delta T = T_h - T_c = 390 - 275$ K. As a resulting illustration we report in Figure 3.9 the electrical current map showing the strong coupling between both intensive variables at the boundaries of the system.

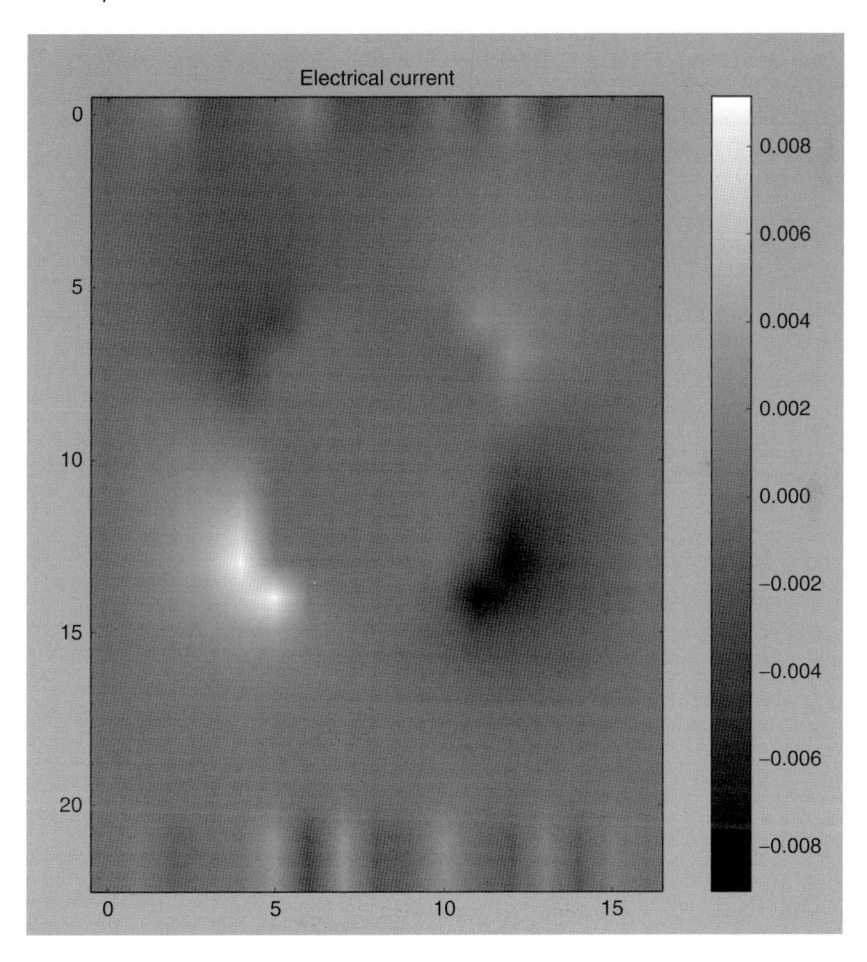

Figure 3.9 Electrical current distribution inside the network. Due to the open-circuit configuration, the current is zero except close to the boundaries between high and low Seebeck coefficients.

References

1. Landauer, R. (1978) Electrical transport and optical properties of inhomogeneous media, in *Proceedings of the 1st Conference on the Electrical Transport and Optical Properties of Inhomogeneous Media*, AIP Conference Proceedings, vol. **40** (eds J.C. Garland and D.B. Tanner), Ohio State University, New York.
2. Zabrocki, K., Müller, E., and Seifert, W. (2010) One-dimensional modeling of thermogenerator elements with linear material profiles. *J. Electron. Mater.*, **39**, 1724–1729.
3. Zabrocki, K., Müller, E., Seifert, W., and Trimper, S. (2011) Performance optimization of a thermoelectric generator element with linear material profiles in a 1d setup. *J. Mater. Res.*, **26** - Focus Issue - Advances in Thermoelectric Material (15), 1963–1974.
4. Müller, E., Karpinski, G., Wu, L.M., Walczak, S., and Seifert, W. (2006) Separated effect of 1D thermoelectric material gradients, in *25th International Conference on Thermoelectrics* (ed. P. Rogl), IEEE, Piscataway, NJ, pp. 204–209.
5. Müller, E., Walczak, S., and Seifert, W. (2006) Optimization strategies for segmented Peltier coolers. *Phys. Status Solidi A*, **203** (8), 2128–2141.
6. Müller, E., Karpinski, G., Wu, L.M., Walczak, S., and Seifert, W. (2006) Numerically based design of thermoelectric elements, in *4th European Conference on Thermoelectrics* (ed. D.M. Rowe), Cardiff University.
7. Domenicali, C.A. (1954) Irreversible thermodynamics of thermoelectricity. *Rev. Mod. Phys.*, **26** (2), 237–275.
8. Müller, E., Zabrocki, K., Goupil, C., Snyder, G.J., and Seifert, W. (2012) Functionally graded thermoelectric generator and cooler elements, in *CRC Handbook of Thermoelectrics: Thermoelectrics and its Energy Harvesting*, Chapter 4 (ed. D.M. Rowe), CRC Press, Boca Raton, FL.

4
Transient Response and Green's Function Technique

Wolfgang Seifert[†], Knud Zabrocki, Steven Achilles, and Steffen Trimper

In the previous chapters, the main objectives in the description of thermoelectric devices were centered around the steady-state operation. In this chapter, we want to highlight the aspects of transient operation and measurements of thermoelectric materials and systems. Both the electric field $\mathbf{E} = \mathbf{E}(\mathbf{r}, t)$ (vector field) and the temperature $T = T(\mathbf{r}, t)$ (scalar field) are, in general, time-dependent quantities. As shown in Chapters 1 and 2, these fields can be determined by the solution of thermal balance equations, which is, in general, a partial differential equation (PDE) derived from the general conservation laws as the conservation of thermal energy and charge conservation combined with constitutive relations such as generalized Fourier's and Ohm's laws in the linear Onsager formalism, which contain information of the material properties, see, for example, Eq. (1.67). In the stationary or steady state, these equations are ordinary differential equations (ODE), whereas in the transient mode, the describing equations are PDEs, for which the mathematical treatment in comparison to ODE is much more complicated.

The generalized heat equation already defined in Eq. (1.89) is

$$\varrho_{\mathrm{md}} c \frac{\partial T}{\partial t} = \nabla(\kappa_J \nabla T) + \frac{j^2}{\sigma_T} - \tau \mathbf{j} \cdot \nabla T - T \mathbf{j} \cdot (\nabla \alpha)_T. \tag{4.1}$$

In the steady-state problems, the material properties such as the Seebeck coefficient α, Thomson coefficient $\tau = T \partial \alpha / \partial T$, electrical conductivity σ_T (or analogously, the electrical resistivity $\varrho = \sigma_T^{-1}$), and the thermal conductivity κ_J especially their influence on the performance of the TE device have been discussed. In transient problems, the mass density ϱ_{md} and the specific heat given in J/(kg K)[1)]

---

1) In Reference [1] the following definition of the specific heat is given: *The specific heat c of a substance at temperature T is defined as $\delta Q/\delta T$ where δQ is the quantity of heat necessary to raise the temperature of unit mass of the substance through the small temperature range from T to $T + \delta T$. It depends on both the temperature and the assumed mode of heating, which is taken here to be at constant strain. ⋯ It should be noted that there is a considerably variety of usage of this matter. Some writers regard the above definition as that of heat capacity, or heat capacity per unit mass, of the substance,*

Continuum Theory and Modelling of Thermoelectric Elements, First Edition. Edited by Christophe Goupil.
© 2016 Wiley-VCH Verlag GmbH & Co. KGaA. Published 2016 by Wiley-VCH Verlag GmbH & Co. KGaA.

c are additional material properties needed to calculate the temperature distribution. The thermal energy balance given by Eq. (4.1) contains different terms, which contribute to the changes in the temperature. The first term on the right-hand side is the Fourier heat, which is often the main heat transport mechanism in solids and is found in the classical heat equation [1–4]. The second term, which is quadratic in the current density **j**, corresponds to the Joule heat. Due to the electrical resistance of the solid, it is heated when an electrical current passes through it. The other two terms are contributions from the coupled thermoelectric effect. Therefore, these terms explicitly contain α and τ, representing the coupling between thermal and electrical effects.

In addition to the thermal energy balance equation, the knowledge of adequate boundary condition (BC) and initial condition (IC) is needed. Initial conditions (ICs) means that, e.g. a temperature distribution $T_0 = T_0(\mathbf{r})$ is given for a certain time t, which in most cases is supposed to be the origin of the time coordinate $t = 0$ without loss of generality.

The boundary conditions capture the properties at certain surfaces of the body. There are different classes of the boundary conditions being prescribed at these surfaces in heat conduction, for example, prescribed temperatures and fluxes as well as heat transfer at the surfaces. This differentiation opens the possibility in classifying the BCs, that is, prescribed temperature is BC of first kind called *Dirichlet* BC, whereas prescribed fluxes are denoted as *von Neumann* BC (BC of second kind). Note that both IC and BC have to be understood as limiting conditions, see [1, Section 1.12, p. 27].

In principle, two kinds of problems that are of importance in the investigation of thermoelectric materials and systems can be distinguished within the given mathematical basis (governing equations, PDE, IC, BC, geometry, internal sources). On the one hand, the *direct* problem is used to determine directly the fields, that is, temperature and electric field, with the knowledge of all parameters such as material properties, IC, and BC, which leads to a so-called well-posed problem. On the other hand, the *inverse* problem is based on the fact that the fields are known at some points and the task is to determine some information of the whole system. Such a "backward calculation" is called *inverse heat conduction problem* (IHCP) [5–13]. These problems can be classified in different ways. One way is following the question: What is the quantity that has to be determined in the calculation? [5]:

- Material properties determination inverse problems
- Boundary value determination inverse problems
- Initial value determination inverse problems
- Source determination inverse problems
- Shape determination inverse problems.

and define the specific heat of a substance as the ratio of its heat capacity per unit mass to that of water. For solids, the effect of the method of heating on the specific heat is usually negligible and c may be replaced by c_p, the specific heat at constant pressure.

Especially, the first point is of major interest in thermoelectricity as it can help to determine the material properties in measurement facilities. In most measuring systems, a calculation routine is needed to evaluate the material properties, for example, dependence on the temperature, starting from the measurement of the temperature or heat flux at certain points in the system. The inverse problem is in the mathematical sense much more complicated in comparison to the direct problem. Often, it is an ill-posed problem, see, for example, [6, 11, 13–15]. Inverse problems are addressed later on in this chapter. By explaining some examples of measuring systems, the general idea of inverse problems is introduced.

The direct problem is mainly used in gaining information about the fields and further on the performance of thermoelectric devices, especially the performance that leads to optimum values. As for the steady state, distinguishing between the working modes should be a criterion of classification.

There are many works concerning transient thermoelectric cooling. Much less is known on the transient effects in thermoelectric generators (TEGs). The next sections are dedicated to review the work concerning the dynamic behavior of thermoelectric devices. Several influential factors lead to dynamic behavior of such a device:

- Varying boundary temperatures $T_h = T_h(t)$ and $T_c = T_c(t)$,
- Varying ambient temperature $T_{amb} = T_{amb}(t)$,
- Changing material properties, for example, the thermal conductivity $\kappa = \kappa[T(\mathbf{r},t), \mathbf{r}(t), t]$,
- Varying load $R_L = R_L(t)$ and/or current $I = I(t)$.

Analytical solutions can be found only in particular cases, for example, with help of integral transformation (e.g., Laplace transformation) or in a (Fourier) series expansion. These techniques will be introduced in connection with the given examples. In general, the use of approximative or numerical solution methods is necessary, which can be shown by the treatment of selected examples. Further details of numerical methods will be presented in Chapter 6.

4.1
Quasi-Stationary Processes

If the timescale of changes, for example, in the working or boundary conditions, is much greater than the response time of the thermoelectric system, a quasi-stationary approach is suitable to describe the transient behavior of the TE system. This should be demonstrated here with an example. The temperatures at the Moon's surface and below the surface are observed during a lunation, the temperature differences are calculated, and the potential of usage of these differences by a TEG is estimated on this basis. A lunation is the variable period of time for one revolution of the Moon around the Earth, which is in average around 29 days.[2]

---

2) 29.530589 days (29 days, 12 h, 44 min, 2.9 s).

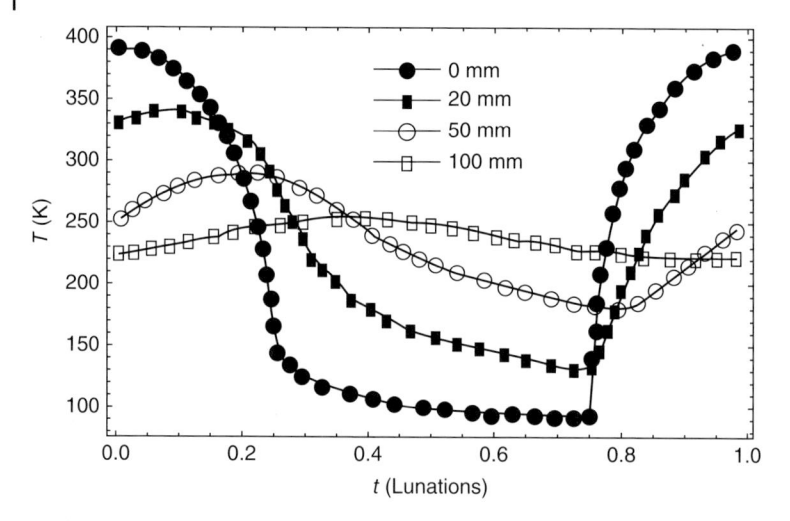

Figure 4.1 Temperatures at the Moon's surface and in different layers during a lunation. Symbols represent measurement points and solid lines represent linear interpolations between these points.

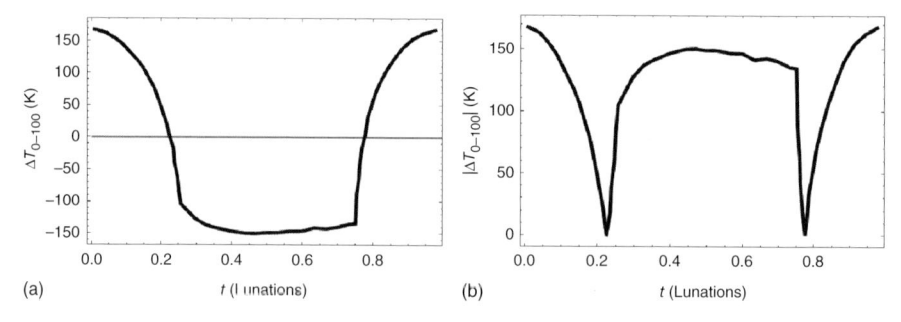

Figure 4.2 Temperature difference and its absolute value during a lunation between a layer in 100 mm depth and the Moon's surface. (a) ΔT_{0-100} (b) $|\Delta T_{0-100}|$

In Figure 4.1, the temperature data over time during a lunation at the Moon's surface and at different layers below the surface are shown. Measurements were taken every h.14–20

In Figure 4.2, the temperature difference and its absolute value during a lunation between a layer in 100 mm depth and the Moon's surface are illustrated. The potential of this temperature difference is valued for the application in a TEG built of a high-performance low-temperature TE material, using the material's figure of merit of $z = 3 \times 10^{-3} \mathrm{K}^{-1}$, which is accessible by classical low-temperature materials as given in, for example, [16].

In Figure 4.3a, the temperatures of the hot side $T_h = T_h(t)$ and the cold side $T_c = T_c(t)$ and the averaged temperature $T_m = T_m(t) = [T_h(t)+T_c(t)]/2$ are shown based on the linear interpolation of the measurement values. The maximum efficiency of a TEG can be determined as a function of time using Ioffe's classical results

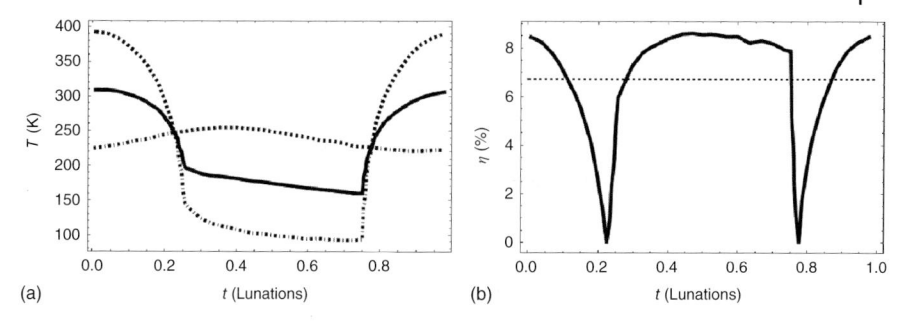

Figure 4.3 (a) Temperatures of the hot (dashed) and cold side (dashed–dotted) of a TEG as well as the averaged (middle) temperature (solid line). (b) The "time-resolved" maximum efficiency based on the temperatures and $z = 3 \times 10^{-3} \text{K}^{-1}$. The dashed line marks the (time) average of the maximum efficiency over one lunation.

by means of these temperatures, see Eq. (2.24) or Eq. (2.57). The time-resolved efficiency is shown in Figure 4.3b. The dashed line marks the time average over one lunation,

$$\bar{\eta} = \int_0^t \eta(t)\,dt,$$

which is in this example around 6.7%. Although the results in Figure 4.3 are illustrated as continuous-time functions, they are in principle only discontinuous-time series of distinct equilibrium states.

To value if it is possible to use a quasi-stationary description or other approximative methods for transient processes, the inherent timescales and response times have to be observed. In Table 4.1, timescales of transient electrical systems are summarized, see [17].

From a thermal point of view, a timescale is manifested by the thermal diffusivity of the material, which contains, besides the thermal conductivity κ, the mass density ϱ_{md} as well as the specific heat c, that is,

$$\lambda = \frac{\kappa}{c\varrho_{md}}. \tag{4.2}$$

Table 4.1 Time duration of transient phenomena in electrical systems.

Nature of the transient phenomena	Time duration
Lightning	0.1 μs–1.0 ms
Switching	10 μs to less than a second
Subsynchronous resonance	0.1 ms–5 s
Transient stability	1 ms–10 s
Dynamic stability, long-term dynamics	0.5–1000 s
Tie line regulation	10–1000 s
Daily load management, operator actions	Up to 24 h

Source: Table 1.1 in Ref. [17].

Table 4.2 Thermal diffusivity of bismuth telluride and the resulting response time.

References		[18]	[19]	[20]	[21]
		\multicolumn{4}{c}{Bismuth telluride Bi_2Te_3}
Density ϱ_{md}	(kg/m³)	—	—	8000	—
Specific heat c	(J/g K)	—	—	0.54	—
Thermal conductivity κ	(W/m K)	—	—	2.01	—
Thermal diffusivity λ	(10^{-6} m²/s)	0.7	0.7–1.0	0.46	1.4
Response time τ_{rs}	(s)	3.6	2.5–3.6	5.5	1.8

The response times refer to the length of TE element of $L = 5$ mm

For the classical thermoelectric material bismuth telluride Bi_2Te_3, the thermal diffusivities and the corresponding response times from different sources are listed in Table 4.2. The response time τ_{rs} depends not only on the material but also on the chosen geometry. For a TE element with constant cross-sectional area A_c and the length L, it is defined as

$$\tau_{rs} = \frac{L^2}{\pi^2 \lambda}. \tag{4.3}$$

The response time is in principle a measure of the thermal relaxation of a body homogeneously made of a distinct material. Note that the thermal relaxation of a TE device is mostly determined by its surroundings. The material in the neighboring components has in most cases a specific heat greater than that of the TE material.

4.2
Supercooling with a Transient Peltier Cooler

The majority of publications of transient effects in thermoelectric devices is dedicated to the phenomenon of *"Peltier supercooling,"* which describes a transient effect in a thermoelectric cooler (TEC). TECs can be used for an effective cooling of on-chip hotspots [22–25]. Going beyond steady-state operation, a dynamically switched TEC can be used to suppress hotspots in electronic chips, see, for example, [25]. Supercooling is reached because there is a temporal and spatial interplay between Peltier cooling (mainly at an interface) and Joule heating as a volume effect, which allows going below minimum cooling temperature observed under steady-state operation. In the following table, which is adopted from Table 1 in Ref. [25], the metrics found under transient response are summarized.

The parameters listed in Table 4.3 are illustrated in Figure 4.4.

To quantify the effect of the dynamic operation of thermoelectric elements, the stationary state is taken as reference quantity; therefore, the basic facts of a TEC in the steady state are repeated here.

Table 4.3 Metrics for the transient behavior of a TEC.

Metric	Notation	Description
Minimum steady-state temperature	$T_{min,ss}$	Steady-state temperature when the optimum current is applied
Minimum transient temperature	$T_{min,tr}$	Minimum temperature after current pulse has been applied
Minimum time	t_{min}	Time to reach the minimum transient temperature
"Overshoot" temperature	T_{os}	Maximum temperature after current pulse has been applied
Transient "supercooling"	ΔT_{sc}	Maximum temperature reduction below the steady-state value of minimum temperature, that is, $\Delta T_{sc} = T_{min,ss} - T_{min,tr}$
Holding time	t_h	Duration for which the transient temperature remains below the steady-state minimum temperature
"Overshoot" temperature difference	ΔT_{ov}	Maximum transient temperature difference above the steady-state minimum temperature, that is, $\Delta T_{ov} = T_{os} - T_{min,ss}$
Settling time	t_{slg}	The time period for the transient temperature to return to the steady-state value under a certain and given accuracy interval
Transient advantage	TA	The integrated area below $T_{min,ss}$ and above the transient temperature curve (during the holding time period)
Transient penalty	TP	The integrated area above $T_{min,ss}$ and below the transient temperature curve (during the settling time period)

Adapted from Table 1 in Ref. [25].

4.2.1
Steady-State Operation of a Thermoelectric Cooler

The Peltier effect is used in a TE device operated as TEC, see Chapters 1 and 2. The TEC is a solid-state active heat pump, which transfers heat from the cold side to the hot side of the element while electrical energy is consumed. Then the heat flux (direction) depends on the direction of the electrical current. For the TEC, the cold side is the absorbing side, whereas the hot side is the sink side. Performance parameters for the TEC are the thermal power at the cold absorbing side $\dot{Q}_c$, the maximum temperature difference ΔT_{max} or minimum cooling temperature $T_{c,min}$, and the coefficient of performance φ.[3]

Again two choices of boundary conditions can be made, see [28]. On the one hand, boundary conditions (BCs) of the first kind (Dirichlet BC), where the junction temperatures are given (T_h and T_c), and on the other hand, mixed BC

---
3) We use here Sherman's notation of the coefficient of performance instead of the term COP [26, 27].

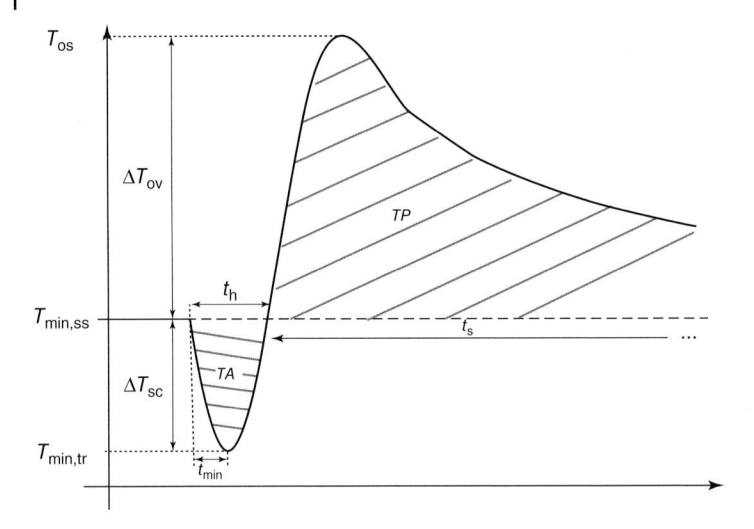

4 Transient Response and Green's Function Technique

Figure 4.4 Schematics of quantifying parameters of transient thermoelectric cooling – Peltier supercooling.

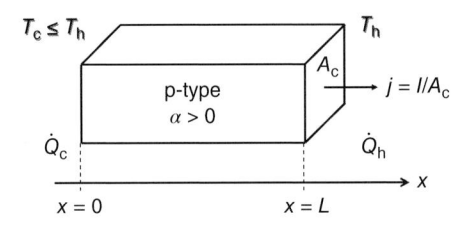

Figure 4.5 Thermoelectric cooler element.

(von Neumann BC), where the heat flux on one side and the temperature on the other side are given [$\dot{Q}_c$ (or $\dot{q}_c = \dot{Q}_c/A_c$) and T_c], can be assumed.

In Figure 4.5, a single TEC element is shown. For the constant properties model (CPM) case, the temperature profile is a parabolic function, which has already been determined in Eq. (2.14) for Dirichlet BC. In general, the solution is

$$T(x) = -\frac{j^2}{2\sigma\kappa}x^2 + C_1 x + C_2. \tag{4.4}$$

For Dirichlet BC, the free constants C_1 and C_2 are determined as follows:

$$C_1 = \frac{1}{L}\left(T_h + \frac{j^2 L^2}{2\sigma\kappa}\right) \quad \text{and} \quad C_2 = T_c, \tag{4.5}$$

whereas for mixed BC, the solution is given in Ref. [28] with

$$C_1 = \frac{j\alpha\left(T_h + \frac{j^2 L^2}{2\sigma\kappa}\right) - \dot{q}_c^{(0)}}{\kappa + j\alpha L} \quad \text{and} \quad C_2 = \frac{\kappa\left(T_h + \frac{j^2 L^2}{2\sigma\kappa}\right) + L\dot{q}_c^{(0)}}{\kappa + j\alpha L}, \tag{4.6}$$

Table 4.4 Material properties of bismuth telluride Bi_2Te_3, see [20].

Parameter	Notation	Unit	Value
Seebeck coefficient	α	μV/K	175
Electrical conductivity	σ	S/m	100,000
Thermal conductivity	κ	W/m K	2.01
Mass density	ϱ_{md}	kg/m³	8000
Specific heat	c	J/kg K	544

where $\dot{q}_c^{(0)}$ is the constant heat flux at the absorbing side. For mixed boundary conditions, the absorbed thermal power is just $\dot{Q}_c = \dot{q}_c^{(0)} A_c$, whereas for fixed junction temperatures, the absorbed thermal power is

$$\dot{Q}_c = -(K + I\alpha)\Delta T + I\alpha T_h - \frac{I^2 R_i}{2}, \quad (4.7)$$

where $\Delta T = T_h - T_c$ is the temperature difference, $K = \kappa A_c/L$ is the thermal conductance, and $R_i = L/(\sigma A_c)$ is the electrical resistance of the element. In terms of local variables, it is

$$\dot{q}_c(j) = \kappa \frac{\Delta T}{L} + j\alpha T_h - \frac{j^2 L}{2\sigma}. \quad (4.8)$$

To illustrate these formulas, an example is given by considering the material properties of a bismuth telluride sample from Ref. [20], which are summarized in Table 4.4, where the data is transformed to SI units.[4]

For this illustration, the geometric parameters of the TE element, see Figure 4.5, the length $L = 1$ mm and the cross-sectional area $A_c = 10$ mm², are chosen similar to those in Ref. [28]. For comparison, another aspect ratio is taken into account, where $L = 5$ mm and $A_c = 1$ mm². The aspect ratio Γ is defined as the quotient

$$\Gamma = \frac{L}{A_c}, \quad (4.9)$$

such that the thermal power for $\Gamma = 100$ m⁻¹ and $\Gamma = 5000$ m⁻¹ (in the steady state) is shown in Figure 4.6.

Two different cases can be identified for a Peltier cooler: on the one hand, the *maximum cooling* case, where $\Delta T \to \Delta T_{max}$ at zero heat absorption $\dot{q}_c \to 0$, and on the other hand, the *pure heat pumping* operation at external isothermal conditions, that is, $\Delta T = 0$. The optimum current for the maximum cooling case can be determined by setting the thermal power in Eq. (4.7) to zero, which results in

$$\dot{Q}_c = 0 \Rightarrow \Delta T(I) = \frac{IL(2A_c \sigma \alpha T_h - IL)}{2A_c \sigma (A_c \kappa + I\alpha L)}. \quad (4.10)$$

The thick contours in Figure 4.6c and d show the found result. To find the optimum current, the derivative $\partial(\Delta T)/\partial I$ has to be set to zero, obtaining the result with the

---
4) For example, the thermal conductivity $\kappa = 4.8 \times 10^{-3}$ cal/s/(K cm).

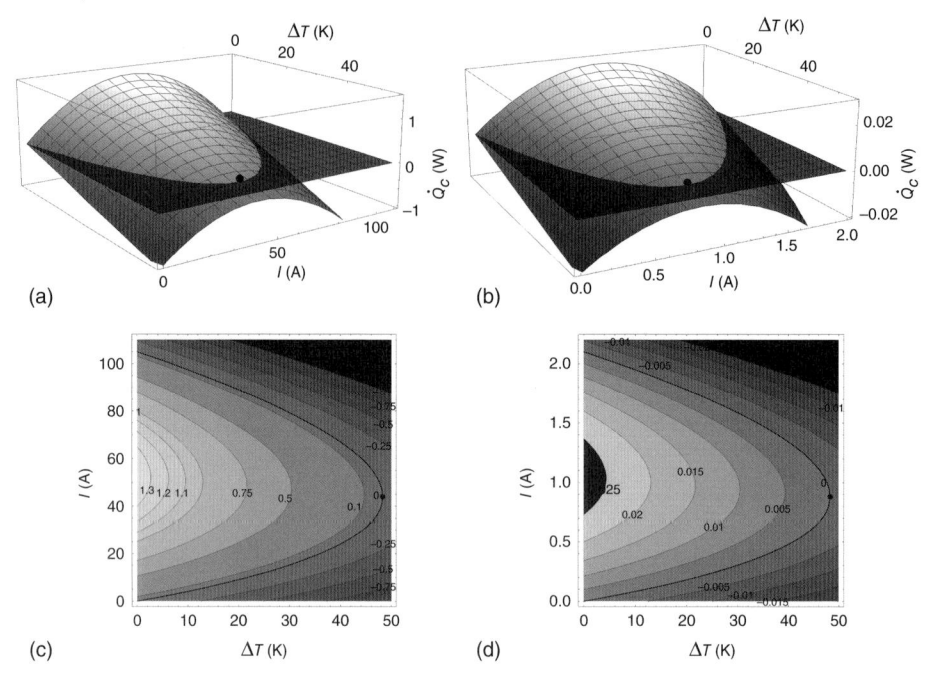

Figure 4.6 Absorbed heat for a single thermoelectric element in the Peltier mode with constant properties, see Table 4.4, $\dot{Q}_c$ in dependence on the temperature difference ΔT and current I for two different aspect ratios. (a) $\Gamma = 100$ m^{-1} (b) $\Gamma = 5000$ m^{-1} (c) $\Gamma = 100$ m^{-1} (d) $\Gamma = 5000$ m^{-1}.

positive root

$$I_{\text{opt}} = \frac{A_c}{L u}\left(-\kappa + \sqrt{\kappa^2 + 2\alpha^2 \sigma \kappa T_h}\right) = \frac{K}{u}(-1 + \sqrt{1 + 2Z T_h}). \quad (4.11)$$

To evaluate the optimum current that maximizes the absorbing heat flux (thermal power) $d\dot{q}_c(j)/dj = 0$ is calculated by means of Eq. (4.8) resulting in

$$j_{\text{opt},\dot{q}_c} = \frac{\sigma \alpha}{L} T_{c,\text{min}} \quad \text{or} \quad I_{\text{opt},\dot{Q}_c} = \frac{\alpha}{R_{\text{in}}} T_{c,\text{min}}. \quad (4.12)$$

On basis of this result, that is, the optimum current (density), the maximum absorbing heat flux/thermal power can be determined

$$\dot{q}_{c,\text{max}} = \frac{\kappa}{L}\left[\Delta T + \frac{1}{2} z T_c^2\right] \quad \text{or} \quad \dot{Q}_{c,\text{max}} = K\left[\Delta T + \frac{1}{2} Z T_c^2\right]. \quad (4.13)$$

To calculate the coefficient of performance φ, the ratio

$$\varphi = \frac{\dot{Q}_c}{P_{\text{el,con}}} \quad (4.14)$$

has to be observed. $\dot{Q}_c$ is known from Eq. (4.7). Furthermore, the thermal power at the sink side is needed and can be determined analogously as

$$\dot{Q}_h = K \Delta T + I \alpha T_h + \frac{I^2 R_{\text{in}}}{2} \quad (4.15)$$

and the consumed electrical power is then

$$P_{el,con} = \dot{Q}_h - \dot{Q}_c = \alpha(T_h - T_c)I + I^2 R_{in}. \quad (4.16)$$

Both Eqs. (4.7) and (4.16) set in Eq. (4.14) give the coefficient of performance depending on the electrical current I

$$\varphi \equiv \varphi(I) = \frac{K\Delta T + I\alpha T_c - \frac{I^2 R_{in}}{2}}{\alpha(T_h - T_c)I + I^2 R_{in}}. \quad (4.17)$$

The optimum current $I_{opt,\varphi}$ is determined with the condition $\partial\varphi(I)/\partial I = 0$ and is

$$I_{opt,\varphi} = -\frac{K}{\alpha}\frac{\Delta T}{T_m}(1 + \sqrt{1 + ZT_m}). \quad (4.18)$$

For this optimum current, a maximum coefficient of performance is reached

$$\varphi_{max} = \frac{T_c}{T_h - T_c}\frac{\sqrt{1 + ZT_m} - T_h/T_c}{\sqrt{1 + ZT_m} + 1}, \quad (4.19)$$

where the first part is a reversible Carnot factor and the second contains the irreversible contributions.

The case of maximum cooling is reached if $\dot{Q}_{c,max} = 0$ and hence $\varphi = 0$ such that

$$\Delta T_{max} = (T_h - T_{c,min}) = \frac{1}{2}ZT_{c,min}^2. \quad (4.20)$$

4.2.2
Important Parameters for the Supercooling Case

Some of the important parameters for supercooling are already given in Table 4.3 and illustrated in Figure 4.4. Here these parameters are explained in more detail.

Usually, the starting point to observe the phenomenon of supercooling in a TEC is the steady state, which is discussed in the previous subsection. From the steady state, a minimum cooling temperature $T_{min,ss}$ or a maximum cooling difference $\Delta T_{max,ss}$ and the corresponding current I_{opt} are known. If a current pulse ($P = I_{pulse}/I_{opt}$) is applied at a certain time for a defined duration, the Peltier junction is supercooled.

In References [19, 29], these experimental and theoretical parameters are investigated in comparison to different approximative solutions from the literature. The supercooling effect can be described as a function of different independent variables, for example, the length of the element L, the hot side temperature T_h, the pulse ratio P, which is also denoted as "normalized pulse magnitude" [30], and the figure of merit of the element Z. The obvious facts that have to be known are then: the amount of increased cooling, the current needed, the amount of time the cooling lasts, and the time between the cooling pulses. The determination of ΔT_{sc} is the focus of most studies on supercooling of a TEC. Obviously, the normalized pulse magnitude P is a parameter that affects the value of $T_{min,tr}$. This can be shown best by an analytical approximation as given in Ref. [31]. In this work, the TE elements are assumed to be infinitely long. This approximation is appropriate for large pulses, that is, $P > 2$. For this case, the times involved are shorter than the characteristic timescale calculated from the thermal diffusivity. For an infinite pulse, a supercooling $\Delta T_{P\infty} \approx \Delta T_{\min,ss}/2$ can be achieved, see, for example, [19, 29, 31].

In general, analytical calculations are only feasible for constant material parameters as already noted for the steady state. In this sense, in most cases, the natural extension of the CPM to the transient state is used to calculate the temperature profile and the performance parameters

$$c\varrho_d \frac{\partial T}{\partial t} = \kappa \frac{\partial^2 T}{\partial x^2} + \frac{I^2}{A_c^2 \sigma}. \tag{4.21}$$

Equation (4.21) is an inhomogeneous heat equation. The inhomogeneity is due to the Joule heat. As this equation is linear in T, classical solution methods such as the Fourier series method can be used to solve the problem analytically. The solution is found with regard to the following boundary conditions

$$T(x = L, t) = T_h \quad \text{and} \quad \left.\frac{\partial T}{\partial x}\right|_{x=0} = \frac{\alpha I T(x = 0, t)}{A_c \kappa} \tag{4.22}$$

and the initial condition, which is the steady-state solution for the maximum temperature difference

$$T_{ss}(x; I_{opt}) = T_c + \Delta T_{\max,ss} \left(\frac{2x}{L} - \frac{x^2}{L^2}\right). \tag{4.23}$$

A solution in the framework of Fourier series is presented in Appendix A in Ref. [18]. A linear approximation of Eq. (4.21) is discussed in Refs [19, 29, 32], where

$$\frac{\partial^2 T}{\partial x^2} + \frac{I^2}{A_c^2 \sigma \kappa} \approx \frac{T - T_{\min,ss}}{\lambda t} \tag{4.24}$$

is the approximation for the equation and the boundary condition at $x = 0$ also has to be approximated, that is,

$$\left.\frac{\partial T}{\partial x}\right|_{x=0} \approx \frac{\alpha I T_c}{A_c \kappa}.$$

The approximate problem can be solved with the following general solution

$$\frac{T(x, t) - T_{\min,ss}}{z T_c^2} = (P^2 - 1)\frac{\lambda t}{L^2} - (P - 1)\frac{\sqrt{\lambda t}}{L} e^{-x/\sqrt{\lambda t}}, \tag{4.25}$$

see Eq. (4) in Refs [19, 29]. This kind of solution results in an algebraic function for ΔT_{sc}, that is,

$$\Delta T_{sc} = \frac{\Delta T_{\max,ss}}{2}\left(\frac{P-1}{P+1}\right). \tag{4.26}$$

The approximation fails to describe the experimentally observed behavior above $P = 3$. Snyder et al. identified an empirically determined formula with only one free parameter to fit to the experimental data

$$\Delta T_{sc} = \Delta T_{P\infty}(1 - e^{1-P}), \tag{4.27}$$

see [19, 29]. Furthermore, it is observed that if P is increased, then the minimum junction temperature decreases, but this state can be sustained for shorter times only. The times t_{min} and t_h are already defined. The difference in both is denoted as *returning time*, that is, $t_{ret} = t_h - t_{min}$. Both are decreased for increasing P. A $(P+1)^{-2}$ functional fit to the experimental data for t_{min} has been applied and proved to fit the data very well, that is,

$$t_{min} = \frac{t_{ret}}{4} = \frac{L^2}{4\lambda(P+1)^2} = \frac{\tau}{(P+1)^2}, \quad \text{where} \quad \tau = \frac{L^2}{4\lambda}. \tag{4.28}$$

After the supercooling period, a rapid heating up of the junction above $T_{min,ss}$ takes place until a maximum of this "overshoot" is reached. ΔT_{ov} depends on the amount of excess Joule heat that is added to the system during the pulse. After reaching T_{os}, the temperature decreases exponentially. The corresponding time constant can be determined from the Fourier series treatment of the heat equation from which a value for $\tau = 4L^2/\pi^2\lambda$ for the fundamental exponential decay $e^{-t/\tau}$ can be found. Higher order terms can be neglected. They have much shorter time constants and smaller amplitudes. Again it can be seen that the characteristic time for this regime is proportional to the square of the TE element length as for the other characteristic times as well.

Further references offer more detailed information and results concerning the supercooling. A first[5] experimental observation was reported by White in 1951 [33]. One of the most cited works is that of Stil'bans and Fedorovich published in 1958 [34, 35]. Two books dedicated to this topic are from Gray [36] and Iordanishvili and Babin [37]. In the 1960s, there was a very active period of investigation of the transient response of Peltier TEC [20, 31, 38–47] followed by some sporadic work [48–58]. A renewed interest has been observed since the 1990s until today [18, 19, 25, 29, 30, 32, 59–78].

4.3
Transient Behavior of a Thermoelectric Generator

As mentioned earlier, there is little information available about the dynamic behavior of TEG, from both analytical and experimental works. This case is analytically more complicated because the electrical current is no longer an independent variable but depends on the junction temperatures. For this, the complete nonlinear problem has to be solved.

In Reference [36], a theoretical analysis of the small-signal dynamic behavior is developed and investigated. In the beginning of the 1960s, Stremler investigated the problem in Refs [79, 80]. The important problem of the device settling time is a main focus of the works besides the description of the transient output. Again the nonlinear character of the describing equations forces to obtain a perturbative

---
5) To the authors' knowledge.

solution more or less by linearization of the PDE. The found analytical results are compared with experimental values.

Shaw presented in his dissertation and subsequent works a derivation with temperature-dependent material properties [81, 82]. It was found in his works that familiar techniques to treat nonlinear problems such as perturbative method and successive approximation cannot be applied straightforwardly. The Seebeck coefficient and the electrical resistivity are assumed as polynomial functions of third order. For the thermal conductivity, a inverse relation is supposed, that is, $\kappa \propto T^{-1}$. The other coefficients such as the specific heat are assumed to be constant. A special Jacobi's expansion is used to perform the analytical calculation.

A first numerical approach to solve the nonlinear PDE was introduced by Baranov *et al.* [83]. In Reference [84], a Fourier series expansion for the PDE of a transient TEG with constant material properties was performed by Cobble. A finite element method was used by El-Genk and Seo to determine the performance parameters of a TEG under transient operation conditions [85].

The application of TEG in vehicles also brought attention to investigate the transient behavior of the TE device, see, for example, [86, 87]. Further, only recent works should be listed here without claim of completeness [61, 73, 88–110].

4.4
Dynamic Measurements of the Thermal Conductivity: Laser Flash Analysis

In the following two sections, two different methods for the dynamical determination of the thermal conductivity are discussed. Although the mathematical treatment of these methods is based on a pure thermal description, the methods are relevant for thermoelectric problems.

The laser flash analysis (LFA) is a measurement method for the determination of the thermal diffusivity λ of a single material but can also be applied to measure the heat transfer through a layered sample. A plane-parallel sample with thickness L is heated at one side by an irradiation pulse (from a laser or flash lamp) and the temperature of the backside $T(x = L, t)$ is observed. The time-dependent temperature rise can be recorded by means of an IR detector. The magnitude of λ determines the rate of the temperature change, that is, the smaller the diffusivity is, the longer it takes until the irradiated heat reaches the backside of the sample. The strongly temperature-dependent thermal diffusivity is measured at different temperatures by placing the sample in a furnace at an adjustable temperature T_{amb}, which is kept constant during one irradiation pulse and its response, see Figure 4.7. The LFA method is reviewed in Refs [111–113]. The LFA principle has rarely been used for the determination of the thermal diffusivity of thermoelectric materials since the late 1960s [114–116], but became quite widespread from the 1990s when it got readily available from commercial suppliers. It became very popular because of its wide temperature range, the simple sample preparation, and its elegant principle

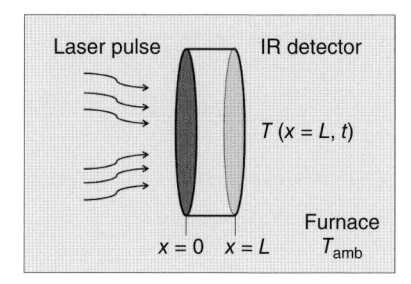

Figure 4.7 The principle of LFA measurement.

eliminating common systematic errors such as heat loss and transfer resistance at the interfaces.

The measurable values of the thermal diffusivity range from 1×10^{-7} to 1×10^{-3} m^2/s in a temperature interval from about 100 to 3300 K [111, p. 281]. The measurement method needs small samples, 6–16 mm in diameter, and a thickness L, which ranges from a fraction to a few millimeters [111, p. 281]. From the thermal diffusivity λ, the thermal conductivity κ can be calculated if the specific heat capacity c and the mass density ϱ_{md} are known. The amount of energy input is only important in so far that it will generate a sufficiently strong rear face signal. But if the irradiated heat flux is known by calibrating the laser and detector, it is feasible to determine the specific heat of an unknown sample relative to a standard reference. The sample's mass density is calculated from two measurements, the weight measurement and the computation of the volume by measuring the geometric dimensions. The density is determined at room temperature, and thermal expansion data is used to calculate the corresponding temperature dependence. Another possibility to estimate the density is achieved by Archimedes' principle.

The mathematical calculation routine of the laser flash method was developed by Parker *et al.* [117]. The basis of the calculation is the solution of the heat equation under the following assumptions, see e.g. Refs. [111, 113]:

- 1D heat flow,
- no heat losses from the sample's surfaces,
- a (chemical) *homogeneous* and *isotropic* sample material (no position dependence),
- a homogeneous/uniform energy absorption over the front face of the sample,
- absorption of the pulse energy in a very thin layer,
- a time-dependent short pulse (infinitesimal short pulse duration) as boundary condition represented by a Dirac delta function,
- property invariance with temperature under experimental conditions.

This solution is found by means of a standard method, that is, Fourier series expansion, which is explained in great detail in Ref. [1]. The temperature at the rear surface of the sample under the above-listed assumptions can be calculated as

$$V(L,t) = \frac{T(L,t)}{T_{max}} = 1 + 2 \sum_{n=1}^{\infty} (-1)^n \exp(-n^2 \omega), \tag{4.29}$$

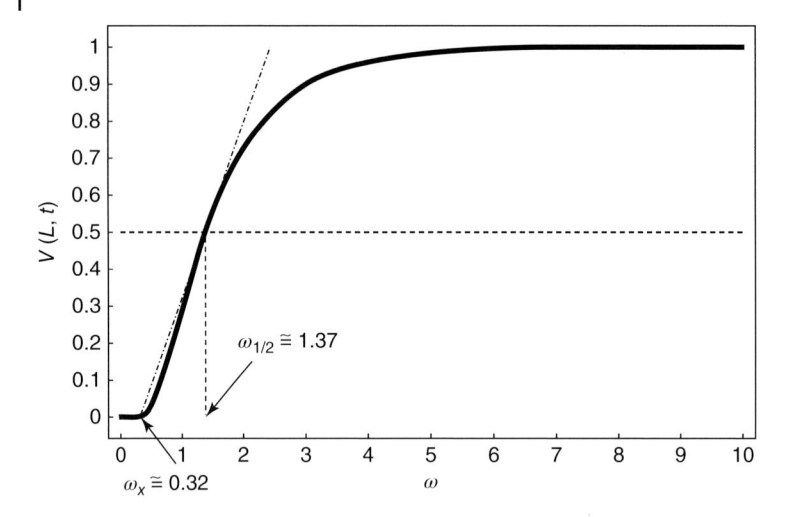

Figure 4.8 Rear surface temperature history in dimensionless variables.

where V is a dimensionless temperature scaled with the maximum temperature T_{max} at the rear surface and $\omega = \pi^2 \lambda t/L^2$ denotes a dimensionless time variable. Equation (4.29) can be expressed as Jacobi's elliptic theta function and evaluated further. It is illustrated in Figure 4.8.

In most cases, the time to half maximum, $t_{1/2}$, where $V = 0.5$ is measured, from which the thermal diffusivity can be determined by

$$\lambda = 1.37 \frac{L^2}{\pi^2 t_{1/2}}. \tag{4.30}$$

Another option is to measure the intercept of the tangent with the horizontal axis as shown in Figure 4.8 to obtain the thermal diffusivity given by

$$\lambda = 0.32 \frac{L^2}{\pi^2 t_x}. \tag{4.31}$$

This ideal solution can be inadequate for particular measurements, because the assumptions made for the calculation can be violated to some extent during an experiment. There are several works aiming at correcting the theory to adjust the theory and experiment. In most cases, only one deviation from the ideal situation is considered in the correction. Corrections have been elaborated for

Finite pulse width: This effect occurs strongly when thin samples of high thermal diffusivity are investigated [113, 118–125].
Radiative heat loss: This effect becomes dominant at high temperatures especially for thick samples [113, 126–134].
Nonuniform heating: This effect can be present at any thermal diffusivity experiment [132, 135].

As an example, the influence of radiation heat losses is illustrated in Figure 4.9, which is adapted from Ref. [132]. Heat losses from the sample are taken into account by the Newtonian cooling law and the dimensionless parameter called

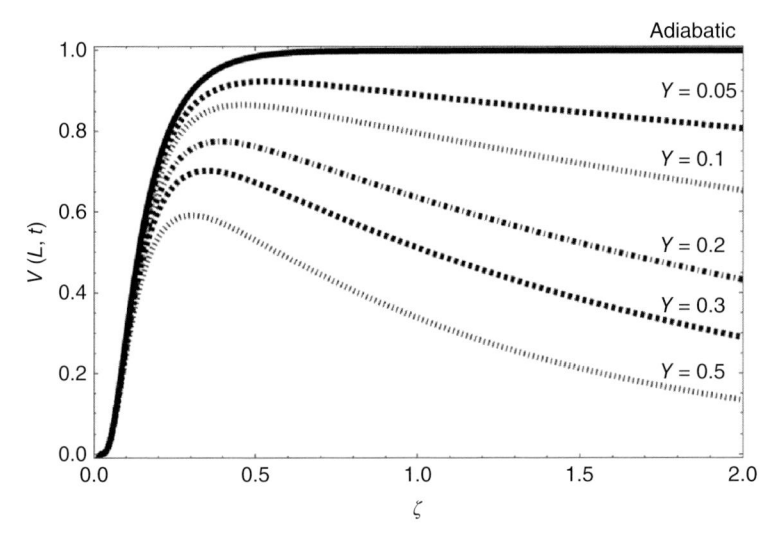

Figure 4.9 Rear surface temperature history in dimensionless variables for various values of the Biot number Y, where $\zeta = t/\tau_0$ with the timescale $\tau_0 = L^2/\lambda$.

Biot number Y. The results of the variation of Y are shown in Figure 4.9 in comparison to the adiabatic case [$Y = 0$, see Eq. (4.29)]. The basis of the solution, which is illustrated, is given by an approximative solution by Josell *et al.* [131], see Eq. (4) in Ref. [132]. The heat losses are supposed to originate from the radiation over the front and rear surfaces, whereas the losses over the lateral, cylindrical side face are neglected, that is,

$$Y = Y_x = 4\epsilon\sigma_{SB} T_{amb}^3 \frac{L}{\kappa} \quad \text{and} \quad Y_r = 0, \tag{4.32}$$

where ϵ is the total hemispherical emissivity of the sample, σ_{SB} is the Stefan–Boltzmann constant, and T_{amb} is the ambient temperature, the steady-state temperature before the pulse heating.

4.5
Dynamic Measurements of the Thermal Conductivity: Classical Ioffe Method

The combined thermoelectric measurement (CTEM) facility[6] is an important method to determine thermoelectric material properties. It is possible to measure the Seebeck coefficient α, the thermal conductivity κ, and the electrical conductivity κ as well as the thermoelectric figure of merit zT in a combined and simultaneous way with this method.

---

6) The CTEM is an in-house measurement facility at the German Aerospace Center (DLR), Institute of Materials Research (Thermoelectric materials and systems), Cologne, Germany, that allows a combined and simultaneous measurement of all important steady-state TE material properties (α, σ, κ) including the Harman zT.

Here the focus lies on the determination of the thermal conductivity and its theoretical background based on the measurement method introduced by Ioffe and Ioffe in 1952, see [136]. This measurement method is a dynamical method to determine the thermal conductivity. In Reference [137], the following statement had been given:

> The flow of heat from one copper block to a second small block through the specimen was measured in intervals of 15–30 sec. Simultaneously the temperature difference between the two blocks was measured.

A few years later, the method had been improved and described, and an error estimation had been given in Ref. [138]. The mathematical description of the problem in this article is rough and is more based on empirical findings. A stricter mathematical description for the special conditions in the Ioffe apparatus had been given by Kaganov [139] in the form of Fourier series. Swann investigated the theoretical basis of the measurement principle in a general form in Refs. [140, 141].

In principle, this problem is nothing but a solution of the (inhomogeneous) Fourier heat equation. In most cases, a one-dimensional transient process is assumed and serves as a basis for the theoretical calculation. This process is described by the heat equation

$$c_p \varrho_{md} \frac{\partial T}{\partial t} = \kappa \frac{\partial^2 T}{\partial x^2} \Rightarrow \frac{\partial T}{\partial t} = \lambda \frac{\partial^2 T}{\partial x^2} \quad \text{with} \quad \lambda = \frac{\kappa}{c_p \varrho_{md}}, \tag{4.33}$$

where it is assumed that the heat capacity c_p, the mass density ϱ_{md}, the thermal conductivity κ, and the combined quantity thermal diffusivity λ are supposed to be independent of temperature, that is, constant.

Stecker and Teubner provided a rigorous mathematical derivation, which is provided here in parts as the original work is in German [142]. Here the focus lies on the idealized, classical Ioffe arrangement for which the principle is explained.

4.5.1
Theoretical Basis of Simple Ioffe Method

The arrangement in the classical Ioffe method is shown in Figure 4.10. A homogeneous, cylindrical sample (body 1) with the length L_1 and the thermal diffusivity $\lambda^{(1)}$ is connected to a (metallic) block (body 2) with the same cross-sectional area

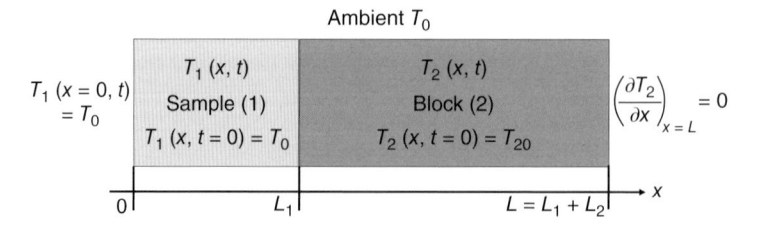

Figure 4.10 The arrangement in the classical Ioffe method.

A_c, length L_2, and thermal diffusivity $\lambda^{(2)}$. It is supposed that the thermal resistance of the sample is large in comparison to the thermal resistance of the metallic block. At $x = 0$ (left end of the sample), the temperature is assumed to be constant as T_0. The same constant temperature is taken to fix the ambient temperature of both the block and the sample. At the beginning of the measurement ($t = 0$), the temperature of the whole block is supposed to be $T_2 = T_{20}$, whereas at this time, the temperature of the whole sample has the same value as it is found at the left boundary and the ambient T_0. The latter assumption is far from experimental initial conditions but will be waived during the first moments of the temporal evolution and is thus of minor importance for the accuracy of the method. Due to the temperature difference, relaxation to an equilibrium temperature will be observed for $t > 0$. The temperature distribution depending on the position and time is sought for times $t > 0$. To realize this temperature relaxation, an electrical analogue can be considered in a first approximation. A capacitor of capacity C (with one side grounded) is discharged via an electrical resistor R beginning at $t = 0$. Then, an analogy can be found between the temperature and the electrical potential. The discharge shows an exponential decay with a time constant $\tau_{\text{rel}} = RC$. As it is possible to determine R by measuring the time constant with the capacity known, the same works as well in the thermal analogue, that is, by measuring the time constant, the thermal resistance of the sample can be calculated if the heat capacity of the block is known. Of course, this analogy is not ideal as in the thermal case, it is practically not possible to separate resistance and capacitance completely. Thus, the analogy is based on the presumption that the heat capacity of the block is much larger than that of the sample and the thermal conductivity of the sample is much smaller than that of the block. Accordingly, an intuitive evaluation of the thermal measurement in analogy to electrical discharge will suffer from and has to be corrected by the small but nonnegligible effects of the nonvanishing heat capacity of the sample and nonvanishing thermal resistivity of the block. The classical Ioffe method has also further disadvantages, which lead to deviation between the experimental situation and the ideal theoretical conditions. These deviations cannot be prevented, for example, $T(0, t) = \text{const.}$ is only possible if the cooling plate, which is used for cooling in the classical method, has an infinitely large heat capacity. Furthermore, the number of measurements in the same temperature range for the same time can be increased by a generalized method where the sample is sandwiched between two blocks and, at the same time, the precision of the measurement is increased. More effects are identified to have an influence on the measurement of τ_{rel}, which can alter the results going beyond the ideal (Fourier) heat conduction. On the one hand, there is a (thermal) contact resistance of the soldered contacts between the sample and the blocks, and on the other hand, thermal radiation to the ambient occurs. Both effects can only be investigated numerically in general because of their nonlinear character, see, for example, [143].

Here the Laplace transformation as analytical tool for the solution of the Ioffe model is presented. Beforehand, the idealized model without consideration of the so-called spurious effects is investigated, although under some approximation, these effects can be included in an analytical calculation. For this purpose, the temperature in the sample is denoted as $T_1 = T_1(x, t)$, whereas the temperature in the block is $T_2 = T_2(x, t)$. The two temperatures are determined by the solution of the following two heat equations:

$$\frac{\partial T_1}{\partial t} = \lambda^{(1)} \frac{\partial^2 T_1}{\partial x^2} \quad \text{and} \quad \frac{\partial T_2}{\partial t} = \lambda^{(2)} \frac{\partial^2 T_2}{\partial x^2} \quad \text{with} \quad \lambda^{(1/2)} = \frac{\kappa^{(1/2)}}{c_p^{(1/2)} \varrho_d^{(1/2)}}. \quad (4.34)$$

The heat equations are solved with regard to suitable initial and boundary conditions. The initial conditions are as already explained

$$T_1(x, t = 0) = T_0 \quad \text{and} \quad T_2(x, t = 0) = T_{20}, \quad (4.35)$$

where T_0 denotes the ambient temperature. For the boundary conditions, the following choice has been made

$$T_1(x = 0, t) = T_0 \quad \text{and} \quad (4.36a)$$

$$\left(\frac{\partial T_2}{\partial x}\right)_{x=L} = 0. \quad (4.36b)$$

To complete the task, the conditions at the interface between the sample and the block at $x = L_1$ are needed, which are

$$T_1(x = L_1-, t) = T_2(x = L_1+, t) \quad \text{and} \quad (4.37a)$$

$$-\kappa_1 \left(\frac{\partial T_1}{\partial x}\right)_{x=L_1-} = -\kappa_2 \left(\frac{\partial T_2}{\partial x}\right)_{x=L_1+}. \quad (4.37b)$$

To simplify the equations, a crossover to dimensionless variables is performed

$$\tilde{x} = \frac{x}{L_1}, \tilde{T} = \frac{T - T_0}{T_{20} - T_0}, \quad \text{and} \quad \tilde{t} = \frac{\lambda^{(1)}}{L_1^2} t. \quad (4.38)$$

For the sake of simplicity, the notations are interchanged after the variable transformation, that is, $\tilde{x} \leftrightarrow x$, $\tilde{t} \leftrightarrow t$, and $\tilde{T} \leftrightarrow T$. In the new variables, the equations have the following form:

$$\frac{\partial T_1}{\partial t} = \frac{\partial^2 T_1}{\partial x^2} \quad \text{and} \quad \frac{\partial T_2}{\partial t} = \lambda^{(2)}/\lambda^{(1)} \frac{\partial^2 T_2}{\partial x^2} = \beta \frac{\partial^2 T_2}{\partial x^2} \quad (4.39)$$

with the definition of the ratio of thermal diffusivities $\beta = \lambda^{(2)}/\lambda^{(1)}$. The initial conditions are changed of course

$$T_1(x, t = 0) = 0 \quad \text{and} \quad T_2(x, t = 0) = 1, \quad (4.40)$$

as well as the boundary conditions

$$T_1(x = 0, t) = 0, \quad \text{and} \quad \left(\frac{\partial T_2}{\partial x}\right)_{x = L/L_1} = 0. \quad (4.41)$$

Analogously, the interface conditions are changed

$$T_1(x = 1-, t) = T_2(x = 1+, t) \quad \text{and} \quad -\kappa_1 \left(\frac{\partial T_1}{\partial x} \right)_{x=1-} = -\kappa_2 \left(\frac{\partial T_2}{\partial x} \right)_{x=1+}, \tag{4.42}$$

where $x = 1-$ and $x = 1+$ mean the left-sided and right-sided limit to $x = 1$, respectively.

A particular operational method, the Laplace transformation technique, is an established method for the calculation of heat equations especially in a composite structure as given here. This method had been introduced and developed by the mathematician Gustav Doetsch (in collaboration with Felix Bernstein) in a series of articles [144–149]. A more general overview especially on the general application of the Laplace transformation on PDE is given in textbooks, for example, [150, 151].

4.5.2
Laplace Transformation and Important Properties

The Laplace transformation (LT) of a function is an integral transformation, which is usually performed with regard to the time variable $t \in [0, \infty)$ and is defined as

$$\mathcal{L}\{f(t)\} = \int_0^\infty e^{-st} f(t) \, dt = \overline{f}(s). \tag{4.43}$$

For PDEs, the functions are functions of more than one variable. Here the target function is the temperature $T = T(x, t)$ depending on the spatial variable x and the time t. For the transformation of $T(x, t)$, the spatial variable is supposed to be fixed, that is, for every fixed x, there is a separate transform, that is,

$$\mathcal{L}\{T(x, t)\} = \int_0^\infty e^{-st} T(x, t) \, dt = \overline{T}(x, s). \tag{4.44}$$

An important property for the application of the LT with regard to PDE is

$$\mathcal{L}\left\{ \frac{\partial T(x, t)}{\partial t} \right\} = s\mathcal{L}\{T(x, t)\} - T(x, t = 0+) \tag{4.45}$$

such that the initial condition (IC) is automatically implemented. For the derivative with regard to the spatial coordinate, the exchangeability of derivative and $\mathcal{L}$-integral is assumed, that is,

$$\mathcal{L}\left\{ \frac{\partial T(x, t)}{\partial x} \right\} = \frac{\partial}{\partial x} \mathcal{L}\{T(x, t)\} = \frac{\partial \overline{T}(x, s)}{\partial x}. \tag{4.46}$$

For the boundary conditions (BCs), a similar exchangeability is supposed. Usually, the BC are given as limiting values, for example,

$$\lim_{x \to a+} T(x, t) = T_0(t) \quad \text{or in short form} \quad T(a + 0, t) = T_0(t). \tag{4.47}$$

```
┌────────────────────────┐
Original space: ┤ Partial differential equation           Solution
                │ + Initial conditions                      ↑
                │ + Boundary conditions                     │
                └────────────────────────┘                  │
                            │                               │
                            ↓                               │
                   ℒ transformation              ℒ⁻¹ transformation
                            │                               │
                            ↓                               │
                ┌────────────────────────┐
Image space:    ┤ Ordinary differential equation  ──→  Solution
                │ + Boundary conditions
                └────────────────────────┘
```

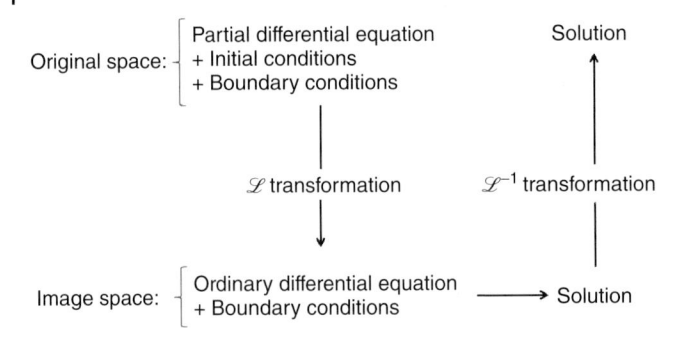

Figure 4.11 Scheme of the application of Laplace transformation for partial differential equations.

The exchangeability means then

$$\mathcal{L}\left\{\lim_{x\to a+} T(x,t)\right\} = \mathcal{L}\{T_0(t)\} \Leftrightarrow \lim_{x\to a+}\mathcal{L}\{T(x,t)\}$$
$$= \mathcal{L}\{T_0(t)\} = \lim_{x\to a+} \overline{T}(x,s) = \overline{T}_0(s). \tag{4.48}$$

By means of the LT, the PDE is transformed into an ODE in the image space. The scheme in Figure 4.11 describes the method briefly. In most cases, the main problem consists in determining the solution function in the t space from the solution in the image space by inverse LT.

4.5.3
Solution of the Classical Ioffe Method

The chosen example is a boundary value problem of a composite system. Such problems have been investigated in the literature with several boundary values/conditions [152–160].

4.5.3.1 LT of Original Equation
The first step is to transform the original equations (4.39) and use the initial conditions (4.40)

$$s\overline{T}_1(x,s) - \underbrace{T_1(x,t=0)}_{=0} = \frac{\partial^2 \overline{T}_1(x,s)}{\partial x^2}, \tag{4.49a}$$

$$s\overline{T}_2(x,s) - \underbrace{T_2(x,t=0)}_{=1} = \beta\,\frac{\partial^2 \overline{T}_2(x,s)}{\partial x^2}. \tag{4.49b}$$

By operating the Laplace transformation on the time derivative, see Eq. (4.45), this derivative vanishes and a system of ODE results, which has to be solved in a first step in the s space and afterward transformed back.

4.5.4
Solution of the Temperatures in the s-Domain

The equation for the temperature in body 1 of the system (sample) in the s-domain is

$$\frac{\partial^2 \overline{T}_1(x,s)}{\partial x^2} - s\overline{T}_1(x,s) = 0. \qquad (4.50)$$

On the other hand, the equation for body 2 (block) is

$$\beta \frac{\partial^2 \overline{T}_2(x,s)}{\partial x^2} - s\overline{T}_2(x,s) + 1 = 0. \qquad (4.51)$$

Both ODEs can be solved using standard techniques, that is, the exponential ansatz leading to

$$\overline{T}_1 = C_1 \exp(\sqrt{s}x) + C_2 \exp(-\sqrt{s}x). \qquad (4.52)$$

Here another fundamental system is used such that the solutions are expressed with hyperbolic functions $[\sinh(x) := (e^x - e^{-x})/2 \text{ and } \cosh(x) := (e^x + e^{-x})/2]$

$$\overline{T}_1 = C_1 \sinh(\sqrt{s}x) + C_2 \cosh(\sqrt{s}x). \qquad (4.53)$$

For the solution of the second part, see Eq. (4.51), the general solution of the homogeneous equation and a particular solution have to be found

$$\overline{T}_2 = C_3 \sinh\left(\sqrt{\frac{s}{\beta}}x\right) + C_4 \cosh\left(\sqrt{\frac{s}{\beta}}x\right) + \frac{1}{s}. \qquad (4.54)$$

The four unknown integration constants C_1, C_2, C_3, and C_4 can be determined by means of the boundary and interface conditions, which have to be transformed as well

$$\overline{T}_1(0,s) = 0 \quad \text{and} \quad \left(\frac{\partial \overline{T}_2(x,s)}{\partial x}\right)_{x=L/L_1} = 0, \qquad (4.55\text{a})$$

$$\overline{T}_1(1,s) = \overline{T}_2(1,s) \quad \text{and} \quad -\kappa_1 \left(\frac{\partial \overline{T}_1(x,s)}{\partial x}\right)_{x=1} = -\kappa_2 \left(\frac{\partial \overline{T}_2(x,s)}{\partial x}\right)_{x=1}. \qquad (4.55\text{b})$$

The left boundary $[x = 0]$ gives

$$\overline{T}_1(0,s) = 0 = C_1 \underbrace{\sinh(0)}_{=0} + C_2 \underbrace{\cosh(0)}_{=1} \Rightarrow C_2 = 0. \qquad (4.56)$$

The other three conditions deliver three equations for three unknown coefficients C_1, C_3, and C_4. First, we determine the derivative of Eq. (4.54)

$$\frac{\partial \overline{T}_2(x,s)}{\partial x} = C_3 \sqrt{\frac{s}{\beta}} \cosh\left(\sqrt{\frac{s}{\beta}}x\right) + C_4 \sqrt{\frac{s}{\beta}} \sinh\left(\sqrt{\frac{s}{\beta}}x\right) \qquad (4.57\text{a})$$

such that the BC at the right boundary $x = {}^L\!/_{L_1}$ is

$$\left(\frac{\partial \overline{T}_2(x,s)}{\partial x}\right)_{x={}^L\!/_{L_1}} = 0 = C_3 \sqrt{\frac{s}{\beta}} \cosh\left(\sqrt{\frac{s}{\beta}\frac{L}{L_1}}\right) + C_4 \sqrt{\frac{s}{\beta}} \sinh\left(\sqrt{\frac{s}{\beta}\frac{L}{L_1}}\right) \quad (4.57\text{b})$$

if $s \neq 0$, then it simplifies to

$$0 = C_3 \cosh\left(\sqrt{\frac{s}{\beta}\frac{L}{L_1}}\right) + C_4 \sinh\left(\sqrt{\frac{s}{\beta}\frac{L}{L_1}}\right). \quad (4.57\text{c})$$

The interface conditions lead to

$$C_1 \sinh(\sqrt{s}) = C_3 \sinh\left(\sqrt{\frac{s}{\beta}}\right) + C_4 \cosh\left(\sqrt{\frac{s}{\beta}}\right) + \frac{1}{s} \quad (4.57\text{d})$$

and

$$-\kappa_1 \sqrt{s} \cosh(\sqrt{s}) C_1 = -\kappa_2 \left[C_3 \sqrt{\frac{s}{\beta}} \cosh\left(\sqrt{\frac{s}{\beta}}\right) + C_4 \sqrt{\frac{s}{\beta}} \sinh\left(\sqrt{\frac{s}{\beta}}\right) \right]. \quad (4.57\text{e})$$

The equation system given by the formulae (4.57a)–(4.57c) can be written in matrix form

$$\begin{pmatrix} \sinh(\sqrt{s}) & -\sinh\left(\sqrt{\frac{s}{\beta}}\right) & -\cosh\left(\sqrt{\frac{s}{\beta}}\right) \\ 0 & \cosh\left(\sqrt{\frac{s}{\beta}\frac{L}{L_1}}\right) & \sinh\left(\sqrt{\frac{s}{\beta}\frac{L}{L_1}}\right) \\ \frac{\kappa_1}{\kappa_2}\cosh(\sqrt{s}) & -\frac{1}{\sqrt{\beta}}\cosh\left(\sqrt{\frac{s}{\beta}}\right) & -\frac{1}{\sqrt{\beta}}\sinh\left(\sqrt{\frac{s}{\beta}}\right) \end{pmatrix} \begin{pmatrix} C_1 \\ C_3 \\ C_4 \end{pmatrix} = \begin{pmatrix} \frac{1}{s} \\ 0 \\ 0 \end{pmatrix}. \quad (4.58)$$

The determinant of the matrix denoted as **A** is

$$\det \mathbf{A}$$
$$= \frac{\kappa_1}{\kappa_2} \cosh(\sqrt{s}) \underbrace{\left[\cosh\left(\sqrt{\frac{s}{\beta}\frac{L}{L_1}}\right)\cosh\left(\sqrt{\frac{s}{\beta}}\right) - \sinh\left(\sqrt{\frac{s}{\beta}\frac{L}{L_1}}\right)\sinh\left(\sqrt{\frac{s}{\beta}}\right)\right]}_{=\cosh\left[\sqrt{\frac{s}{\beta}}\left(\frac{L}{L_1}-1\right)\right]=\cosh\left(\sqrt{\frac{s}{\beta}}\frac{L_2}{L_1}\right)}$$
$$+ \frac{1}{\sqrt{\beta}} \sinh(\sqrt{s}) \underbrace{\left[\sinh\left(\sqrt{\frac{s}{\beta}\frac{L}{L_1}}\right)\cosh\left(\sqrt{\frac{s}{\beta}}\right) - \cosh\left(\sqrt{\frac{s}{\beta}\frac{L}{L_1}}\right)\sinh\left(\sqrt{\frac{s}{\beta}}\right)\right]}_{=\sinh\left[\sqrt{\frac{s}{\beta}}\left(\frac{L}{L_1}-1\right)\right]=\sinh\left(\sqrt{\frac{s}{\beta}}\frac{L_2}{L_1}\right)}$$
$$= \frac{\kappa_1}{\kappa_2} \cosh(\sqrt{s}) \cosh\left(\sqrt{\frac{s}{\beta}\frac{L_2}{L_1}}\right) + \frac{1}{\sqrt{\beta}} \sinh(\sqrt{s}) \sinh\left(\sqrt{\frac{s}{\beta}\frac{L_2}{L_1}}\right).$$

4.5 Dynamic Measurements of the Thermal Conductivity: Classical Ioffe Method

To determine the unknown coefficients, further determinants have to be calculated

$$C_1 = \frac{1}{\det \mathbf{A}} \begin{vmatrix} \frac{1}{s} & -\sinh\left(\sqrt{\frac{s}{\beta}}\right) & -\cosh\left(\sqrt{\frac{s}{\beta}}\right) \\ 0 & \cosh\left(\sqrt{\frac{s}{\beta}}\frac{L}{L_1}\right) & \sinh\left(\sqrt{\frac{s}{\beta}}\frac{L}{L_1}\right) \\ 0 & -\frac{1}{\sqrt{\beta}}\cosh\left(\sqrt{\frac{s}{\beta}}\right) & -\frac{1}{\sqrt{\beta}}\sinh\left(\sqrt{\frac{s}{\beta}}\right) \end{vmatrix} = \frac{\sinh\left(\sqrt{\frac{s}{\beta}}\frac{L_2}{L_1}\right)}{s\sqrt{\beta}\,\det \mathbf{A}},$$

(4.59a)

$$C_3 = \frac{1}{\det \mathbf{A}} \begin{vmatrix} \sinh\left(\sqrt{s}\right) & \frac{1}{s} & -\cosh\left(\sqrt{\frac{s}{\beta}}\right) \\ 0 & 0 & \sinh\left(\sqrt{\frac{s}{\beta}}\frac{L}{L_1}\right) \\ \frac{\kappa_1}{\kappa_2}\cosh\left(\sqrt{s}\right) & 0 & -\frac{1}{\sqrt{\beta}}\sinh\left(\sqrt{\frac{s}{\beta}}\right) \end{vmatrix}$$

$$= \frac{\kappa_1 \cosh\left(\sqrt{s}\right)\sinh\left(\sqrt{\frac{s}{\beta}}\frac{L}{L_1}\right)}{s\kappa_2 \det \mathbf{A}},$$

(4.59b)

$$C_4 = \frac{1}{\det \mathbf{A}} \begin{vmatrix} \sinh\left(\sqrt{s}\right) & -\sinh\left(\sqrt{\frac{s}{\beta}}\right) & \frac{1}{s} \\ 0 & \cosh\left(\sqrt{\frac{s}{\beta}}\frac{L}{L_1}\right) & 0 \\ \frac{\kappa_1}{\kappa_2}\cosh\left(\sqrt{s}\right) & -\frac{1}{\sqrt{\beta}}\cosh\left(\sqrt{\frac{s}{\beta}}\right) & 0 \end{vmatrix}$$

$$= -\frac{\kappa_1 \cosh\left(\sqrt{s}\right)\cosh\left(\sqrt{\frac{s}{\beta}}\frac{L}{L_1}\right)}{s\kappa_2 \det \mathbf{A}},$$

(4.59c)

The results of the coefficients, see Eq. (4.59), are set into Eqs. (4.53) and (4.54) to obtain the results of the temperature profiles in the s-domain:

$$\overline{T}_1(x,s) = \frac{1}{s} \frac{\sinh(\sqrt{s}x)\sinh\left(\sqrt{\frac{s}{\beta}}\frac{L_2}{L_1}\right)}{\zeta \cosh(\sqrt{s})\cosh\left(\sqrt{\frac{s}{\beta}}\frac{L_2}{L_1}\right) + \sinh(\sqrt{s})\sinh\left(\sqrt{\frac{s}{\beta}}\frac{L_2}{L_1}\right)},$$

(4.60a)

$$\overline{T}_2(x,s) = \frac{1}{s} - \zeta\frac{1}{s}\frac{\cosh\left[\sqrt{\frac{s}{\beta}}\left(\frac{L}{L_1}-x\right)\right]}{\zeta \cosh(\sqrt{s})\cosh\left(\sqrt{\frac{s}{\beta}}\frac{L_2}{L_1}\right) + \sinh(\sqrt{s})\sinh\left(\sqrt{\frac{s}{\beta}}\frac{L_2}{L_1}\right)}$$

(4.60b)

with the abbreviation $\zeta = \sqrt{\beta}\frac{\kappa_1}{\kappa_2}$.

4.5.5
Inverse Laplace Transformation

To find the inverse Laplace transformation, it is convenient to change the results in the s-space from hyperbolic functions to circular functions with the relations

$$\sinh(x) = -i\sin(ix) \quad \text{and} \quad \cosh(x) = \cos(ix).$$

By using these relations, the solutions of the temperatures change to

$$\overline{T}_1(x,s) = \frac{1}{s} \frac{(-i)\sin(i\sqrt{s}x)(-i)\sin\left(i\sqrt{\frac{s}{\beta}\frac{L_2}{L_1}}\right)}{\zeta\cos(i\sqrt{s})\cos\left(i\sqrt{\frac{s}{\beta}\frac{L_2}{L_1}}\right) - i\sin(i\sqrt{s})(-i)\sin\left(i\sqrt{\frac{s}{\beta}\frac{L_2}{L_1}}\right)}$$

$$= \frac{1}{s} \frac{-\sin(\sqrt{-s}x)\sin\left(\sqrt{-\frac{s}{\beta}\frac{L_2}{L_1}}\right)}{\zeta\cos(\sqrt{-s})\cos\left(\sqrt{-\frac{s}{\beta}\frac{L_2}{L_1}}\right) - \sin(\sqrt{-s})\sin\left(\sqrt{-\frac{s}{\beta}\frac{L_2}{L_1}}\right)},$$

(4.61a)

$$\overline{T}_2(x,s) = \frac{1}{s} - \zeta\frac{1}{s}\frac{\cos\left[\sqrt{-\frac{s}{\beta}}\left(\frac{L}{L_1} - x\right)\right]}{\zeta\cos(\sqrt{-s})\cos\left(\sqrt{-\frac{s}{\beta}\frac{L_2}{L_1}}\right) - \sin(\sqrt{-s})\sin\left(\sqrt{-\frac{s}{\beta}\frac{L_2}{L_1}}\right)}.$$

(4.61b)

For the inverse transformation, the zeros of the denominator have to be known or determined

$$F(\xi) = \zeta\cos(\xi)\cos(\mu\xi) - \sin(\xi)\sin(\mu\xi) \Rightarrow F(\xi) \equiv 0 \quad (4.62)$$

with $\xi = \sqrt{-s}$ and $\mu = 1/\sqrt{\beta}L_2/L_1$. For a similar problem, it has been shown that the roots of the transcendental Eq. (4.62) are real and simple, see [1, Section 12.8 "Composite solids" IV p. 324].

Only in particular cases, the zeros can be determined analytically. In general, the determination has to be performed with numerical or graphical methods. We demonstrate this procedure by an example. In Table 4.5, the thermal properties of the TE sample as well as of the copper block are summarized. By means of those values, the parameters introduced in the previous calculation can be deduced. Instead of Eq. (4.62)

$$G(\xi) = \tan(\xi)\tan(\mu\xi) = \zeta \quad (4.63)$$

is used.

Table 4.5 Material data – Ioffe method.

	κ (W/(m K))	c (J/(kg K))	ρ_d (kg/m³)
TE sample (1)	1.1	302	6050
Copper block (2)	398	385	8930

Table 4.6 Geometric data and derived quantities.

	L (mm)	A_c (mm²)	λ (10^{-6} m²/s)	$\beta = \lambda^{(2)}/\lambda^{(1)}$	$\zeta = \sqrt{\beta}\frac{\kappa_1}{\kappa_2}$	$\mu = \frac{1}{\sqrt{\beta}}\frac{L_2}{L_1}$
(1)	1	1	0.60205	192.283	0.0383248	0.360578
(2)	5	1	115.763			

Table 4.7 First zeros of Eq. (4.63).

k	ξ_k	$s_k = -\xi_k^2$
1	0.319667	−0.102187
2	3.15924	−9.98082
3	6.25198	−39.0872
4	8.61206	−74.1676
5	9.54759	−91.1564
6	12.5733	−158.088
7	15.6563	−245.12

By substituting the values from Table 4.6 in Eq. (4.63), the zeros can be determined numerically (see Table 4.7). The zeros are determined by means of a computer algebra system, for example, MATHEMATICA or MAPLE, after specifying the rough position graphically, see Figure 4.12.

4.5.6
Inversion Theorem for the Laplace Transformation

For the back or inverse transformation, some standard theorems are used, which often can be applied in the framework of the parabolic transport equations as given here especially in the context of heat transfer. One straightforward property that has not been mentioned yet is the linearity of the transformation, that is,

$$\mathcal{L}\{f_1 + f_2\} = \mathcal{L}\{f_1\} + \mathcal{L}\{f_2\}. \tag{4.64}$$

Another theorem often used is the convolution or superposition theorem, also known as Duhamel's theorem

$$\mathcal{L}\left\{\int_0^t f_1(t')f_2(t-t')\,dt'\right\} = \mathcal{L}\left\{\int_0^t f_2(t')f_1(t-t')\,dt'\right\} = \mathcal{L}\{f_1(t)\}\,\mathcal{L}\{f_2(t)\}. \tag{4.65}$$

A consequence and particular case of this is the following relation:

$$\mathcal{L}\left\{\int_0^t f(t')\,dt'\right\} = \frac{1}{s}\mathcal{L}\{f(t)\}. \tag{4.66}$$

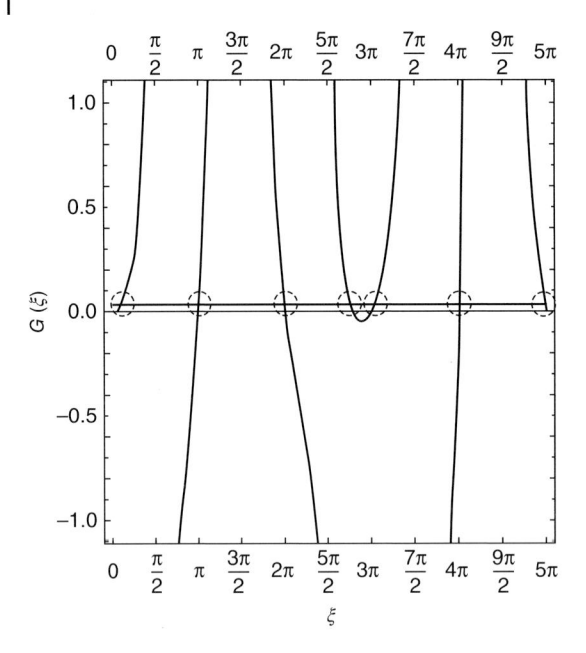

Figure 4.12 Zeros of the function $G(\xi) = \zeta$.

The most important theorem for the back transformation is the "Inversion theorem." Although (mathematical) details cannot be given here, for those details, see [1, 150, 151], the formal calculation should be presented. The integral equation for the inverse Laplace transformation

$$f(t) = \mathcal{L}^{-1}\{\overline{f}(s)\} = \frac{1}{2\pi i}\lim_{c\to\infty}\int_{\gamma-ic}^{\gamma+ic} e^{st}\overline{f}(s)\,ds \Rightarrow f(t) = \frac{1}{2\pi i}\oint_C e^{st}\overline{f}(s)\,ds \quad (4.67)$$

is called the Fourier–Mellin integral, Mellin's inverse formula, or the Bromwich integral.

In application, especially in the theory of boundary value problems, the most often case to be found is that the Laplace transform, for example, $\overline{T}(x,s)$, has an infinite number of poles or isolated essential singular points in the left half space and for those types where the functions are *meromorphic*. A practical calculation of the integral equation (4.67) can be performed by means of Cauchy's residue theorem from complex analysis. The contour integral can be split into two parts. The curve or contour C is chosen as illustrated in Figure 4.13. All poles lie on the left of γ. A series of curves C_n has to be chosen such that the points $\gamma \pm ic_n$ are connected and the curve between these points contains the poles $s_0, s_1, \ldots, s_n$ but does not pass through these poles. Then, Cauchy's residue theorem gives

$$\frac{1}{2\pi i}\int_{\gamma-ic_n}^{\gamma+ic_n} e^{st}\overline{f}(s)\,ds + \frac{1}{2\pi i}\oint_{\mathscr{C}_n} e^{st}\overline{f}(s)\,ds = \sum_{k=0}^{n} r_k(t), \quad (4.68)$$

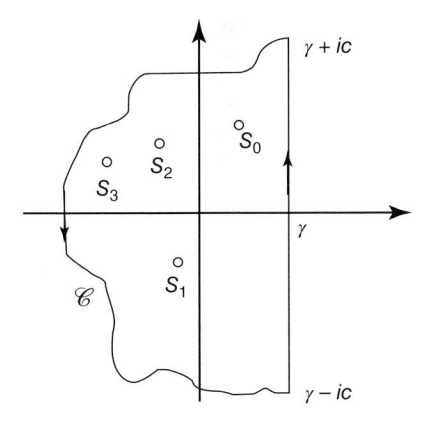

4.5 Dynamic Measurements of the Thermal Conductivity: Classical Ioffe Method

Figure 4.13 Complex plane with poles.

where $r_k(t)$ is the residue of $e^{st}\overline{f}(s)$ in s_k. In the limit $n \to \infty$, it can be shown that the second part of the left-hand side of Eq. (4.68) vanishes and

$$f(t) = \sum_{k=0}^{\infty} r_k(t). \tag{4.69}$$

If $\overline{f}(s)$ is a quotient of two holomorphic functions $p(s)$ and $q(s)$ and has only single poles, then

$$r_k = \frac{p(s_k)}{q'(s_k)} e^{s_k t}. \tag{4.70}$$

4.5.7
Inversion of the Temperature Profiles

In Eq. (4.61), the temperature profiles are given in the s-domain. The main task is then the back transformation (inversion) of these results in the t-domain. For this, the first step is to define the following auxiliary functions $\overline{y}_1$ and $\overline{y}_2$

$$\overline{T}_1(x, s) = \frac{1}{s}\overline{y}_1(x, s), \tag{4.71a}$$

$$\overline{T}_2(x, s) = \frac{1}{s} - \frac{1}{s}\overline{y}_2(x, s) \tag{4.71b}$$

and use the properties defined in the previous subsection such that

$$\mathcal{L}^{-1}\{\overline{T}_1(x, s)\} = T_1(x, t) = \mathcal{L}^{-1}\left\{\frac{1}{s}\overline{y}_1(x, s)\right\} = \int_0^t y_1(x, t')\,dt', \tag{4.72a}$$

$$\mathcal{L}^{-1}\{\overline{T}_2(x, s)\} = T_2(x, t) = \mathcal{L}^{-1}\left\{\frac{1}{s} - \frac{1}{s}\overline{y}_2(x, s)\right\} = 1 - \int_0^t y_2(x, t')\,dt', \tag{4.72b}$$

where

$$y_1(x,t) = \mathcal{L}^{-1}\{\bar{y}_1(x,s)\} \quad \text{and} \quad y_2(x,t) = \mathcal{L}^{-1}\{\bar{y}_2(x,s)\}. \tag{4.72c}$$

The two auxiliary functions can be expressed as a fraction of two holomorphic functions, each of which has only single poles

$$\bar{y}_1(x,s) = \frac{p_1(x,s)}{q_1(x,s)} \quad \text{and} \quad \bar{y}_2(x,s) = \frac{p_2(x,s)}{q_2(x,s)}, \tag{4.73}$$

where $q_1(x,s) = q_2(x,s)$, which are identical to $F[\xi(s)]$ from Eq. (4.62). The same relation can be used for the determination of the zeros of the denominator

$$F'(s) = \frac{dF}{ds} = \frac{d\xi}{ds}\frac{dF}{d\xi} = \frac{d\xi}{ds}F'(\xi). \tag{4.74}$$

As a result, the derivative of F with respect to ξ, see Eq. (4.62) is given by

$$F'(\xi) = -[\mu + \zeta]\sin(\xi)\cos(\mu\xi) - [1 + \mu\zeta]\cos(\xi)\sin(\mu\xi). \tag{4.75}$$

From the definition of $\xi = \sqrt{-s}$, the relation $d\xi/ds = -1/2\sqrt{-s} = -1/2\xi$ is calculated.

From the inversion, the following relations are found:

$$T_1(x,t) = 2\sum_{k=1}^{\infty}\frac{\sin(\xi_k x)\sin(\mu\xi_k)\left[1 - e^{-\xi_k^2 t}\right]}{\{[\mu + \zeta]\sin(\xi_k)\cos(\mu\xi_k) + [1 + \mu\zeta]\cos(\xi_k)\sin(\mu\xi_k)\}\xi_k}, \tag{4.76a}$$

$$T_2(x,t) = 1 - 2\sum_{k=1}^{\infty}\frac{\cos\left[\xi_k\frac{1}{\sqrt{\beta}}\left(\frac{L}{L_1} - x\right)\right]\left[1 - e^{-\xi_k^2 t}\right]}{\{[\mu + \zeta]\sin(\xi_k)\cos(\mu\xi_k) + [1 + \mu\zeta]\cos(\xi_k)\sin(\mu\xi_k)\}\xi_k} \tag{4.76b}$$

4.6
Green's Function Approach in Thermoelectricity

In view of the huge number of papers applying Green's function (GF) methods in solid-state physics, quantum field theory, statistical physics, and so on (for an overview, see, e.g., [161–166]) we may share the reader's skepticism concerning the relevance of this section. However, the usage of GF in classical nonequilibrium situations is much less developed,[7] and there is a great effort in studying those problems under the aspect of a better understanding of the underlying physical processes and as an example aimed at reaching a maximum performance in thermoelectric elements. For the researchers within the TE community, the

---

7) In particular, the discussions are mostly restricted to diffusion and heat conductivity [1, 167].

application of classical GF method is a quite new aspect, which has, to the best of our knowledge, never been discussed before in detail.

With regard to this situation, the intention of this section is to bridge the gap between the general method used in theoretical physics and the classical theory of thermoelectricity.

Great effort has been made to study such complex problems as thermoelectricity, thermodiffusion, and electrodiffusion as well as thermogalvanic/-magnetic effects on a microscopic level. In particular, a significant progress is achieved in the observation of the spin Seebeck effect [168, 169].

Recently, the thermoelectric (TE) properties in carbon nanotubes have been studied using an atomistic nonequilibrium Green's function approach (NEGF) [170]; for the quantum mechanical model to study the coupled electron–phonon transport in one-dimensional systems, see also [171].

However, due to the broad applicability of TE effects, for instance, in thermoelectric power generators and for temperature sensing, the usage of the powerful apparatus of the conventional Green's function technique is still required. It seems to us that the GF method is not sufficiently applied within the phenomenological approach based on the linear Onsager theory. In particular, the GF method allows in a very transparent manner the inclusion of a broad class of boundary conditions relevant for a diversity of applications. With regard to the specific material properties, one is often confronted with graded or composite materials, which are identified by nonconstant material parameters such as thermal conductivity and Seebeck or Peltier coefficients. Moreover, for a precise characterization of materials, very detailed temperature distributions, diffusion profiles, or current distributions within the material are necessary to understand experimental realizations. It is therefore the aim of this section to demonstrate the great advantages in applying GF for that kind of problem.

The basic concept of GF consists of determining a response to a point-like source. The knowledge of the corresponding response function or GF allows then the solution of the underlying differential equation under inclusion of a diverse class of BCs or ICs. So, the GF, calculated as response to a point-like source, for instance, enables to find a closed representation of the total evolution equation; this solution is simply given by a convolution of the GF and the inhomogeneity of the differential equations. Generally, the GF depends on the differential equation, the body shape, and the type of BCs used.

In the next subsections, we demonstrate how the GF technique works in the classical theory of thermoelectricity. The procedure is illustrated by the generalized heat conduction equation

$$\nabla \cdot (-\kappa \nabla T) = \varrho j^2 - \tau \mathbf{j} \cdot \nabla T. \tag{4.77}$$

The analysis concerns primarily the 1D steady state as well as an extension to time evolution. Let us emphasize that the method is not restricted to that problem but can be easily extended to other classes of models.

4.6.1
Continuity Equations

The basic equations for a generalized description of thermoelectric effects are given by a transient set of continuity equations,

$$\frac{\partial \rho_{el}}{\partial t} + \nabla \cdot \mathbf{j} = 0, \tag{4.78}$$

$$\frac{\partial u}{\partial t} + \nabla \cdot \mathbf{q} = \mathbf{j} \cdot \mathbf{E}, \tag{4.79}$$

suggesting that changes in the charge density ρ_{el} are related to the electrical currents while the changes of the energy density u are accompanied by changes in the heat current and may originate by a source given by the Joule heat.

The energy density can be expressed by the specific heat c, mass density ϱ_{md}, and temperature T according to

$$u = \varrho_{md} c T. \tag{4.80}$$

Then the second continuity equation can be rewritten as

$$\varrho_{md} c \frac{\partial T}{\partial t} + \nabla \cdot \mathbf{q} = \mathbf{j} \cdot \mathbf{E}. \tag{4.81}$$

Equations (4.78) and (4.81) are the most general relations between currents, sources, and field variables. However, this set of equations is quite complicated and before we discuss a transient analysis of thermoelectric transport, we will focus on steady-state solutions of those equations. In particular, the steady-state behavior enables us to introduce Green's function technique.

4.6.2
Green's Function Approach in the Steady State

The steady state of thermoelectric devices is characterized by the conservation of charge and energy. The related Eqs. (4.78) and (4.81) are reduced to

$$\nabla \cdot \mathbf{j} = 0, \; \nabla \cdot \mathbf{q} = \mathbf{j} \cdot \mathbf{E},$$

with the Joule heat representing a source of heating. Under isotropic conditions, the constitutive relations for the (total) heat and electrical current density are given by

$$\mathbf{q} = -\kappa \nabla T + \Pi \mathbf{j}, \; \mathbf{j} = \sigma \mathbf{E} - \Theta \nabla T, \tag{4.82}$$

with Peltier coefficient $\Pi = \alpha T$ and Seebeck conductivity $\Theta = \sigma \alpha$.

Since the steady state means a *constant* (total) electrical current density, the thermal energy balance [Eq. (4.77)] leads to

$$\nabla \cdot (-\kappa \nabla T) + T \mathbf{j} \cdot \nabla \alpha = \mathbf{j}^2 / \sigma. \tag{4.83}$$

The last equation represents a Poisson-like differential equation. This equation is nonlinear in temperature T as soon as one of the parameters κ, α, or σ depends on T.

Green's function, $G(\mathbf{r}, \mathbf{r}')$, of a linear differential operator $\hat{L}_\mathbf{r} = \hat{L}(\mathbf{r})$ acting on distributions over a subset of the Euclidean space (here $\mathbb{R}^3$) at a point $\mathbf{r}'$ is any solution of

$$\hat{L}_\mathbf{r} G(\mathbf{r}, \mathbf{r}') = \delta(\mathbf{r} - \mathbf{r}'), \tag{4.84}$$

where δ is the Dirac delta distribution and

$$\hat{L}_\mathbf{r} = \nabla \cdot (-\kappa(\mathbf{r})\nabla) + (\mathbf{j} \cdot \nabla \alpha(\mathbf{r})) \tag{4.85}$$

is the operator defined in Eq. (4.83).

In this notation, $\mathbf{r}'$ denotes the position of a thermoelectric point source. Because the differentiation is carried out with respect to $\mathbf{r}$, the operator $\hat{L}_\mathbf{r}$ is characterized by an index as well. The GF itself has the dimension of a thermal resistance and is introduced to describe the temperature distribution. For details, we refer to [172].

Green's function and the temperature are related via an integral expression,

$$T(\mathbf{r}) = \int_\Omega \left\{ \frac{j^2}{\sigma(\mathbf{r}')} G(\mathbf{r}, \mathbf{r}') \right\} d^3 r' - \oint_{d\Omega} \left\{ \kappa(\mathbf{r}') T(\mathbf{r}') \nabla_{\mathbf{r}'} G(\mathbf{r}, \mathbf{r}') \right\} \cdot d\mathbf{f}' \tag{4.86}$$

involving information about volume properties and heat flow through the surface of the device.

By using Snyder's TE potential, the electric field $\mathbf{E}$ and the temperature are related via

$$\mathbf{E}(\mathbf{r}) = \nabla \Phi(\mathbf{r}) = \frac{\mathbf{j}}{\sigma(\mathbf{r})} + \alpha(\mathbf{r}) \nabla T(\mathbf{r}). \tag{4.87}$$

That relation allows for the determination of the electrical quantities in terms of a path-independent line integral

$$\Phi(\mathbf{r}) = \int_{\mathbf{r}_o}^{\mathbf{r}} \left(\frac{\mathbf{j}}{\sigma(\mathbf{r}')} + \alpha(\mathbf{r}') \nabla_{\mathbf{r}'} T(\mathbf{r}') \right) \cdot d\mathbf{r}'. \tag{4.88}$$

4.6.3
One-Dimensional Green's Functions in the Steady State

The one-dimensional (1D) version of Eq. (4.83) with $\mathbf{j} = j\mathbf{e}_x$, j=const., is given by

$$-\frac{d}{dx}\left(\kappa(x)\frac{d}{dx}T(x)\right) + \left(j\frac{d\alpha(x)}{dx}\right) T(x) = \frac{j^2}{\sigma(x)}. \tag{4.89}$$

The corresponding GF obeys the equation

$$-\frac{d}{dx}\left(\kappa(x)\frac{d}{dx}G(x, x')\right) + \left(j\frac{d\alpha(x)}{dx}\right) G(x, x') = \delta(x - x'). \tag{4.90}$$

By knowing Green's function, the temperature profile can be obtained after evaluating the subsequent integral [Eq. (4.86)],

$$T(x) = \int \frac{j^2}{\sigma(x')} G(x, x') dx' - \left[T(x)\kappa(x) \frac{dG(x, x')}{dx} \right]_{x=0}^{x=L}. \tag{4.91}$$

The result suggests that the influence of the electrical conductivity can be separated from the thermoelectric coupling. In this respect, the GF method allows a systematic variation of $\sigma(x)$ for given $\kappa(x)$ and $\alpha(x)$.

The electrical potential Φ can be calculated using Eq. (4.88),

$$\Phi(x) = \int_0^x \left(\frac{j}{\sigma(x')} + \alpha(x') \frac{dT(x')}{dx'} \right) dx'. \tag{4.92}$$

Notice that in addition to the differential equation (4.89), some BCs have to be specified. In the case of Dirichlet conditions, that is, the given temperatures at the left and right end of the device, the temperature profile $T(x)$ can be transformed to an auxiliary temperature profile $\tilde{T}(x)$, satisfying the homogeneous BCs. The transformation reads

$$\tilde{T}(x) = T(x) - \frac{T_2 - T_1}{L} x - T_1, \; x \in [0, L]. \tag{4.93}$$

In terms of $\tilde{T}(x)$, the one-dimensional thermal energy balance takes the form

$$-\frac{d}{dx}\left(\kappa(x) \frac{d}{dx} \tilde{T}(x) \right) + \left(j \frac{d\alpha(x)}{dx} \right) \tilde{T}(x) = J(x) \tag{4.94}$$

with an modified source term $J(x)$. The Green's function still satisfies the same differential equation but because of homogeneous BCs, the relation to the auxiliary temperature profile is modified according to

$$\tilde{T}(x) = \int_0^L J(x) G(x, x') dx'. \tag{4.95}$$

From that equation, one finds the physical temperature distribution

$$T(x) = \int_0^L J(x) G(x, x') dx' + \frac{T_2 - T_1}{L} x + T_1. \tag{4.96}$$

In Table 4.8, we have listed different realizations of the source field $J(x)$ and related Green's functions. Note that the shorthand notations $x_> = \max(x, x')$ and $x_< = \min(x, x')$ are used. For the general case of spatially varying transport coefficients α, κ, and σ, there is still no general solution available. Green's function for such problems depends sensitively on the exact realization of the transport coefficients. An example has been discussed in Ref. [172].

With regard to a more systematic discussion, it seems to be appropriate to focus on TE applications. On the one hand, this concerns the discussion of the efficiency η of a thermogenerator (see, e.g., [173–175]), on the other hand, the coefficient of performance of a Peltier cooler [28, 176], which can be expressed in terms of the temperature profile and thus by means of Green's functions.

4.6.4
Perturbative Approach to a Full Description (1D)

Green's functions for special cases regarding the transport coefficients are listed in Table 4.8. However, in case of a quite general spatially dependent thermopower

Table 4.8 Summary of Green's functions (see also [172]).

Scenario	Source field $J(x)$	Green function $G(x,x')$
General	$\dfrac{j^2}{\sigma(x)} - j\dfrac{d\alpha(x)}{dx}\left(\dfrac{T_2 - T_1}{L}x + T_1\right) + \dfrac{d\kappa(x)}{dx}\dfrac{T_2 - T_1}{L}$	—
CPM	$\dfrac{j^2}{\sigma}$	$\dfrac{1}{\kappa}\left(1 - \dfrac{x_>}{L}\right)x_<$
$\sigma(x)$	$\dfrac{j^2}{\sigma(x)}$	$\dfrac{1}{\kappa}\left(1 - \dfrac{x_>}{L}\right)x_<$
$\kappa(x)$	$\dfrac{j^2}{\sigma(x)} - \dfrac{d\kappa(x)}{dx}\dfrac{T_2 - T_1}{L}$	$\left(1 - \dfrac{\int_0^{x_>}\frac{d\xi}{\kappa(\xi)}}{\int_0^{L}\frac{d\xi}{\kappa(\xi)}}\right)\int_0^{x_<}\dfrac{d\xi}{\kappa(\xi)}$

$\alpha(x)$, a compact expression for corresponding Green's function is still open. In Reference [172], a particular form function for the thermopower was chosen and an analytic expression for Green's function was derived. For a more complicated realization of spatially dependent transport coefficient, one can use a perturbative approach leading to a hierarchy of Green's functions.

Such a description is based on a Dyson equation, which offers a relation between the GF of a solvable problem and the total GF under the influence of a perturbation. The total system consists of a reference system, characterized by an operator denoted by $\hat{L}^0$, and a perturbative part $\Delta\hat{L}$. The reference system fulfills the relation

$$\hat{L}_\mathbf{r}^0 \, G^0(\mathbf{r},\mathbf{r}') = \delta(\mathbf{r} - \mathbf{r}'), \tag{4.97}$$

while the total operator includes the perturbation $\hat{L}_{\Delta,\mathbf{r}}$:

$$\hat{L}_\mathbf{r} \, G(\mathbf{r},\mathbf{r}') = (\hat{L}_\mathbf{r}^0 - \hat{L}_{\Delta,\mathbf{r}}) \, G(\mathbf{r},\mathbf{r}') = \delta(\mathbf{r} - \mathbf{r}'). \tag{4.98}$$

The relation between both Green's functions G and G^0 is given by the Dyson equation

$$G(\mathbf{r},\mathbf{r}') = G^0(\mathbf{r},\mathbf{r}') + \int_\Omega G^0(\mathbf{r},\mathbf{r}_1) \hat{L}_{\Delta,\mathbf{r}_1} \, G(\mathbf{r},\mathbf{r}') d^3 r_1. \tag{4.99}$$

To be more specific, let us consider a 1D system with constant gradient of the thermopower but spatially dependent $\kappa(x)$. Then the GF $G^0 \equiv G_\kappa(x,x')$ of the reference system is analytically known (see Table 4.8). If the gradient of the thermopower can be assumed to be a small perturbation, then the Dyson equation reads

$$G(x,x') = G_\kappa(x,x') - \int_0^L G_\kappa(x,x_1) j \dfrac{d\alpha(x_1)}{dx_1} G(x_1,x') dx_1. \tag{4.100}$$

This equation is solvable either by means of orthogonal basis functions or by using an iterative approach. As a result, one finds Green's function $G(x, x')$ of the complete thermoelectric problem. Knowing this the Green's function, the convolution integral [Eq. (4.96)] leads to the uniquely determined temperature profile. If the gradient of the thermopower cannot be considered as a small perturbation, then it might be favorable to assume a constant thermal conductivity but spatially dependent thermopower and search for an analytic expression for that problem. Here the reference system is given by the GF $G^0 \equiv G_\alpha(x, x')$ and the complete GF $G(x, x')$ fulfills the Dyson equation

$$G(x, x') = G_\alpha(x, x')$$
$$+ \int_0^L dx_1 \, G_\alpha(x, x_1) \left(\frac{d\kappa(x_1)}{dx_1} \frac{dG(x_1, x')}{dx_1} + (\kappa(x_1) - \kappa_o) \frac{d^2 G(x_1, x')}{dx_1^2} \right) \quad (4.101)$$

or equivalently by

$$G(x, x') = \frac{\kappa_o}{\kappa(x')} G_\alpha(x, x')$$
$$+ \int_0^L dx_1 \, G_\alpha(x, x_1) \left(1 - \frac{\kappa_o}{\kappa(x_1)} \right) j \frac{d\alpha(x_1)}{dx_1} G(x_1, x')$$
$$+ \int_0^L dx_1 \, G_\alpha(x, x_1) \left(1 - \frac{\kappa_o}{\kappa(x_1)} \right) \frac{d\kappa(x_1)}{dx_1} \frac{dG(x_1, x')}{dx_1}. \quad (4.102)$$

For more details, we refer to [172].

4.7 Linear Transient Approach

4.7.1 Relaxation Time

By using dimensional analysis, only two possible variants of a relaxation time τ_{rel} can be detected when excluding absolute temperature T as a parameter:

A) τ_{rel} with Seebeck coefficient α and electrical conductivity σ, but without κ:
$$\tau_{rel} = \frac{\alpha^2 \sigma}{\varrho_{md} c^2},$$
where c is the specific heat (at constant volume) measured in J/(kg K).

B) τ_{rel} with thermal conductivity κ but without α and σ, and additionally containing a typical length scale L such as the length of the device:
$$\tau_{rel} = \frac{\varrho_{md} c L^2}{\kappa} = \frac{L^2}{\lambda}.$$

We add here the dimensions for control:
$[\varrho_{md}] = $ kg/m^3, $[c] = $ J/(kg K), $[\kappa] = $ W/(m K), $[\lambda] = [\kappa/(\varrho_{md} c)] = $ m²/s.

Let us give some arguments favoring variant B[8]:

1) The smaller the body is, the faster the temperature gradient in a material is subsided. Therefore, a length scale should be included in the related relaxation time.
2) The time behavior should not controlled by the Seebeck coefficient. Instead of that the amplitude of the Peltier heat generated is determined thereby.
3) Variant B contains the reciprocal thermal diffusivity ($1/\lambda$), which can be measured through a temporal decay process.

So far, we cannot see that the time behavior of thermoelectric systems is explicitly determined by the thermoelectric properties and not solely by the thermal properties. It seems plausible to use a purely thermal model, which is based on the assumption that the transient behavior of thermoelectric systems is carrying the conventional thermal character. In the subsequent section, let us verify this idea in more detail.

4.7.2
Transient Field Equations

Within Onsager's approach of a linear response theory, the constitutive relations are under isotropic conditions, see already Eq. (4.82),

$$\mathbf{j} = \sigma \mathbf{E} - \sigma \alpha \nabla T, \quad \mathbf{q} = -\kappa \nabla T + \alpha T \mathbf{j} \tag{4.103}$$

with the electric field vector $\mathbf{E}$, electrical current density $\mathbf{j}$, temperature T, total heat flux $\mathbf{q}$, the nabla operator ∇, and the material properties given by the isothermal electrical conductivity σ, the thermal conductivity κ under zero current, and the Seebeck coefficient α.

Inserting the relations (4.103) (for $\mathbf{q}$ and $\mathbf{E}$) into Eq. (4.81) and using nabla calculus, the transient thermal energy balance can be rewritten as

$$\varrho_{md} c \frac{\partial T}{\partial t} + \nabla \cdot (-\kappa \nabla T) = \frac{j^2}{\sigma} - T \mathbf{j} \cdot \nabla \alpha - \alpha T \nabla \cdot \mathbf{j}. \tag{4.104}$$

The set of coupled equations (4.78) and (4.104) is the starting point describing the time-dependent thermoelectric processes. By combining both equations, we get

$$\varrho_{md} c \frac{\partial T}{\partial t} + \nabla \cdot (-\kappa \nabla T) + T \mathbf{j} \cdot \nabla \alpha = \frac{j^2}{\sigma} + \alpha T \frac{\partial \rho_{el}}{\partial t}, \tag{4.105}$$

which is the transient analog to Eq. (4.83).

We further assume that the equilibrium distribution of the temperature $T^{(eq)}(\vec{r})$ is reached at a constant electrical current density $\mathbf{j} = \mathbf{j}_o$. This equilibrium, which can be an optimal state, is governed by the spatial field equation

$$\kappa \Delta T^{(eq)}(\vec{r}) = -\frac{j_o^2}{\sigma} + T^{(eq)}(\vec{r}) \mathbf{j}_o \cdot \nabla \alpha. \tag{4.106}$$

---
8) The question of whether variant A is a pure mathematical result or can be used to explain any transient thermoelectric process remains unanswered.

Note that Eq. (4.106) results from Eq. (4.104) when $\kappa =$ const., $\partial/\partial t = 0$ and $\nabla \cdot \mathbf{j}_o = 0$, respectively. Thus, in the steady state, the Joule heat and a gradient in Seebeck (distributed Peltier heat and Thomson heat) appear as sources of the Fourier heat flux, $\mathbf{q}_\kappa = -\kappa \nabla T$.

4.7.3
Transient Linear Response Approximation

Let us now consider small deviations from the thermodynamic equilibrium and weak electrical currents. Under these conditions, one can derive an appropriate approximation of Eq. (4.104). By neglecting all quadratic terms as j^2, $(\nabla T)^2$, $\mathbf{E}^2$, as well as, $T\mathbf{E}$, one can de facto use the approximation

$$\mathbf{j} \approx -\alpha \sigma \nabla T. \tag{4.107}$$

Doing so, we discuss the last two terms on the right-hand side of Eq. (4.104)

$$-T\mathbf{j}\cdot\nabla\alpha - -T\sigma(\mathbf{E}\quad \alpha\nabla T)\cdot\nabla\alpha \approx T\alpha\sigma\nabla T\cdot\nabla\alpha = T\sigma\nabla T\cdot\nabla(\alpha^2/2)\sim 0. \tag{4.108}$$

Since $\nabla(\alpha^2/2)$ is very small, this term can be neglected. The other term gives

$$-\alpha T \nabla \cdot \mathbf{j} = -\alpha T \nabla \cdot (\sigma \mathbf{E} - \sigma \alpha \nabla T) \approx \alpha T \nabla \cdot (\alpha \sigma \nabla T). \tag{4.109}$$

By neglecting the variations in σ and assuming only small deviations from the equilibrium distribution of the temperature $T^{(\text{eq})}$, we further find using Eq. (4.108)

$$-\alpha T \nabla \cdot \mathbf{j} \approx \alpha^2 \sigma T^{(\text{eq})} \Delta T. \tag{4.110}$$

Hence, a first-order approximation of Eq. (4.104) is

$$\varrho_{\text{md}} c \frac{\partial T}{\partial t} = \kappa \Delta T + \alpha^2 \sigma T^{(\text{eq})} \Delta T = \kappa (1 + z T^{(\text{eq})}) \Delta T, \tag{4.111}$$

which is a heat equation with an effective thermal conductivity, which contains the power factor $\alpha^2 \sigma$ implicitly. This result underlines once again the importance of the dimensionless figure of merit.

The further consideration is based on the (static) expression of the thermal conductivities given by [177]

$$\kappa_E = \kappa_J + \alpha^2 \sigma T \quad \text{or} \quad \frac{\kappa_E}{\kappa_J} = 1 + zT, \tag{4.112}$$

where $\kappa \equiv \kappa_J$ corresponds to the open-circuit configuration (zero electrical current) while κ_E corresponds to the short-circuit configuration. Thus, a transient linear response approximation of Eq. (4.104) is given by

$$\varrho_{\text{md}} c \frac{\partial T}{\partial t} = \kappa_E \Delta T \quad \Longleftrightarrow \quad \frac{\partial T}{\partial t} = \lambda_E \Delta T \tag{4.113}$$

where $\lambda_E = \kappa_E/(\varrho_{\text{md}} c) = \lambda(1 + zT)$ can be introduced as thermal diffusivity (of a short-circuit configuration), which is, in general, a temperature-dependent material property.

The rough approximation leads to a heat conduction equation with modified parameters due to the thermoelectric coupling. Damping processes due to the

Joule heat are not taken into account. Nevertheless, this simple model offers the possibility of approximately describing the transient behavior of TE systems in a conventional way.

4.8 Time-Dependent Green's Function Approach

The transient heat conduction equation is of fundamental importance in diverse scientific fields, and its solution has been sufficiently investigated, for an overview see, for example, [1, 162, 178, 179]. There are different solution techniques depending on the dimension of the problem, on the coordinate system to be used, and on the area to be investigated: If the medium is not the whole space, then we need to specify boundary and ICs for $T(\mathbf{r}, t)$ in order to solve the heat equation uniquely. Depending on the problem considered, Eq. (4.113) can be solved analytically or by means of available FEM codes (e. g., ANSYS).

The Green's function technique can be used to develop an improved solution variant of the transient thermoelectric problem. We therefore focus on the solution of an inhomogeneous heat equation; for a constant thermal conductivity κ, it is given by

$$\left(\varrho_{\mathrm{md}} c \frac{\partial}{\partial t} - \kappa \Delta + \mathbf{j} \cdot \nabla \alpha \right) T(\mathbf{r}, t) = f(\mathbf{r}, t) \tag{4.114}$$

together with boundary and ICs yet to be specified.

From Eq. (4.105), we conclude that the inhomogeneity $f(\mathbf{r}, t)$ is given by

$$f(\mathbf{r}, t) = \frac{j^2}{\sigma} + \alpha T \frac{\partial \rho_{\mathrm{el}}}{\partial t} = \varrho \mathbf{j} \cdot \mathbf{j} - \alpha T \nabla \cdot \mathbf{j}, \tag{4.115}$$

where the electrical current density $\mathbf{j}(t)$ and the charge density $\rho_{\mathrm{el}}(\mathbf{r}, t)$ are coupled via Eq. (4.78).

However, when referring to the transient linear approximation derived in Section 4.7.3, the operation of thermoelectric devices under nonstationary conditions (e.g., the beginning of a cooling regime at constant electrical current) can be described by the heat conduction equation

$$\left(\varrho_{\mathrm{md}} c \frac{\partial}{\partial t} - \kappa_E \Delta \right) T(\mathbf{r}, t) = \varrho j^2 \quad \Rightarrow \quad \left(\frac{\partial}{\partial t} - \lambda_E \Delta \right) T(\mathbf{r}, t) = \frac{\varrho j^2}{\varrho_{\mathrm{md}} c}. \tag{4.116}$$

Then, the GF technique provides the following results: Starting with an infinite domain, the corresponding Green's function is defined by the equation

$$\left(\frac{\partial}{\partial t} - \lambda_E \Delta \right) G(\mathbf{r}, t; \mathbf{r}', t') = \delta(\mathbf{r} - \mathbf{r}')\delta(t - t') \tag{4.117}$$

together with the boundary condition

$$\lim_{|\mathbf{r}| \to \infty} G(\mathbf{r}, t; \mathbf{r}', t') < \infty \tag{4.118}$$

and the initial condition

$$G(\mathbf{r}, t = 0; \mathbf{r}', t') = 0. \tag{4.119}$$

The *fundamental solution* can be obtained, for instance, by the Laplace transform with respect to the time variables and the Fourier transform with respect to the spatial variables. The fundamental solution, the *free-space* Green's function, in three dimensions is given by [162]

$$G(\mathbf{r}, t; \mathbf{r}', t') = \frac{H(t - t')}{(4\pi \lambda_E (t - t'))^{3/2}} \, e^{-\frac{(\mathbf{r} - \mathbf{r}')^2}{4\lambda_E (t - t')}} \qquad (4.120)$$

with $H(t - t')$ being the Heavyside step function. For problems over finite domains, the solution may be obtained analytically, that is, by explicitly defining the volume and boundary conditions or alternatively using the free-space Green's function and combining the fundamental and their particular solutions. The second method is successfully applied to problems in electrostatics and sometimes referred as the method of mirror charges. Nevertheless, we should bear in mind that the second method can be become tedious, in particular if the domain has a very complicated shape.

Finally, the solution to the general formulation of the time-dependent heat equation using the *reciprocity property* and *adjoint Green's function* (see [162]) is given by

$$\begin{aligned}
T(\mathbf{r}, t) = & \int_0^{t^+} dt' \int_\Omega d^3 r' \, G(\mathbf{r}, t; \mathbf{r}', t') f(\mathbf{r}', t') \\
& + \lambda_E \int_0^{t^+} dt' \int_{\partial\Omega} dS' \cdot \mathbf{n} \Big(G(\mathbf{r}, t; \mathbf{r}', t') \nabla' T(\mathbf{r}', t') \\
& \quad - T(\mathbf{r}', t') \nabla' G(\mathbf{r}, t; \mathbf{r}', t') \Big) \\
& + \int_\Omega d^3 r' \, G(\mathbf{r}, t; \mathbf{r}', t' = 0) \, T(\mathbf{r}', t' = 0).
\end{aligned}$$

The first term represents the effects of volume sources, the second accounts for the boundary contributions as well as for contributions at the surfaces ($\partial\Omega$), and the last term includes the initial data.

In the following, we construct a Green's function for one-dimensional problems, that is, we are looking for a solution to the general problem

$$\left(\frac{\partial}{\partial t} - \lambda_E \frac{\partial^2}{\partial x^2} \right) T(x, t) = f(x, t), \quad t > 0, \quad 0 < x < L. \qquad (4.121)$$

Similarly, Green's function satisfies

$$\left(\frac{\partial}{\partial t} - \lambda_E \frac{\partial^2}{\partial x^2} \right) G(x, t; x', t') = \delta(x - x') \delta(t - t'). \qquad (4.122)$$

In order to keep the things as general as possible, we assume that Green's function should obey boundary conditions of Robinson type at $x = 0$ and $x = L$,

$$\alpha_1 G(x = 0, t; x', t') + \beta_1 \nabla_x G(x = 0, t; x', t') = 0, \quad t < 0, \qquad (4.123)$$
$$\alpha_2 G(x = L, t; x', t') + \beta_2 \nabla_x G(x = L, t; x', t') = 0, \quad t < 0, \qquad (4.124)$$

and the initial condition

$$G(x, 0; x', t') = 0. \tag{4.125}$$

Again by using the Laplace transformation, the time dependence can be suitably handled, and the remaining problem is finding an appropriate system of *orthonormal* eigenfunctions $\phi_n(\mathbf{r})$ of a Sturm–Liouville problem,

$$(\nabla^2 + k^2)\phi_n(x) = 0 \tag{4.126}$$

together with the boundary conditions

$$\alpha_1 \phi_n(x = 0) + \beta_1 \nabla_x \phi_n(x = 0) = 0, \tag{4.127}$$

$$\alpha_2 \phi_n(x = L) + \beta_2 \nabla_x \phi_n(x = L) = 0, \tag{4.128}$$

which can be used to find the (discrete or continuous) eigenvalues k_n of Eq. (4.126).

After performing all back transformations, we obtain Green's function for the 1D transient problem

$$G(x, t; x', t') = H(t - t') \sum_{n=1}^{\infty} \phi_n(x) \phi_n(x') e^{-k_n^2 \lambda_E (t - t')}, \tag{4.129}$$

where the sum belongs to a summation over all (discrete or continuous) eigenvalues and corresponding eigenfunctions. It is worth noting that this form of Green's function belongs to the spectral (or Lehmann) representation. Please note that the special functional dependence of the conductivities defines the eigenfunctions and related eigenvalues.

References

1. Carslaw, H.S. and Jaeger, J.C. (1959) *Conduction of Heat in Solids*, 2nd edn, Oxford at the Calerendon Press.
2. Fourier, J.B.J. (1822) *Théorie Analytique de la Chaleur*, F. Didot.
3. Narasimhan, T.N. (1999) Fourier's heat conduction equation: history, influence, and connections. *Rev. Geophys.*, **37** (1), 151–172.
4. Narasimhan, T. (1999) Fourier's heat conduction equation: history, influence, and connections. *J. Earth Syst. Sci.*, **108**, 117–148.
5. Grysa, K. (2011) *Heat Conduction - Basic Research - Inverse Heat Conduction Problems*, Chapter 1, InTech, pp. 3–36.
6. Beck, J.V., Blackwell, B., and St. Clair, C.R. Jr. (1985) *Inverse Heat Conduction - Ill-Posed Problems*, John Wiley & Sons, Inc.
7. Alifanov, O.M. (1994) *Inverse Heat Transfer Problems, International Series in Heat and Mass Transfer*, Springer-Verlag, Berlin, Heidelberg.
8. Özişik, M.N. (2000) *Inverse Heat Transfer: Fundamentals and Applications*, CRC Press.
9. Ang, D.D., Gorenflo, R., Le, V.K., and Trong, D.D. (2002) *Moment Theory and Some Inverse Problems in Potential Theory and Heat Conduction*, Lecture Notes in Mathematics, vol. **1792**, Springer-Verlag.
10. Taler, J. and Duda, P. (2006) *Solving Direct and Inverse Heat Conduction Problems*, Springer-Verlag.
11. Samarskii, A.A. and Vabishchevich, P.N. (2007) *Numerical Methods for Solving Inverse Problems of Mathematical Physics, Inverse and Ill-Posed Problems*, Walter de Gruyter.

12. Orlande, H.R.B., Fudym, O., Maillet, D., and Cotta, R.M. (2011) *Thermal Measurements and Inverse Techniques*, CRC Press.
13. Kabanikhin, S.I. (2012) Inverse and Ill-Posed Problems: Theory and Applications, *Inverse and Ill-Posed Problems Series*, vol. **55**, Walter de Gruyter.
14. Lavrent'ev, M.M., Romanov, V.G., and Shishatskiĭ, S.P. (1986) *Ill-Posed Problems of Mathematical Physics and Analysis*, Translation of Mathematical Monographs, vol. **64**, American Mathematical Society.
15. Petrov, Yu.P. and Sizikov, V.S. (2005) *Well-Posed, Ill-Posed, and Intermediate Problems with Applications*, Inverse and Ill-Posed Problems Series, VSP.
16. Kutasov, V.A., Lukyanova, L.N., and Vedernıkov, M.V. (2006) *Thermoelectric Handbook - Macro to Nano - Shifting the Maximum Figure-of-Merit of $(Bi, Sb)_2(Te, Se)_3$ Thermoelectrics to Lower Temperatures*, Chapter 37, CRC Press - Taylor & Francis Group, LLC, pages 37-1–37-18.
17. Das, J.C. (2010) *Transients in Electrical Systems: Analysis, Recognition, and Migration*, McGraw-Hill Professional.
18. Zhou, Q., Bian, Z., and Shakouri, A. (2007) Pulsed cooling of inhomogeneous thermoelectric materials. *J. Phys. D: Appl. Phys.*, **40** (14), 4376–4381.
19. Snyder, G.J., Fleurial, J.-P., Caillat, T., Yang, R., and Chen, G. (2002) Supercooling of Peltier cooler using a current pulse. *J. Appl. Phys.*, **92** (3), 1564–1569.
20. Landecker, K. and Findlay, A.W. (1961) Study of the fast transient behaviour of Peltier junctions. *Solid-State Electron.*, **3** (3–4), 239–260.
21. Green, A. and Cowles, L.E.J. (1960) Measurement of thermal diffusivity of semiconductors by ångström's method. *J. Sci. Instrum.*, **37** (9), 349–351.
22. Zhang, Y., Shakouri, A., and Zeng, G. (2004) High-power-density spot cooling using bulk thermoelectrics. *Appl. Phys. Lett.*, **85** (14), 2977–2979.
23. Wang, P., Bar-Cohen, A., Yang, B., Solbrekken, G.L., and Shakouri, A. (2006) Analytical modeling of silicon thermoelectric microcooler. *J. Appl. Phys.*, **100** (1), 014501.
24. Litvinovitch, V., Wang, P., and Bar-Cohen, A. (2008) Impact of integrated superlattice μTEC structures on hot spot remediation. Thermal and Thermomechanical Phenomena in Electronic Systems, 2008. ITHERM 2008. 11th Intersociety Conference on, pp. 1231–1241.
25. Manno, M., Wang, P., and Bar-Cohen, A. (2014) Pulsed thermoelectric cooling for improved suppression of a germanium hotspot. *IEEE Trans. Compon. Packag. Manuf. Technol.*, **4** (4), 602–611.
26. Sherman, B., Heikes, R.R., and Ure, R.W. Jr. (1960) Calculation of efficiency of thermoelectric devices. *J. Appl. Phys.*, **31** (1), 1–16.
27. Sherman, B., Heikes, R.R., and Ure, R.W. Jr. (1960) Calculation of efficiency of thermoelectric devices, in *Thermoelectric Materials and Devices*, Materials Technology Series, Chapter 15 (eds I.B. Cadoff and E. Miller), Reinhold Publishing Cooperation, New York, pp. 199–226.
28. Seifert, W., Ueltzen, M., and Müller, E. (2002) One-dimensional modelling of thermoelectric cooling. *Phys. Status Solidi A*, **1** (194), 277–290.
29. Snyder, G.J., Fleurial, J.-P., Caillat, T., Chen, G., and Yang, R.G. (2004) Current Pulses Momentarily Enhance Thermoelectric Cooling. Technical report, NPO-30553 NASA Tech Briefs.
30. Yang, R., Chen, G., Ravi Kumar, A., Snyder, G.J., and Fleurial, J.-P. (2005) Transient cooling of thermoelectric coolers and its applications for microdevices. *Energy Convers. Manage.*, **46** (9–10), 1407–1421.
31. Babin, V.P. and Iordanishvili, E.K. (1969) Enhancement of thermoelectric cooling in nonstationary operation. *Sov. Phys.-Tech. Phys.*, **14** (2), 293–298. Translated from Zhurnal Tekhnicheskoi Fiziki, **39** (2), 399–406, (1969).
32. Miner, A., Majumdar, A., and Ghoshal, U. (1999) Thermoelectromechanical refrigeration based on transient thermoelectric effects. *Appl. Phys. Lett.*, **75** (8), 1176–1178.

33. White, W.C. (1951) Some experiments with Peltier effect. *Electr. Eng.*, **70** (7), 589–591.
34. Stil'bans, L.S. and Fedorovich, N.A. (1958) The operation of refrigerating thermoelectric elements in nonstationary conditions. *Sov. Phys.-Tech. Phys.*, **3** (1), 460–463.
35. Stil'bans, L.S. and Fedorovich, N.A. (1958) O rabote ohlazhdajushhih termoelementov v nestacionarnom rezhume. *Zh. Tekh. Fiz.*, **28**, 489–492. (in Russian).
36. Gray, P.E. (1960) *The Dynamic Behavior of Thermoelectric Devices*, Technology Press of the Massachusetts Institute of Technology, Cambridge, MA.
37. Iordanishvili, E.K. and Babin, V.P. (1983) *Non-Stationary Processes in Thermoelectric and Thermomagnetic Energy Conversion Systems*, Nauka, Moskva. (in Russian).
38. Alfonso, N. and Milnes, A.G. (1960) Transient response and ripple effects in thermoelectric cooling cells. *Electr. Eng.*, **79** (6), 443–449.
39. Parrott, J.E. (1960) The interpretation of the stationary and transient behaviour of refrigerating thermocouples. *Solid-State Electron.*, **1** (2), 135–143.
40. Reich, A.D. and Madigan, J.R. (1961) Transient response of a thermocouple circuit under steady currents. *J. Appl. Phys.*, **32** (2), 294–301.
41. Parrott, J.E. (1962) The stationary and transient characteristics of refrigerating thermocouples. *Adv. Energy Convers.*, **2**, 141–152.
42. Idnurm, M. and Landecker, K. (1963) Experiments with Peltier junctions pulsed with high transient currents. *J. Appl. Phys.*, **34** (6), 1806–1810.
43. Horvay, J.B. (1963) Thermoelectric transients. *IEEE Trans. Appl. Ind.*, **82** (66), 111–115.
44. Horvay, J.B. (1963) Method of operating thermoelectric cooling unit.
45. Naer, V.A. (1965) Transient regimes of thermoelectric cooling and heating units. *J. Eng. Phys. Thermophys.*, **8** (4), 340–344.
46. Balko, B. and Berger, R.L. (1968) Measurement and computation of thermojunction response times in the submillisecond range. *Rev. Sci. Instrum.*, **39** (4), 498–503.
47. Grinberg, G.A. (1968) Transient operation of thermoelectric cooling devices. *Sov. Phys.-Tech. Phys.*, **13** (3), 308–312. Translated from Zhurnal Tekhnicheskoi Fiziki, **38** (3), 418–424, (1968).
48. Bywaters, R.P. and Blum, H.A. (1970) The transient behavior of cascade thermoelectric heat pumps. *Energy Convers.*, **10**, 193–200.
49. Field, R.L. (1971) A study of the fast transient behavior of pulsed thermoelectric coolers. PhD thesis. Institute of Technology - Southern Methodist University.
50. Anatychuk, L.I., Dimitrashchuk, V.T., Luste, O.Ya., and Tsyganyuk, Yu.S. (1972) Turbulent thermoelectric current in a nonstationary temperature field. *Sov. Phys. J.*, **15** (3), 325–330.
51. Iordanishvili, E.K. and Malkovich, B.E.Sh. (1972) Experimental study of transient thermoelectric cooling. *J. Eng. Phys. Thermophys.*, **23** (3), 1158–1163.
52. Ivanova, K.F., Kaganov, M.A., and Rivkin, A.S. (1977) Control of a nonstationary process of thermoelectric cooling by variation of the geometric shape of the branches of a thermoelement. *J. Eng. Phys.*, **32** (3), 295–298.
53. Hoyos, G.E., Rao, K.R., and Jaeger, D. (1977) Fast transient response of novel Peltier junctions. *Energy Convers.*, **17**, 45–54.
54. Hoyos, G.E. (1977) Fast transient response of novel Peltier junctions. PhD thesis. The University of Texas at Arlington.
55. Hoyos, G.E., Rao, K.R., and Jerger, D. (1977) Numerical analysis of transient behavior of thermoelectric coolers. *Energy Convers.*, **17** (1), 23–29.
56. Woodbridge, K. and Ertl, M.E. (1977) Pulse refrigeration in Bi_2Te_3. *Phys. Status Solidi A*, **44** (2), K123–K125.
57. Woodbridge, K. and Ertl, M.E. (1978) Pulsed Ettingshausen cooling in bismuth. *J. Phys. F: Met. Phys.*, **8** (9), 1941.
58. Field, R.L. and Blum, H.A. (1979) Fast transient behavior of thermoelectric coolers with high current pulse and

finite cold junction. *Energy Convers.*, **19**, 159–165.
59. Nagy, M.J. and Buist, R.J. (1996) Transient analysis of thermal junctions within a thermoelectric cooling assembly. 15th International Conference on Thermoelectrics, pp. 288–292.
60. Taylor, P.J., Jesser, W.A., Rosi, F.D., and Derzko, Z. (1997) A model for the nonsteady-state temperature behaviour of thermoelectric cooling semiconductor devices. *Semicond. Sci. Technol.*, **12** (4), 443.
61. Dudarev, Yu. and Maksimov, M. (1998) Mathematical modeling of the nonstationary operation of thermoelectric current sources. *Tech. Phys.*, **43**, 737–738, doi: 10.1134/1.1259064.
62. Miner, A., Majumdar, A., and Ghoshal, U. (1999) Thermo-electro-mechanical refrigeration based on transient thermoelectric effects. Thermoelectrics, 1999. 18th International Conference on, pp. 27–30.
63. Huang, B.J. and Duang, C.L. (2000) System dynamic model and temperature control of a thermoelectric cooler. *Int. J. Refrig.*, **23** (3), 197–207.
64. Naji, M., Alata, M., and Al-Nimr, M.A. (2003) Transient behaviour of a thermoelectric device. *Proc. Inst. Mech. Eng., Part A: J. Power Energy*, **217**, 615–621.
65. Thonhauser, T., Mahan, G.D., Zikatanov, L., and Roe, J. (2004) Improved supercooling in transient thermoelectrics. *Appl. Phys. Lett.*, **85** (15), 3247–3249.
66. Makino, Y. and Maruyama, T. (2006) Transient cooling of heat-generating materials with thermoelectric coolers. *Chem. Eng. Technol.*, **29** (6), 711–715.
67. Chakraborty, A. and Ng, K.C. (2006) Thermodynamic formulation of temperature-entropy diagram for the transient operation of a pulsed thermoelectric cooler. *Int. J. Heat Mass Transfer*, **49** (11–12), 1845–1850.
68. Ravich, Yu.I. and Gordienko, A.N. (2007) A method for calculating the transient time of a multi-stage thermoelectric cooler. *Semiconductors*, **41** (1), 110–114.
69. Ju, Y.S. (2008) Impact of interface resistance on pulsed thermoelectric cooling. *J. Heat Transfer*, **130**, 014502.
70. Mitrani, D., Salazar, J., Miguel, A.T., García, J., and Chávez, J.A. (2009) Transient distributed parameter electrical analogous model of TE devices. *Microelectron. J.*, **40** (9), 1406–1410. Quality in Electronic Design 2nd IEEE International Workshop on Advances in Sensors and Interfaces Thermal Investigations of ICs and Systems.
71. Cheng, C.-H., Huang, S.-Y., and Cheng, T.-C. (2010) A three-dimensional theoretical model for predicting transient thermal behavior of thermoelectric coolers. *Int. J. Heat Mass Transfer*, **53** (910), 2001–2011.
72. Mao, J.N., Chen, H.X., Jia, H., and Qian, X.L. (2012) The transient behavior of Peltier junctions pulsed with supercooling. *J. Appl. Phys.*, **112** (1), 014514 (9 pages).
73. Sullivan, O.A. (2012) Embedded thermoelectric devices for on-chip cooling and power generation. Master's thesis. Georgia Institute of Technology.
74. Meng, J.-H., Wang, X.-D., and Zhang, X.-X. (2013) Transient modeling and dynamic characteristics of thermoelectric cooler. *Appl. Energy*, **108**, 340–348.
75. De Aloysio, G. and de Monte, F. (2014) Thermal characterization of micro thermoelectric coolers: an analytical study. *J. Phys. Conf. Ser.*, **547** (1), 012007.
76. Ma, M. and Yu, J. (2014) A numerical study on the temperature overshoot characteristic of a realistic thermoelectric module under a current pulse operation. *Int. J. Heat Mass Transfer*, **72**, 234–241.
77. Mannella, G.A., Carrubba, V.L., and Brucato, V. (2014) Peltier cells as temperature control elements: experimental characterization and modeling. *Appl. Therm. Eng.*, **63** (1), 234–245.
78. Shen, L., Chen, H., Xiao, F., Yang, Y., and Wang, S. (2014) The step-change cooling performance of miniature thermoelectric module for pulse laser. *Energy Convers. Manage.*, **80**, 39–45.
79. Stremler, F.G. (1960) The initial transient behavior of a thermoelectric

generator. Master's thesis. Massachusetts Institute of Technology.
80. Stremler, F.G. and Gray, P.E. (1961) The initial transient behavior of a thermoelectric generator. *Trans. Am. Inst. Electr. Eng., Part I: Commun. Electron.*, **80** (4), 367–372.
81. Shaw, D.T. (1964) A non-linear analysis of the transient behavior of thermoelectric generators with temperature dependent physical parameters. PhD thesis. Purdue University.
82. Shaw, D.T.-K. (1966) The start-up transient behavior of thermoelectric generators with temperature-dependent physical parameters. *Nucl. Sci. Eng.*, **24**, 227–238.
83. Baranov, A.P., Manasyan, Yu.G., and Solov'ev, A.E. (1969) Digital-computer theoretical investigations of nonstationary processes in thermoelectric generators. *J. Eng. Phys. Thermophys.*, **17**, 974–979.
84. Cobble, M.H. (1982) Transient analysis of a thermoelectric generator. Proceedings of the 4th International Conference on Thermoelectric Energy Conversion, Arlington, TX, pp. 52–55.
85. El-Genk, M. and Seo, J.T. (1987) *Analysis of the Transient Behavior of Thermoelectric Generators*, Space Nuclear Power Systems 1986 CONF-860102, Orbit Book Company.
86. Anatychuk, L.I., Luste, O.J., and Kuz, R.V. (2011) Theoretical and experimental study of thermoelectric generators for vehicles. *J. Electron. Mater.*, **40** (5), 1326–1331.
87. Altstedde, M.K., Rinderknecht, F., and Friedrich, H. (2014) Integrating phase-change materials into automotive thermoelectric generators. *J. Electron. Mater.*, **43** (6), 2134–2140.
88. Schacht, R. (2002) Entwurf und Simulation von Makromodellen zur transienten Simulation von thermo-elektrischen Kopplungen in einem Netzwerksimulator. PhD thesis. Technischen Universität Berlin. (in German).
89. Alata, M., Al-Nimr, M.A., and Naji, M. (2003) Transient behavior of a thermoelectric device under the hyperbolic heat conduction model. *Int. J. Thermophys.*, **24** (6), 1753–1768.
90. Bechtold, T., Rudnyi, E.B., and Korvink, J.G. (2005) Dynamic electro-thermal simulation of microsystems - a review. *J. Micromech. Microeng.*, **15** (11), R17–R31.
91. Chen, M., Rosendahl, L.A., Bach, I., Condra, T., and Pedersen, J.K. (2006) Transient behavior study of thermoelectric generators through an electro-thermal model using SPICE. Thermoelectrics, 2006. ICT '06. 25th International Conference on, pp. 214–219.
92. McCarty, R., Hallinan, K.P., Sanders, B., and Somphone, T. (2006) Enhancing thermoelectric energy recovery via modulations of source temperature for cyclical heat loadings. *J. Heat Transfer*, **129** (6), 749–755.
93. McCarty, R., Monaghan, D., Hallinan, K., and Sanders, B. (2007) Experimental verification of thermal switch effectiveness in thermoelectric energy harvesting. *J. Thermophys. Heat Transfer*, **21** (3), 505–511.
94. Apostol, M. and Nedelcu, M. (2010) Pulsed thermoelectricity. *J. Appl. Phys.*, **108** (2), 023702 (7 pages).
95. Moser, A., Rendler, L., Kratschmer, M., and Woias, P. (2010) Transient model for thermoelectric generator systems harvesting from the natural ambient temperature cycle. Proceedings of PowerMEMS 2010, Leuven, Belgium, pp. 431–434.
96. Crane, D.T. (2011) An introduction to system-level, steady-state and transient modeling and optimization of high-power-density thermoelectric generator devices made of segmented thermoelectric elements. *J. Electron. Mater.*, **40** (5), 561–569.
97. Yan, D. (2011) Modeling and application of a thermoelectric generator. Master's thesis. University of Toronto.
98. Crane, D.T., Koripella, C.R., and Jovovic, V. (2012) Validating steady-state and transient modeling tools for high-power-density thermoelectric generators. *J. Electron. Mater.*, **41** (6), 1524–1534.
99. Montecucco, A., Buckle, J.R., and Knox, A.R. (2012) Solution to the 1-d unsteady heat conduction equation

with internal joule heat generation for thermoelectric devices. *Appl. Therm. Eng.*, **35**, 177–184.
100. Moser, A., Erd, M., Kostic, M., Cobry, K., Kroener, M., and Woias, P. (2012) Thermoelectric energy harvesting from transient ambient temperature gradients. *J. Electron. Mater.*, **41** (6), 1653–1661.
101. Yan, D., Dawson, F.P., Pugh, M.C., and El-Deib, A. (2012) Time dependent finite volume model of thermoelectric devices. Energy Conversion Congress and Exposition (ECCE), 2012 IEEE, pp. 105–112.
102. Nguyen, N.Q. and Pochiraju, K.V. (2013) Behavior of thermoelectric generators exposed to transient heat sources. *Appl. Therm. Eng.*, **51** (1–2), 1–9.
103. Yan, Y. and Malen, J.A. (2013) Periodic heating amplifies the efficiency of thermoelectric energy conversion. *Energy Environ. Sci.*, **6**, 1267–1273.
104. Yusop, A.Md., Mohamed, R., and Ayob, A. (2013) Model building of thermoelectric generator exposed to dynamic transient sources. *IOP Conf. Ser. Mater. Sci. Eng.*, **53** (1), 012015.
105. Felgner, F., Exel, L., Nesarajah, M., and Frey, G. (2014) Component-oriented modeling of thermoelectric devices for energy system design. *IEEE Trans. Ind. Electron.*, **61** (3), 1301–1310.
106. Fergus, J.W., Yerkes, K., and Yost, K. (2014) Numerical modeling of multi-material thermoelectric devices under static and cyclic thermal loading. *J. Electron. Mater.*, **43** (2), 393–403.
107. Meng, J.-H., Zhang, X.-X., and Wang, X.-D. (2014) Dynamic response characteristics of thermoelectric generator predicted by a three-dimensional heat-electricity coupled model. *J. Power Sources*, **245**, 262–269.
108. Montecucco, A. (2014) Efficiently maximising power generation from thermoelectric generators. PhD thesis. University of Glasgow.
109. Yan, D., Dawson, F.P., Pugh, M., and El-Deib, A.A. (2014) Time-dependent finite-volume model of thermoelectric devices. *IEEE Trans. Ind. Appl.*, **50** (1), 600–608.
110. Wu, G. and Yu, X. (2015) A comprehensive 3D finite element model of a thermoelectric module used in a power generator: a transient performance perspective. *J. Electron. Mater.*, t.b.a, 1–9.
111. Maglić, K.D. and Taylor, R.E. (1992) The apparatus for thermal diffusivity measurement by the laser pulse method, in *Compendium of Thermophysical Property Measurement Methods - Recommended Measurement Techniques and Practices*, vol. **2**, Chapter 10, Springer Science+Business Media, pp. 281–314.
112. Taylor, R. (1995) *CRC Handbook of Thermoelectrics - Measurement of Thermal Properties*, Chapter 16, CRC Press, pp. 165–180.
113. ASTM Standard E1461-01 (2001) Standard Test Method for Thermal Diffusivity by the Flash Method, ASTM International, doi: 10.1520/E1461, http://www.astm.org.
114. Rowe, D.M. and Shukla, V.S. (1981) The effect of phonon grain boundary scattering on the lattice thermal conductivity and thermoelectric conversion efficiency of heavily doped fine grained, hot pressed silicon germanium alloy. *J. Appl. Phys.*, **52** (12), 7421–7426.
115. Wood, C. and Zoltan, A. (1984) Simple high temperature thermal diffusivity apparatus. *Rev. Sci. Instrum.*, **55** (2), 235–237.
116. Vining, C.B., Zoltan, A., and Vandersande, J.W. (1989) Determination of the thermal diffusivity and specific heat using an exponential heat pulse, including heat-loss effects. *Int. J. Thermophys.*, **10** (1), 259–268.
117. Parker, W.J., Jenkins, R.J., Butler, C.P., and Abbott, G.L. (1961) Method of determining thermal diffusivity, heat capacity and thermal conductivity. *J. Appl. Phys.*, **32** (9), 1679.
118. Cape, J.A. and Lehman, G.W. (1963) Temperature and finite pulse time effects in the flash method for measuring thermal diffusivity. *J. Appl. Phys.*, **34** (7), 1909–1913.
119. Taylor, R.E. and Cape, J.A. (1964) Finite pulse time effects in the flash diffusivity

technique. *Appl. Phys. Lett.*, **5** (10), 212–213.

120. Watt, D.A. (1966) Theory of thermal diffusivity by pulse technique. *Br. J. Appl. Phys.*, **17** (2), 231.

121. Larson, K.B. and Koyama, K. (1967) Correction for finite-pulse time effects in very thin samples using the flash method of measuring thermal diffusivity. *J. Appl. Phys.*, **38** (2), 465–474.

122. Heckman, R.C. (1973) Finite pulse time and heat loss effects in pulse thermal diffusivity measurements. *J. Appl. Phys.*, **44** (4), 1455–1460.

123. Etori, K. (1975) Remarks on shifts of thermal diffusivity of a solid by finite width of heat pulse. *Jpn. J. Appl. Phys.*, **14** (9), 1345.

124. Azumi, T. and Takahashi, Y. (1981) Novel finite pulse width correction in flash thermal diffusivity measurement. *Rev. Sci. Instrum.*, **52** (9), 1411–1413.

125. Xue, J., Liu, X., Lian, Y., and Taylor, R. (1993) The effects of a finite pulse time in the flash thermal diffusivity method. *Int. J. Thermophys.*, **14** (1), 123–133.

126. Cowan, R.D. (1961) Proposed method of measuring thermal diffusivity at high temperatures. *J. Appl. Phys.*, **32** (7), 1363–1370.

127. Cowan, R.D. (1963) Pulse method of measuring thermal diffusivity at high temperatures. *J. Appl. Phys.*, **34** (4), 926–927.

128. Clark, L.M. III and Taylor, R.E. (1975) Radiation loss in the flash method for thermal diffusivity. *J. Appl. Phys.*, **46** (2), 714–719.

129. Balageas, D.L. (1982) Nouvelle méthode d'interprétation des thermogrammes pour la détermination de la diffusivité thermique par la méthode impulsionelle (méthode "flash"). *Rev. Phys. Appl.*, **17**, 227–237.

130. Srinivasan, N.S., Xiao, X.G., and Seetharaman, S. (1994) Radiation effects in high temperature thermal diffusion measurements using the laser flash method. *J. Appl. Phys.*, **75** (5), 2325–2331.

131. Josell, D., Warren, J., and Cezairliyan, A. (1995) Comment on "Analysis for determining thermal diffusivity from thermal pulse experiments". *J. Appl. Phys.*, **78** (11), 6867–6869.

132. Baba, T. and Ono, A. (2001) Improvement of the laser flash method to reduce uncertainty in thermal diffusivity measurements. *Meas. Sci. Technol.*, **12** (12), 2046.

133. Blumm, J. and Opfermann, J. (2002) Improvement of the mathematical modeling of flash measurements. *High Temp. High Press.*, **34** (5), 515–521.

134. Opfermann, J. and Blumm, J. (2006). Device for detecting thermal conductivity by means of optical pulses. US Patent 7,038,209 B2, filed Sept. 4, 2003, NETZSCH-Geraetebau GmbH, Selb, DE.

135. McKay, J.A. and Schriempf, J.T. (1976) Corrections for nonuniform surface heating errors in flash method thermal diffusivity measurements. *J. Appl. Phys.*, **47** (4), 1668–1671.

136. Ioffe, A.V. and Ioffe, A.F. (1952) Prostoy metod izmereniya teploprovodnosti. *Zh. Tekh. Fiz.*, **22**, 2005–2013. engl. A simple method for measuring the thermal conductivity.

137. Joffe, A.F. (1956) Heat transfer in semiconductors. *Can. J. Phys.*, **34** (12A), 1342–1355.

138. Ioffe, A.V. and Ioffe, A.F. (1958) Measurement of the thermal conductivity of semiconductors proposed by A.V. Ioffe and A. F. Ioffe. *Sov. Phys.-Tech. Phys.*, **3** (11), 2163–2168.

139. Kaganov, M.A. (1958) A theoretical analysis of the method of measuring thermal conductivity of semiconductors proposed by A. V. Ioffe and A. F. Ioffe. *Sov. Phys.-Tech. Phys.*, **3**, 2169–2172.

140. Swann, W.F.G. (1959) Theory of the A. F. Joffe method for rapid measurement of the thermal conductivity of solids. *J. Franklin Inst.*, **267** (5), 363–380.

141. Swann, W.F.G. (1959) Concerning thermal junction resistance in the A. F. Joffe method for measurement of thermal conductivity. *J. Franklin Inst.*, **268** (4), 294–296.

142. Stecker, K. and Teubner, M. (1967) Untersuchung verschiedener Methoden der Wärmeleitfähigkeitsmessung an Halbleitern. *Wiss. Z. Univ. Halle-Wittenberg*, **XVI** (1), 1–26. engl.

Investigations on different methods for the measurement of the thermal conductivity of semiconductors.

143. Zabrocki, K., Ziolkowski, P., Dasgupta, T., de Boor, J., and Müller, E. (2013) Simulations for the development of thermoelectric measurements. *J. Electron. Mater.*, **42** (7), 2402–2408.

144. Bernstein, F. and Doetsch, G. (1925) Probleme aus der Theorie der Wärmeleitung - I.Mitteilung. Eine neue Methode zur Integration partieller Differentialgleichungen. Der lineare Wärmeleiter mit verschwindender Anfangstemperatur. *Math. Z.*, **22** (1), 285–292. (in German).

145. Doetsch, G. (1925) Probleme aus der Theorie der Wärmeleitung - II. Mitteilung. Der lineare Wärmeleiter mit verschwindender Anfangstemperatur. Die allgemeinste Lösung und die Frage der Eindeutigkeit. *Math. Z.*, **22** (1), 292–306. (in German).

146. Doetsch, G. (1925) Über das Problem der Wärmeleitung, in *Jahresbericht der Deutschen Mathematiker-Vereinigung*, vol. **33**, B.G. Teubner, pp. 45–52. (in German).

147. Doetsch, G. (1926) Probleme aus der Theorie der Wärmeleitung - III. Mitteilung. Der lineare Wärmeleiter mit beliebiger Anfangstemperatur. Die zeitliche Fortsetzung des Wärmezustandes. *Math. Z.*, **25**, 608–626. (in German).

148. Bernstein, F. and Doetsch, G. (1927) Probleme aus der Theorie der Wärmeleitung - IV.Mitteilung. Die räumliche Fortsetzung des Temperaturablaufs. Bolometerproblem. *Math. Z.*, **26**, 89–98. (in German).

149. Doetsch, G. (1928) Probleme aus der Theorie der Wärmeleitung - V. Mitteilung. Explizite Lösung des Bolometerproblems. *Math. Z.*, **28**, 567–578. (in German).

150. Doetsch, G. (1950) *Handbuch der Laplace-Transformation*, Verlag Birkhäuser, Basel. (in German).

151. Doetsch, G. (1956) *Anleitung zum praktischen Gebrauch der Laplace-Transformation*, R. Oldenbourg, München. (in German).

152. Lowan, A.N. (1935) Heat conduction in a semi-infinite solid of two different materials. *Duke Math. J.*, **1** (1), 94–102.

153. Churchill, R.V. (1936) Temperature distribution in a slab of two layers. *Duke Math. J.*, **2** (2), 405–414.

154. Newcomb, T.P. (1959) Flow of heat in a composite solid. *Br. J. Appl. Phys.*, **10** (5), 204–206.

155. Giere, A.C. (1965) Transient heat flow in a composite slab - constant flux, zero flux boundary conditions. *Appl. Sci. Res., Sect. A*, **14** (1), 191–198.

156. Swan, G.W. and Dick, J.J. (1974) Analytic solution for transient heat flow in a shocked composite slab. *J. Appl. Phys.*, **45** (9), 3851–3855.

157. Al-Mujahid, A. and Zedan, M.F. (1990) Transient heat-conduction response of a composite plane wall. *Wärme - und Stoffübertragung*, **26** (1), 33–39.

158. de Monte, F. (2000) Transient heat conduction in one-dimensional composite slab. a 'natural' analytic approach. *Int. J. Heat Mass Transfer*, **43** (19), 3607–3619.

159. de Monte, F. (2002) An analytic approach to the unsteady heat conduction processes in one-dimensional composite media. *Int. J. Heat Mass Transfer*, **45** (6), 1333–1343.

160. Sun, Y. and Wichman, I.S. (2004) On transient heat conduction in a one-dimensional composite slab. *Int. J. Heat Mass Transfer*, **47** (6-7), 1555–1559.

161. Polyanin, A.D. (2002) *Handbook of Linear Partial Differential Equations for Engineers and Scientists*, Chapman & Hall/CRC Press, Boca Raton, FL.

162. Duffy, D.G. (2001) *Green's Function with Applications*, Chapman & Hall / CRC Press, Boca Raton, FL.

163. Economou, E.N. (2005) *Green's Functions in Quantum Physics*, Springer Series in Solid-State Sciences, 3rd edn, Springer-Verlag.

164. Datta, S. (2005) *Quantum Transport: Atom to Transistor*, Cambridge University Press, Cambridge, New York.

165. Datta, S. (1995) *Electronic Transport in Mesoscopic Systems*, Cambridge University Press, Cambridge, New York.

166. Nolting, W. (2009) *Fundamentals of Many-Body Physics*, Springer-Verlag, Berlin.
167. Cole, K.D., Beck, J.V., Haji-Sheikh, A., and Litkouhi, B. (2011) *Heat Conduction Using Green's Functions*, 2nd edn, CRC/Taylor and Francis, Boca Raton, FL.
168. Uchida, K., Takahashi, S., Harii, K., Ieda, J., Koshibae, W., Ando, K., Maekawa, S., and Saitoh, E. (2008) Observation of the spin Seebeck effect. *Nature*, **455**, 778.
169. Uchida, K., Adachi, H., Ota, T., Nakayama, H., Maekawa, S., and Saitoh, E. (2010) Observation of the longitudinal spin-Seebeck effect in magnetic insulators. *Appl. Phys. Lett.*, **97**, 172505.
170. Jiang, J.-W., Wang, J.-S., and Li, B. (2011) A nonequilibrium Green's function study of thermoelectric properties in single-walled carbon nanotubes. *J. Appl. Phys.*, **109**, 014326.
171. Lü, J.T. and Wang, J.-S. (2007) Coupled electron and phonon transport in one-dimensional atomic junctions. *Phys. Rev. B*, **76**, 165418.
172. Achilles, S., Seifert, W., and Trimper, S. (2011) Green's function approach in the classical theory of thermoelectricity. *Phys. Status Solidi B*, **248** (12), 2821–2833.
173. Snyder, G.J. and Ursell, T.S. (2003) Thermoelectric efficiency and compatibility. *Phys. Rev. Lett.*, **91** (14), 148301.
174. Snyder, G.J. (2006) Thermoelectric power generation: efficiency and compatibility, in *CRC Handbook of Thermoelectrics: Macro to Nano*, Chapter 9 (ed. D.M. Rowe), Taylor and Francis, Boca Raton, FL.
175. Seifert, W., Zabrocki, K., Snyder, G.J., and Müller, E. (2010) The compatibility approach in the classical theory of thermoelectricity seen from the perspective of variational calculus. *Phys. Status Solidi A*, **207** (3), 760–765.
176. Müller, E., Walczak, S., and Seifert, W. (2006) Optimization strategies for segmented Peltier coolers. *Phys. Status Solidi A*, **203** (8), 2128–2141.
177. Goupil, C., Seifert, W., Zabrocki, K., Müller, E., and Snyder, G.J. (2011) Thermodynamics of thermoelectric phenomena and applications. *Entropy*, **13** (8), 1481–1517.
178. Cannon, J.R. (1984) *The One–Dimensional Heat Equation*, Encyclopedia of Mathematics and its Applications, vol. 23, 1st edn, Addison-Wesley Publishing Company/Cambridge University Press.
179. Evans, L.C. (1998) *Partial Differential Equations*, American Mathematical Society.

5
Compatibility

Wolfgang Seifert†, G. Jeffrey Snyder, Eric S. Toberer, Volker Pluschke, Eckhard Müller, and Christophe Goupil

The compatibility approach introduced in 2002/2003 by Snyder and Ursell [1, 2] evolves from the analysis of the thermal and electric transport equations. Being a temperature-based method, the compatibility approach focuses on the local performance of a TE material and its optimization. In the successive works [3–5], this concept has been further developed on the basis of a one-dimensional, stationary, and unifying model with material grading for the thermoelectric generator (TEG) and cooler (TEC). This work builds on earlier investigations dating from the 1960s [6–14].

In this chapter, we analyze the potential and the limitations of the compatibility approach, and we discuss the new Thomson cooler concept.

5.1
Relative Current Density and Compatibility Factors

In TE systems, the transport behavior is often examined in one-dimensional models by assuming that the heat flux and the electrical current are parallel (or antiparallel).

The relative current density u is defined as the ratio of electrical current density $\mathbf{j} = j\mathbf{n}$ to Fourier heat flux $\mathbf{q}_\kappa = -\kappa \nabla T$ with respect to the flow direction $\mathbf{n}$,

$$u = \frac{j^2}{\mathbf{q}_\kappa \cdot \mathbf{j}} = \frac{j}{\mathbf{q}_\kappa \cdot \mathbf{n}} = \frac{-j}{\kappa \nabla T \cdot \mathbf{n}}. \tag{5.1}$$

In the general case of a TE device in nonstationary operation, u depends (as a local flow) on both "potentials" T and μ where $\mu(T)$ is given by the equation of state.[1] In today's available bulk TE materials, the coupling between μ and T is weak

---

1) Note that here and further in this chapter $\mu(T)$ means the chemical potential $\mu_c(T)$. The subscript is omitted for the sake of simplicity.

Continuum Theory and Modelling of Thermoelectric Elements, First Edition. Edited by Christophe Goupil.
© 2016 Wiley-VCH Verlag GmbH & Co. KGaA. Published 2016 by Wiley-VCH Verlag GmbH & Co. KGaA.

[15, 16].[2] Under this assumption, we characterize the stationary operation mode of a TE device by two equations for $u(T)$ and $\mu(T)$. This is particularly possible when fixed boundary temperatures are used as the natural boundary conditions of a temperature-based method. (Then, μ is also fixed at both ends of the device and the variation of μ is small but nonzero.) Under these preconditions, the differential equation for $u(T)$ derives naturally from the thermal energy balance[1].[3]

$$\frac{d}{dT}\left(\frac{1}{u}\right) = -T\frac{d\alpha}{dT} - u\rho\kappa \quad \text{or, alternatively,} \quad u'(T) = \tau u^2 + \rho\kappa u^3. \tag{5.2}$$

Furthermore, we assume that the error is negligible when integrating over T.[4] Using fixed boundary temperatures T_a and T_s, the electric current density (which is necessary for a full description of the TE system and often assumed to be a constant) is obtained by the scaling integral [1].

$$j = -\frac{1}{L}\int_{T_a}^{T_s} u\kappa \, dT. \tag{5.3}$$

This integral can easily be proved in a 1D approach using $dT = T'(x)dx = -j/(\kappa u)dx$. For the notation of the boundary temperatures, see [17] and Section 2.3.1 in this book.

The natural field of applications of the compatibility method includes

- temperature-dependent material or segmentation in "ideal" devices without parasitic losses with fixed temperatures at both ends of the device, and
- material optimization with respect to temperature; this can be transformed to one spatial dimension (x-axis) as long as there is a monotonic temperature profile $T(x)$.

The mathematical analysis shows that maximization of the global performance of a TE leg is based on integral formulations that can be optimized with respect to the relative current density u. This value of u where the TE element reaches the highest performance is called *compatibility factor s* and was introduced by Snyder and Ursell [1, 2] as a second characteristic for material optimization along with the maximization of zT. Currently, compatibility factors s have been proposed for such performance parameters arising from the local contribution to the thermoelectric material:

2) If we neglect heat transport due to phonons, the equation of state of the electron gas in a nondegenerate semiconductor is usually considered. Referring even to an ideal Fermi gas (with Fermi energy ϵ_F), the temperature dependence of the chemical potential is given by

$$\mu(T) = \epsilon_F\left[1 - \frac{\pi^2}{12}\left(\frac{k_B T}{\epsilon_F}\right)^2 - \mathcal{O}(T^4)\right].$$

Observe that the temperature-dependent effect is proportional to $(T/T_F)^2$ with a very high temperature equivalent $T_F = \epsilon_F/k_B$ (1 eV≙11600 K). So, the second term is of order 10^{-4}; for more information, see Section 1.5.5.
3) To show this, introduce $u(T)$ in the heat equation [Eq. (2.5)] and eliminate the spatial derivative using $\nabla(1/u) = d/dT(1/u)\nabla T$. Note that $\mathbf{j}$ and ∇T are vectors. A detailed derivation is given in Section 6.5.1.
4) When T and μ cannot be separated, we have $u(T, \mu)$ and any change in T leads to a change in μ.

- TEG: compatibility factor for maximal electric power (s_P or $s_{g,P}$) [4]: $s_P = \frac{z}{2\alpha}$

- TEG: compatibility factor for maximal η (s_g) [1]: $s_g = \frac{\sqrt{1+zT}-1}{\alpha T}$

- TEC: compatibility factor for maximal φ (s_c) [3]: $s_c = \frac{-\sqrt{1+zT}-1}{\alpha T}$

In this chapter, we derive and discuss these factors again.

The advantage of using the relative current density is that the multidimensional thermoelectric problem can be reduced (in a wider range of applications) to a one-dimensional heat flow problem formulated in $u(T)$ (or $u(x)$). This reduces numerical complexity, for example, in comparison to finite element calculations.

However, we wish to emphasize at this point that another equation of state[5], complex boundary conditions of real applications or deviations from the model of an ideal device, may lead to erroneous results if the compatibility method is applied outside its range of validity. Nevertheless, the compatibility method has proved to be a useful instrument for material optimization, as an important aspect of device optimization. It has been found that sufficient compatibility is – besides an increase in the figure of merit – essential for efficient operation of a thermoelectric device and that compatibility will facilitate rational material selection and the engineering of functionally graded materials (FGM).[6]

In the following, the authors give an overview on the fundamental results of the thermoelectric generator and cooler, including a discussion on compatibility from the perspective of variational calculus. A particular focus was on the role of ideal self-compatibility, that is, adjusting compatibility locally at any position along a thermoelectric leg to achieve maximum efficiency of a TEG and maximum COP of a TEC. Further, we reconsider maximum power output from a TEG in connection with power-related compatibility [4], and we discuss the new Thomson cooler concept [19]. Also, noncontinuously graded (i.e., segmented) elements are summarized under the topic of FGM since they lead to the same functional effect as continuously graded elements do. The importance of the compatibility approach has been made apparent first for a segmented TEG [1, 2, 20].

5.2
Compatibility and Segmented Thermogenerators

If the compatibility factor s of one part of the thermoelectric element is significantly different from the s of another part, there will be no suitable current in a TEG where both parts are operating close to maximum efficiency. This is the physical basis for thermoelectric compatibility and is most apparent for segmented generators. As this subsection strictly focuses on TEG, we refer to s_g as simply s.

---

5) We refer here to C.B. Vining's conclusions in Ref. [15] that in systems near an appropriate electronic phase transition, large zT values can be expected, which may lead to improved properties of the TE material. "Standard" equations of state with a low coupling between μ and T are discussed in Section 1.5.5.
6) An analysis of FGM problems by the Anatychuk group is given in Ref. [18].

To achieve high efficiency, segmented generators use large temperature differences to increase the Carnot efficiency $\eta_C = \Delta T/T_h$. Since the material thermoelectric properties (α, σ, κ) vary with temperature, it is not desirable or even possible (most have maximum operating temperature where they may melt or otherwise decompose) to use the same material throughout an entire, large temperature drop. Ideally, different materials can be combined such that a material with high efficiency at high temperature is *segmented* (Figure 5.1) with a different material with high efficiency at low temperature [21]. In this way, both the materials operate only in their most efficient temperature range.

If u could be constrained to be always equal to s, then the most efficient material to choose for a segment would be that with the highest thermoelectric figure of merit z. In this case, known as *infinite staging* [13] (or upper limit of efficiency [8]), the interface temperature between the segments would ideally be the temperature at which the z values of both materials cross. For example, according to Figure 5.3, the best, infinitely staged p-leg in the temperature range from 0 °C to 1000 °C would contain $(Bi, Sb)_2Te_3$, Zn_4Sb_3, TAGS, $CeFe_4Sb_{12}$, and Si Ge with interfaces of about 200 °C, 400 °C, 550 °C, and 700 °C.

Unfortunately, in a generator made of homogeneous temperature-dependent material, $u = s$ is generally not achieved exactly at more than one location, so a compromise value for u must be selected. If the compatibility factors s of the segmented materials differ substantially, all segments cannot be simultaneously operating efficiently, and the overall efficiency may actually decrease as compared to a single segment alone. Figure 5.2 shows graphically that a suitable average value for u can be found for the three materials $(Bi, Sb)_2Te_3$, Zn_4Sb_3, and $CeFe_4Sb_{12}$, which have compatibility factors within about a factor of 2. The reduced efficiency at this average u is not far from the maximum reduced efficiency. Si Ge, on the other hand, has a much lower value for s, such that if the u shown in Figure 5.2 is used, a large negative efficiency will result for the Si Ge segment and the overall efficiency will decrease. If a smaller value of u is used, so that positive efficiency will result from the Si Ge segment, the efficiency of the other segments will have deteriorated more than the efficiency increase from the Si Ge segment. Thus, despite

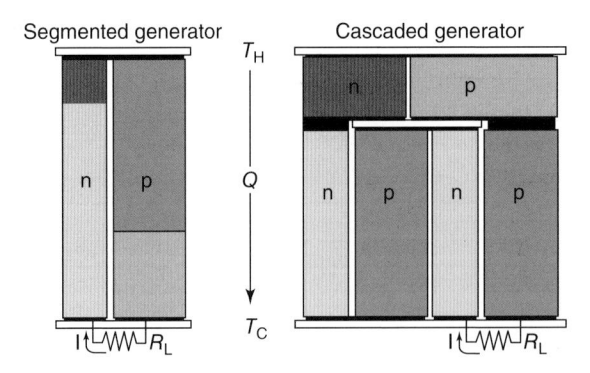

Figure 5.1 Schematic diagram comparing segmented and cascaded TEGs. The cascaded generator has a cascading ratio of 3. (Reprint of [20, Figure 9.6] with kind permission).

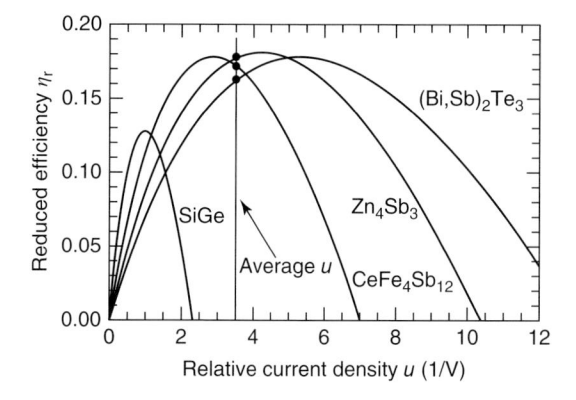

Figure 5.2 Comparison of reduced efficiency as relative current density, u, varies for different p-type thermoelectric materials. An average value for u can be found for $(Bi, Sb)_2Te_3$ (125 °C), Zn_4Sb_3 (300 °C), and $CeFe_4Sb_{12}$ (550 °C), which achieves a reduced efficiency (indicated with a dot) near the maximum efficiency. Si Ge (800 °C), on the other hand, has such a low compatibility factor s that using a u value appropriate for the other materials would result in a negative reduced efficiency for Si Ge. This makes Si Ge incompatible for segmentation with the other thermoelectric materials. (Reprint of [20, Figure 9.7] with kind permission).

having a reasonably high value of z for good efficiency, Si Ge cannot be segmented with the other materials shown in Figure 5.2 because of different compatibility factors.

As a rule of thumb, the compatibility factors of segmented materials should be within about a factor of 2. Within this range, a suitable average u can be used, which will allow an efficiency close to that determined by z. Outside this range of s, are materials that are incompatible, and the efficiency will be substantially less than that expected from z. The compatibility factor is, therefore, similar to z, a thermoelectric property essential for designing an efficient segmented thermoelectric device.

For segmented generators, materials with high z value that have similar compatibility factors, $s = s_g$, need to be selected. Other factors (not considered here) may also affect the selection such as: thermal and chemical stability, heat losses, coefficient of thermal expansion, processing requirements, availability, and cost [22]. The compatibility factor (Figure 5.3) can be used to explain why segmentation of $(AgSbTe_2)_{0.15}(GeTe)_{0.85}$ (TAGS) with Sn Te or Pb Te has produced little extra power [23], but using filled skutterudite would increase the efficiency from 10.5% to 13.6% [24]. Very high efficiency segmented generators to 1000 °C could be designed with skutterudites or Pb Te/TAGS as long as compatible, high temperature materials are used [24]. The compatible, high zT n-type material La_2Te_3 [25] would be ideal as long as a compatible p-type material is found. For the high temperature p-type element, a material with high zT value that is also compatible with Pb Te, TAGS, or skutterudite has been identified in the $Yb_{14}Mn_{1-x}Al_xSb_{11}$ based material [26] with a maximum zT of approximately 1.0. For a material with

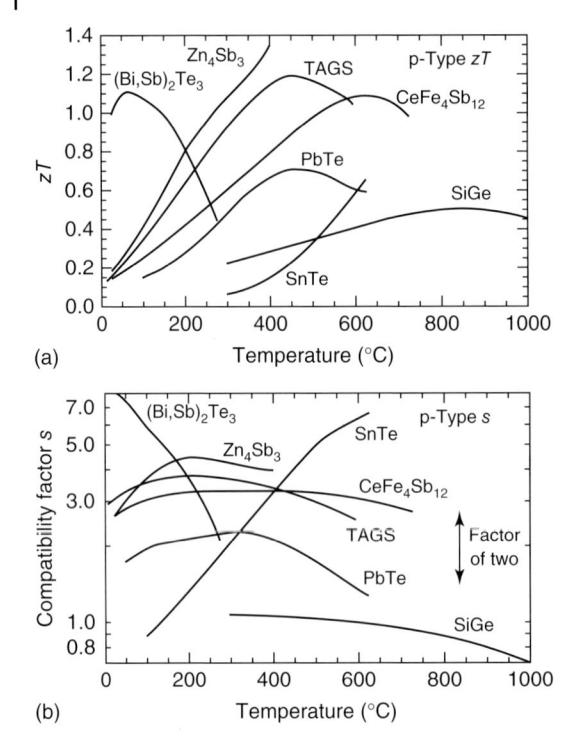

Figure 5.3 (a) Figure of merit (zT) and (b) compatibility factor (s) for p-type materials. (Reprint of [20, Figure 9.8(a) and (b)] with kind permission.).

a low zT value to be compatible with $PbTe$, TAGS, or skutterudite, it must have $s > 1.5\text{V}^{-1}$, ideally $s \approx 3\text{V}^{-1}$. Since $s \approx z/(2\alpha)$, see Eq. (5.13), the material with $zT \approx 0.5$ cannot be a high Seebeck coefficient band or polaron semiconductor. Materials with high z and s values exhibit thermoelectric properties typical of metals (Figure 5.4) with high α value. In a metal, the thermal conductivity is dominated by the electronic contribution by the Wiedemann–Franz law $\kappa = LT/\varrho$ where $L \approx 2.44 \cdot 10^{-8} \text{V}^2/\text{K}^2$. The compatibility factor $s \approx \alpha/(2\kappa\varrho) \approx \alpha/(2LT)$ would then be appropriate if $\alpha > 100\,\mu\text{V/K}$ at 1000 K, see [24]. For example, a candidate for such a refractory p-type metal is $Cu_4Mo_6Se_8$, see [27].

An overview of the theoretical efficiency of the best performing unicouples designed from segmenting the state-of-the-art TE materials is given in Ref. [28].

5.3
Reduced Efficiencies and Self-Compatible Performance

The analysis of segmented TEGs strikingly demonstrates that increasing the average zT does not always lead to an increase in the overall TE efficiency, and so, an understanding of the compatibility factor is needed to explain device performance.

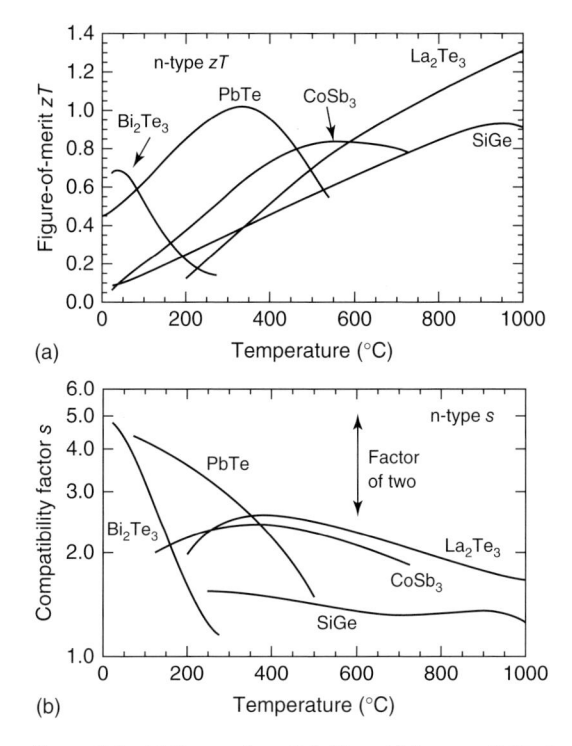

Figure 5.4 (a) Figure of merit (zT) and (b) compatibility factor (s) for n-type materials. (Reprint of [20, Figure 9.9(a) and (b)] with kind permission.).

This also applies for continuously graded materials. We call the consideration of compatibility in the same homogeneous material, which is operated in a temperature gradient and thus shows a spatial variation of $u(x)$ and $s(x)$ in a dissimilar manner, as an issue of self-compatibility. Next we will show that self-compatibility locally maximizes the device efficiency for a given zT and can be achieved by adjusting the relative ratio of the TE material parameters that make up zT. Optimally graded elements (or legs) are called self-compatible elements.

5.3.1
Performance Integrals for Efficiency and COP

The exact performance of a TE leg with temperature-dependent material properties can be computed straightforward at a local scale using reduced variables. The summation metric for a continuous system in one dimension is based on Zener [29] and similar derivations given in Refs [3, 6, 8, 19, 30]. The model is based on an ideal single-element device (prismatic TE element of length L and fixed boundary temperatures) without parasitic losses; for more information, see [17, 31] and in Section 2.3.1. Then, the device figure of merit is equal to the material's figure of merit, $z = \alpha^2/(\rho\kappa)$ (see also Section 2.3). In terms of the new local variables

u and η_r, the efficiency η of a TEG leg and the coefficient of performance (COP)[7] φ of a Peltier cooler leg are given by Seifert et al. [3, 6, 8]

$$\text{TEG } (T_s \leq T \leq T_a) \qquad \eta = 1 - \exp\left(-\int_{T_s}^{T_a} \frac{1}{T} \eta_r(u, T) \, dT\right) \qquad (5.4)$$

$$\text{TEC } (T_a \leq T \leq T_s) \qquad -\frac{1}{\varphi} = 1 - \exp\left(\int_{T_a}^{T_s} \frac{1}{T} \frac{1}{\varphi_r(u, T)} \, dT\right) \qquad (5.5)$$

with reduced efficiency η_r for TEG and reduced COP φ_r for TEC[8], respectively,

$$\eta_r(u, T) = \frac{u \frac{\alpha}{z}\left(1 - u \frac{\alpha}{z}\right)}{u \frac{\alpha}{z} + \frac{1}{zT}} = \frac{1 - \frac{\alpha}{z} u}{1 + \frac{1}{u \alpha T}} \qquad (5.6)$$

and

$$\varphi_r(u, T) = \frac{1 + \frac{1}{u \alpha T}}{1 - \frac{\alpha}{z} u} = \frac{1}{\eta_r(u, T)}. \qquad (5.7)$$

Note that the TEG and TEC cases are formally distinguished by the sign of $u(T)$ because of the reversed current direction in relation to the orientation of the temperature gradient. In fact, the reduced efficiencies η_r and φ_r introduced at a local level are formally reciprocal to each other [3]. This is simply the consequence of the formally reciprocal definition of efficiency η (TEG) and φ (TEC).

The value of u, which maximizes the reduced efficiency, is defined as compatibility factor s [1, 2]. The necessary condition for an extreme value

$$\frac{\partial \eta_r(u, T)}{\partial u} = 0 \quad \Longrightarrow \quad u_{\text{opt}} =: s$$

leads to different compatibility factors for maximum efficiency of a TEG (s_g) and maximum COP of a TEC (s_c)

$$s_g(T) = \frac{-1 + \sqrt{1 + zT}}{\alpha T} \quad \text{and} \quad s_c(T) = \frac{-1 - \sqrt{1 + zT}}{\alpha T}. \qquad (5.8)$$

These compatibility factors are, similar to the material's figure of merit z, temperature-dependent material properties. If we assume the feasibility to achieve complete self-compatibility (by infinite staging), we can apply $u_{\text{opt}} = s_g$ for TEG and $u_{\text{opt}} = s_c$ for TEC to the integrals (5.4), (5.5), so that they take their maximal value with the optimal "reduced efficiencies" [1, 3].

$$\eta_{r,\text{opt}} = \varphi_{r,\text{opt}} = \frac{\sqrt{1 + zT} - 1}{\sqrt{1 + zT} + 1}. \qquad (5.9)$$

---

7) We follow here Sherman's notation and use φ instead of COP in TEC formulae.
8) The minus sign in Eq. (24) in Ref. [3] is a misprint.

Then, the integrals for maximum performance of a self-compatible element or leg are given by

$$\ln(1 - \eta_{sc}) = \int_{T_a}^{T_s} \frac{\eta_{r,opt}}{T} \, dT = \int_{T_a}^{T_s} \frac{1}{T} \frac{\sqrt{1+zT}-1}{\sqrt{1+zT}+1} \, dT, \quad (5.10)$$

$$\ln\left(1 + \frac{1}{\varphi_{sc}}\right) = \int_{T_a}^{T_s} \frac{1}{T \varphi_{r,opt}} \, dT = \int_{T_a}^{T_s} \frac{1}{T} \frac{\sqrt{1+zT}+1}{\sqrt{1+zT}-1} \, dT, \quad (5.11)$$

with integrands varying monotonically with zT. Alternatively, these integrals read in Sherman's notation[9] [6]

$$\eta_{sc} = 1 - \exp\left(-\int_{T_s}^{T_a} \frac{1}{T} \frac{\sqrt{1+zT}-1}{\sqrt{1+zT}+1} \, dT\right) \quad (5.12a)$$

and

$$\varphi_{sc} = \left[\exp\left(\int_{T_a}^{T_s} \frac{1}{T} \frac{\sqrt{1+zT}+1}{\sqrt{1+zT}-1} \, dT\right) - 1\right]^{-1}. \quad (5.12b)$$

While the integrals (5.10) and (5.11) do not exhibit extremal properties with respect to the zT value [31], a constraint variational problem can be solved for the figure of merit $z(T)$ [32], see this chapter, Section 5.9.2.

Analytical expressions of the integrals (5.10), (5.11) can be found for CPM as well as for self-compatible elements ($u = s$ throughout) under particular assumptions, see Section 5.3.1 and the appendix of [33].

A comment on the influence of different constraints had been given by Ybarrondo [9]. He compared the results for device $Z(T) \, T = \text{const.}$ and $Z = \text{const.}$ and pointed out that $Z = \text{const.}$ can be an assumption that more closely approximates the actual temperature dependence of Z. By relating to the TE material itself, this appears convincing since the real temperature dependence of the transport parameters is moderate at moderate temperature ranges; therefore, it is usual and sufficient to estimate the device performance approximately with constant material properties, that is, with a constant figure of merit.

5.3.2
Local Efficiency Dependence on Current (TEG)

Whether in power generation or Peltier cooling mode, the reversible, useful thermoelectric effects compete with the irreversible Joule heating. Because the linear effects are directly proportional to the electric current while the irreversible Joule heating is proportional to the square of the current, there is necessarily an

---
9) Note that Sherman et al. derived the same maximum performance for infinitely staged devices being optimized with respect to the load ratio, which is in principle identical to the variation of j or u.

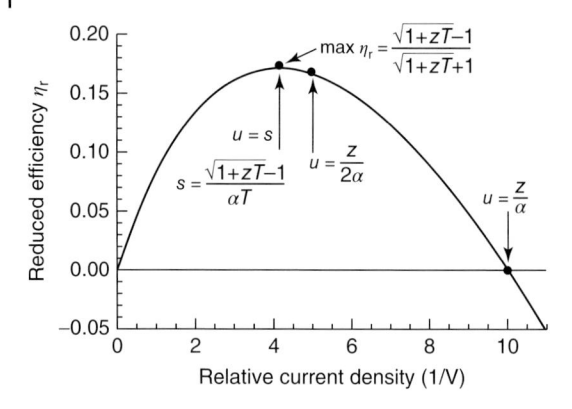

Figure 5.5 Variation of reduced efficiency [Eq. (5.6)] with relative current density u. The maximum efficiency is achieved at the compatibility factor, $u = s$. For the plot, $zT = 1, \alpha T = 0.1$ V similar to the values for (Bi, Sb)$_2$Te$_3$. (Reprint of [20, Figure 9.2] with kind permission).

optimum operating current to achieve the optimum efficiency.[10] The variation of reduced efficiency with u current [Figure 5.5, Eq. (5.6)] is analogous to the variation of the power output to the electric current: At zero u current, there is voltage produced but neither power nor efficiency. As u increases, the efficiency increases to a maximum value and then decreases through zero. Past this zero-efficiency crossing where $u = z/\alpha$, the Ohmic voltage drop is greater than the Seebeck voltage produced, and thus, the power output and efficiency are negative.

The value of u, which achieves the largest reduced efficiency for a TEG [Eq. (5.6)], is thermoelectric power generation compatibility factor s_g, see Eq. (5.8). For small zT value, this can be approximated by

$$s_g \approx \frac{z}{2\alpha}. \tag{5.13}$$

This approximation is reconsidered in the next section. The largest reduced efficiency $\eta_{r,opt} = \max \eta_r$ is given by Eq. (5.9).

In the general case (α, σ, κ spatially dependent), the thermoelectric figure of merit z is the material property that determines the maximum local efficiency. From Eq. (5.8), it is clear that the compatibility factor is, similar to z, a temperature-dependent material property. Thus, s cannot be changed with device geometry or the alteration of electric or thermal currents.

10) This can be shown exactly for constant material properties where we find the following optimal resp. maximum values (see also [5] and Ch. 2 of this book):

TEG: $j_{opt,\eta} = \frac{\kappa \Delta T}{L \alpha T_m}(-1+\sqrt{1+z T_m})$, $\eta_{max} = \frac{\Delta T}{T_h} \frac{\sqrt{1+z T_m}-1}{\sqrt{1+z T_m}+T_c/T_h}$

TEC: $j_{opt,\varphi} = \frac{\kappa \Delta T}{L \alpha T_m}(1+\sqrt{1+z T_m})$, $\frac{1}{\varphi_{max}} = \frac{\Delta T}{T_c} \frac{\sqrt{1+z T_m}+1}{\sqrt{1+z T_m}-T_h/T_c}$

with the mean temperature $T_m = (T_c+T_h)/2$ and $\Delta T = T_h - T_c$.

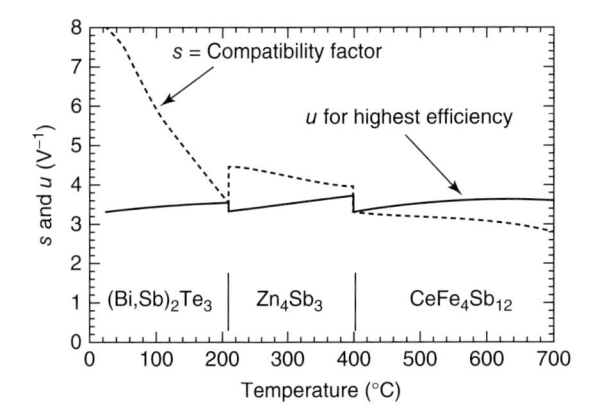

Figure 5.6 Variation of relative current density, u, with temperature for a typical thermoelectric generator. The total variation of u within a material and the change at the segment interfaces are less than 20%. The u shown is that which achieves the highest overall efficiency. For $(Bi, Sb)_2Te_3$ and Zn_4Sb_3, u is less than the compatibility factor s, while for the $CeFe_4Sb_{12}$, segment u is greater than s. (Reprint of [20, Figure 9.3] with kind permission).

If $u \neq s$, then the efficiency is less than the maximum efficiency of Eq. (5.9). Since $u = -j^2/(\kappa \nabla T \cdot \mathbf{j})$, there is some control over u from the applied current density j. However, once u is selected at one point, it cannot be adjusted in a thermoelectric leg (with given material) to follow the temperature variation of s (Figure 5.6), because the variation of u is fixed by the governing equation (5.2).

An alternative method is to use a functionally graded thermoelectric device where α, ϱ, and κ are adjusted by doping or otherwise changing the material as a function of position. The characteristics of optimally graded material are considered as follows.

Conveniently, the variation of u within a thermogenerator segment is typically small.[11] Since all segments in a TE element are electrically and thermally in series, the same current $I = A_c j$ and similar conduction heat $A_c \kappa \nabla T$ flows through each segment. When the electric current is close to zero ($j \approx 0$), the heat flow is very uniform, so u is nearly constant. For $j \neq 0$, the Fourier heat is only slightly modified by the change of the temperature gradient due to the Thomson and Joule heat, see Eq. (2.5). TEGs operating at peak efficiency typically have u that varies by less than 20% within all thermoelectric materials in the entire element, see Fig 5.6.

The actual reduced efficiency of a material depends not only on the maximum reduced efficiency [Eq. (5.9)] determined by z but also on how close u is to s (Figure 5.5). The actual reduced efficiency, Eq. (5.6), is always less than the maximum reduced efficiency [Eq. (5.9)], because u, as determined by the transformed heat equation (5.2), varies differently from the material property s, so they cannot

---

11) Thermoelectric coolers, on the other hand, are typically driven with much higher $|u|$ value; for more information, see, for example, [19] and Section 5.10.

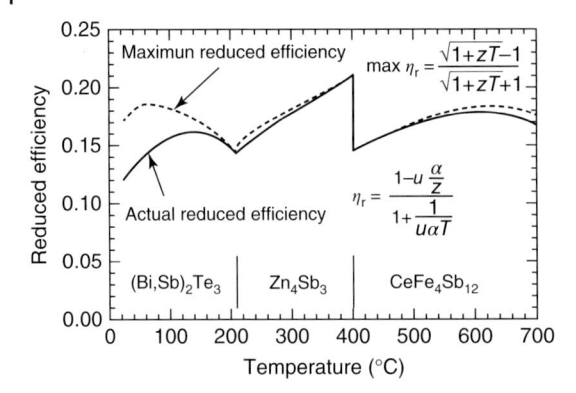

Figure 5.7 Local, reduced efficiency (using optimized u from Figure 5.6) compared to the maximum reduced efficiency [if $u = s$ for all temperatures, Eq. (5.9)]. The difference is most substantial in the regions where u is most distant from s (Figure 5.6). (Reprint of [20, Figure 9.4] with kind permission).

be equal at more than one or a few isolated points (one intersection point for CPM). The difference between the maximum and actual reduced efficiency is the largest for large differences between u and s. This can be seen graphically in Figures 5.6 and 5.7.

5.4
Power-Related Compatibility

We consider here the electrical power output from a TEG with fixed geometry (fixed length in a 1D model) and proceed in the next section with a discussion of the power generation under further global constraints.

We begin with some formal mathematical considerations using Eq. (5.6). Since the local reduced efficiency η_r is obtained when scaling the efficiency of a TEG (as the ratio of electrical power output to the thermal energy supplied) to local Carnot efficiency $\eta_C = dT/T$ [3], we need to differentiate only the numerator of η_r in Eq. (5.6) to obtain the TEG's power compatibility factor s_P [4]

$$\frac{d}{du}\left[u\frac{\alpha}{z}\left(1 - u\frac{\alpha}{z}\right)\right] = 0 \quad \Longrightarrow \quad u_{\text{opt},P} \equiv s_P = \frac{z}{2\alpha}. \tag{5.14}$$

Note that s_P also results when expanding the square root in s_g (see Eq. (5.8)) up to the first order. This means that optimization strategies on the basis of comparable constraints for power output and efficiency will lead (apart from minor changes) to similar results when z is small.

In addition to efficiency, an integral expression can also be found for power generation[12] of a TEG, $-P$. From the results published in Refs [4, 5], we find for the

---

12) Power output is defined here according to thermodynamic rules: quantities put into the system are positive, see also [5, 17].

5.4 Power-Related Compatibility

net electrical power output density p (see also Section 2.3.2)

$$p_{net}(j) = -P(j)/A_c = -\int_0^L \pi_{el}(x)dx = -\int_0^L \left(\frac{j^2}{\sigma(x)} + j\,\alpha(x)\,T'(x)\right) dx, \quad (5.15)$$

where π_{el} is the differential electrical power. As the temperature gradient is negative for TEG, the net electrical power output is derived from the difference between the thermoelectric voltage and the Ohmic voltage drop.

Equation (5.15) implies that the power output increases with increasing electrical current and increasing temperature gradient at large values of α; the local relationship between α and σ does not matter. Although the maximum power value does not depend on κ_{av} or z_{av} explicitly, κ should be large so that –when z is limited – σ should be as large as possible; in other words, the lowest possible total resistance of the device is crucial for maximum power output.

We remark that there are different mathematical representations of the differential electrical power, π_{el}. With $j = -u\,\kappa\,T'(x)$ we have, for example,

$$\pi_{el}(x) = \frac{j^2}{\sigma(x)} + j\,\alpha(x)\,T'(x) = j\,\alpha(x)\left(1 - u(x)\frac{\alpha(x)}{z(x)}\right)T'(x). \quad (5.16)$$

By applying the transcribed variant of the constitutive relation for the total heat flux q, $T' = (j\,\alpha T - q)/\kappa$, we further obtain π_{el} as a function not purely depending on zT (see also [34])

$$\pi_{el} = \frac{j^2}{\sigma}\left(1 + zT - \frac{z}{\alpha}\frac{q}{j}\right). \quad (5.17)$$

With $q = j\,(1/u + \alpha T)$, we finally have

$$\pi_{el}(x) = \frac{j^2}{\sigma(x)}\left(1 - \frac{z(x)}{\alpha(x)\,u(x)}\right). \quad (5.18)$$

Because $\pi_{out}(x) = -\pi_{el}$ depends on u and j, the target of an optimization procedure for the power output of a TEG of fixed length is to find not only the optimal u but also explicitly the optimal electrical current density j_{opt}. The reason is that the power output is a pure electrical quantity whereas the relative current density u (1D definition next) mirrors the ratio of thermal to electric quantities at a local level.

Power-related self-compatible elements (index sc) can be constructed when $u = s_p$, see [4, 5]. Then, we find from Eq. (5.18) for the differential electrical power output

$$\pi_{out}^{(sc)} = -\pi_{el}^{(sc)} = j_{opt}^2/\sigma(x), \quad (5.19)$$

with $\sigma(x)$ being part of the compatible material data set. Properties and specifics of power-related self-compatible elements together with the results of an example calculation have been discussed in Ref. [35].

Here we focus on the integral for the power output produced: By using the criterion $u = s_p = \alpha\,\sigma/(2\kappa)$ and the 1D definition of the relative current density,

$$u(x) = -\frac{j}{\kappa\,T'(x)},$$

we find for the optimal temperature profile

$$\frac{dT}{dx} = -\frac{2 j_{opt}}{\sigma(x)\, \alpha(x)}. \qquad (5.20)$$

Plugging j_{opt} from Eq. (5.20) into Eq. (5.19), we obtain for the power output for self-compatible material [5]

$$p_{net,max}^{(sc)} = \int_0^L \pi_{out}^{(sc)}(x)\, dx = \frac{1}{4}\int_0^L \left(\frac{dT}{dx}\right)^2 \alpha^2(x)\, \sigma(x)\, dx. \qquad (5.21)$$

The integral in (5.21) contains the power factor $\alpha^2 \sigma$ and additionally the optimal temperature profile. The latter is correlated to $\kappa(x)$ via the heat equation; the (optimal) thermal conductivity is needed for the construction of any self-compatible TE element. An example based on two given constant material properties $\alpha, \sigma =$ const. is discussed in Ref. [35].

Another interesting example is the consideration of two spatial material profiles $\alpha(x)$ and $\sigma(x)$. Proof is given in Ref. [35] that

$$p_{net,max}^{(sc)} \leq p_{net,max}^{cpm} \qquad (5.22)$$

for any combination of profiles $\alpha(x), \sigma(x) > 0$ when $u = s_p$. This equation expresses an essential feature of power-related self-compatible TE elements, which are constructed based on the profiles $\alpha(x)$ and $\sigma(x)$: for such elements, the maximum power output is almost optimal at CPM, whereby CPM is strictly related to individual averages of the spatial profiles. Thus, $p_{net,max}^{cpm}$ arises as an upper bound. This also holds for the particular case of a constant material's figure of merit or, equivalently, for the case of a constant power factor together with $\kappa =$ const. if $u = s_p$. In other words, the criterion $u = s_p$ is not sufficient for maximizing the electrical power output.

In Reference [34] (see Eq. (11)), the local ratio $\eta_{loc}(x) = \pi_{out}(x)/q(x)$ has been suggested as a local criterion for maximum power output. However, this criterion is "efficiency-orientated" because we find (see also [17, Secton 4.6.1.])

$$\eta_{loc}(x) = \frac{\pi_{out}(x)}{q(x)} = \frac{u\alpha(z - u\alpha)}{z(1 + u\alpha T)}\, T'(x), \qquad (5.23)$$

and with the integral kernel $K(u, T)$ defined in Ref. [33], see Eq. (9)

$$\int_0^L \eta_{loc}(x)\, dx = \int_{T_a}^{T_s} K(u, T)\, dT = \ln(1 - \eta). \qquad (5.24)$$

This result is essential when there are only marginal differences in the optimal grading for maximum P and maximum η. Then the "local efficiency" defined in Eq. (5.23) – which is nothing more than the TEG's reduced efficiency per temperature formulated in the x-domain – could be a suitable approximation also for power output optimization; $\eta_{loc}(x)$ is optimal if $u(x) = s_g(x) = \left[1-\sqrt{1+z(x)\, T(x)}\right]/[\alpha(x)\, T(x)]$.

A comparison with the results of power generation published by Zabrocki et al. [36, 37][13] clearly shows that one has to distinguish carefully between two issues:

13) Zabrocki et al. used a model assuming linear spatial material profiles in a 1D setup.

self-compatibility as a physical state in TE elements and self-compatibility as a local criterion for performance maximization; the latter does only exist for η and φ because variational formulations can be found for these two performance parameters based on the relative current density u alone, see next section. For maximum power, the relation between the electrical and thermal fluxes is not relevant, j and q must be as large as possible. Consequently, an optimal relative current u alone is not a relevant parameter for power output optimization. Thus, in the strict sense, there is no local selection criterion for the composition of a graded thermogenerator to achieve maximum power output; this also applies for the power factor $\alpha^2 \sigma$. The fact that $\alpha^2 \sigma$ has been considered for a long time as a seemingly local criterion for maximum power output proves the effectiveness of the Constant Properties Model (CPM); there is no doubt CPM can be considered in practice as a reliable orientation especially for weak and moderate gradients.

A novel method to optimizing the material grading in thermoelectric converters was published in 2012 by Gerstenmaier and Wachutka [38]. An adaptation of this approach to optimizing the spatial thermoelectric material profiles for maximum electrical power output of a graded thermogenerator with fixed length is presented in Ref. [39] (see also Section 5.4 in this book). However, the recent discussion on the maximum power from a TEG [40, 41] has shown that even such a well-established problem may still raise further interesting questions.

5.5
Optimal Material Grading for Maximum Power Output

To illustrate the principal difficulties in an attempt to find the optimal material gradients, which provide maximum electrical power from a segmented TE element of fixed length, a simple but instructive example shall be given, which makes evident that the formal constraint of fixed length is losing its physical sense in a problem where arbitrary material gradients are considered. The analytical finding that a local formulation of the produced power can be provided but cannot be traced back to material properties only, rather also containing the carrier density explicitly, has striking consequences to the choice of gradient of segmentation schemes, which lead to high performance. The explicit occurrence of j leads to a strong interaction between the segments in the process of compatibility but means in particular that theoretically no upper bound for the produced power can be formulated for an individual segment.

Imagine a TE element of fixed overall length L consisting of two CPM segments, a short one consisting of a usual TE material with the Seebeck coefficient α_{TE}, electrical and thermal conductivity σ_{TE}, κ_{TE}, respectively, and a long one consisting of an extreme metallic material with low Seebeck coefficient but extremely high electrical and thermal conductivity, such that its resistance $R_m \ll R_{TE}$ is much smaller than that of the TE segment whereas its thermal conductance $K_m \gg K_{TE}$ is much larger than that of the TE segment. Then, the maximum power production of the element is mainly linked to the production of the TE segment $P_{el,max} = (\alpha \Delta T)^2 / 4R$.

where also almost the full temperature difference is concentrated whereas the metallic segment does hardly contribute to any power production or losses. Thus, also, the optimal current is almost exclusively determined by the TE segment and will increase when the length of the TE segment is chosen shorter. Thus, with choosing its length fraction shorter and shorter (while keeping the relations of K and R between both segments), R_{TE} will fall, j_{opt} will rise, and thus the produced power will exceed any limit (even with mediocre TE properties of TE segment) while formally keeping the length of the overall element fixed.

This obvious case not only clarifies that theoretical gradient recipes of the TE material properties that maximize the electrical output power cannot be provided, based on the idealizing boundary conditions of fixed hot and cold side temperature, but also makes evident that this choice of boundary conditions does not lead to practically useful results if extreme properties of the element are assumed.

Thus, it is clear that the consideration of local optimization of power output makes sense only, if ever, under additional global constraints. Another simple example shall be added to show that even fixing the device resistance R is not sufficient as a constraint. This is a practically relevant constraint since a real system will always work in an electric circuit containing parasitic losses (contacts, bridges, lead wires), in addition to the "useful" load. R has to be kept large in relation to the parasitic losses. Anyway, also, the constraint of fixed R is not sufficient to avoid divergence of power production in the element since a high Seebeck coefficient can be chosen arbitrarily without violating a zT limit since a high κ value can be chosen as well.

Only if, additionally, a limit of the thermal conductance K of the element is introduced as a global constraint, power will be limited if we assume that also zT shall be finite. Then K and zT together limit the heat flux, zT limits the efficiency and, hence, the power. Also, the constraint of limited K is practically relevant as explained in Chapter 1 since also parasitic thermal resistances cannot be excluded from a real system.

The problem is thus reduced to the maximization of the thermoelectric voltage $\int \alpha(x) T'(x) dx$. For purely T-dependent α, $\int \alpha(T) dT$ has to be maximized, that is, a material with the highest Seebeck curve in the framework of the zT limit is best. With a temperature-dependent limit $z(T)T$, $\sigma(T)/\kappa(T)$ has to be minimized under the constraint of fixed R, K, leaving little flexibility for tuning of the most suitable temperature profile, just by the interplay of the differences between spatial and temperature averages. If the Seebeck coefficient depends explicitly on the location, then the coincidence of high Seebeck coefficient and large temperature gradient has to be achieved. Mainly, this means that high Seebeck coefficients and low thermal conductivities should coincide; the zT limit requires that at these positions, the electrical conductivity should be low; hence, in a first approximation, both the thermal and the electrical resistance should concentrate where the Seebeck is high, leaving other parts of the elements more or less inactive. Thus, although strongly varying profiles of temperature and properties could be involved, the maximum power, coming from the active part with a combination of properties as linked by the figure of merit, cannot be highly increased above what a CPM with a comparable ZT would produce. Secondary effects, local Joule and Thomson heat and the spatial distribution of these with their influence on the temperature profile, have to be taken into consideration to adjust the optimal properties' profiles for maximum power.

5.6
The Criterion "u = s" and Calculus of Variations

Now, the results in Section 5.3.1 shall be reformulated from the perspective of calculus of variations [33] where the Euler–Lagrange differential equation is a fundamental equation. It states that if I_F is defined by an integral of the form

$$I_F = \int_{x_1}^{x_2} F\left[x, y(x), y'(x)\right] dx, \quad x_1 \leq x \leq x_2 \tag{5.25}$$

then I_F has a stationary value if the Euler–Lagrange differential equation is satisfied:

$$\frac{\partial F}{\partial y} - \frac{d}{dx}\left(\frac{\partial F}{\partial y'}\right) = 0. \tag{5.26}$$

By taking this as a basis, we point out that a well-defined variational problem has been solved in the last section; the target is to search for an extreme value (maximum) of the integral

$$\int_{T_a}^{T_s} K(u(T), T)\, dT \rightarrow \text{Max.} \iff \delta \int_{T_a}^{T_s} K(u(T), T)\, dT = 0, \tag{5.27}$$

where we have the same kernel K for integrals of both generator and cooler [see Eqs. (5.4)–(5.6)]

$$K(u, T) = \frac{1}{T} \eta_r(u, T) = \frac{1}{T}\frac{1}{\varphi_r(u, T)}$$

$$= \frac{1}{T}\frac{u\frac{\alpha}{z}(1 - u\frac{\alpha}{z})}{u\frac{\alpha}{z} + \frac{1}{zT}} = \frac{u\alpha(z - u\alpha)}{z(1 + u\alpha T)}. \tag{5.28}$$

The symbol δ in Eq. (5.27) denotes the first variation of the functional (here an integral), which has to vanish to be an extreme value just as the first derivative has to vanish as a necessary condition for finding an extreme value of a function. As the integral kernel K does not depend on the derivative $u'(T)$, the Euler–Lagrange differential equation reduces to the necessary condition

$$\frac{\partial K(u, T)}{\partial u} = 0. \tag{5.29}$$

The roots of the equation $\partial K(u,T)/\partial u = 0$ are the compatibility factors (positive for TEG and negative for TEC) given in Eq. (5.8).

Note that the kernel $K(u, T)$ for both applications, TEG and TEC, can be written in various ways (see also [17, Section 4.6.1]). By using the thermoelectric potential

Φ, see Eq. (1.92), we also find

$$K(\Phi, T) = \frac{1}{T} \frac{1 - \frac{u\alpha}{z}}{1 + \frac{1}{uT\alpha}} = \frac{1}{T} \frac{1 - \frac{\alpha}{z(\Phi - T\alpha)}}{1 + \frac{z(\Phi - T\alpha)}{zT\alpha}} = \frac{\alpha}{\Phi} \left[1 - \frac{1}{zT\left(\frac{\Phi}{\alpha T} - 1\right)} \right]. \quad (5.30)$$

For this case, Snyder [1, 20] has shown that the global efficiency η is simply given by the relative change of the thermoelectric potential with temperature variation; an analogous relation can be found for the COP (see [3, 31])

$$\eta = 1 - \frac{\Phi(T_s)}{\Phi(T_a)} \quad \text{and} \quad \varphi = \left(\frac{\Phi(T_s)}{\Phi(T_a)} - 1 \right)^{-1}. \quad (5.31)$$

This result points to the importance of the TE potential as a *function of state*; for details, see also [42]. The optimal TE potential for both TEG and TEC is given in Section 1.10.2 in this book.

To sum up, we obtain an optimum kernel and, thus, maximum performance parameters if $u(T) - s(T)$ is strictly fulfilled for any temperature in the given interval. Thus, by using a variational formulation, maximization of the global parameters has been traced back to local optimization. The power output of a TEG, $-P$, is (unlike η and φ) a pure electrical quantity. Therefore, the integral also depends on the electric current density j; from Eq. (5.16), we find for the electrical power output per unit cross-sectional area A_c (see also [4, 5])

$$-\frac{P}{A_c} = \int_{T_s}^{T_a} K_P[j, u(T)] dT \quad \text{with} \quad K_P(j, u) = j \alpha \left(1 - u \frac{\alpha}{z}\right). \quad (5.32)$$

The target of the optimization procedure for the power output of a TEG of fixed length is to find not only the optimum u but also explicitly the optimum current density j_{opt}. Since the integral kernel depends not only on u but also on the electric current density, a direct application of Eq. (5.29) will not match here. However, proof can be given from Eq. (5.32) for the power compatibility factor s_P if we evaluate the derivation of the kernel K_P with respect to j:

$$\frac{\partial K_P}{\partial j} = \alpha \left(1 - u \frac{\alpha}{z}\right) - j \alpha \frac{\alpha}{z} \frac{\partial u}{\partial j} = \alpha - 2u \frac{\alpha^2}{z} = 0$$

$$\Rightarrow u_{opt,P} = \frac{z}{2\alpha} = s_P, \quad (5.33)$$

where $\partial u/\partial j = u/j$ has been used, which follows from the definition of u.

The spatial coordinate has been established as the independent coordinate for an empirical approach to 1D steady-state problems with graded materials. Strictly speaking, u and s are function of both temperature and position, but in many case one can consider two separate approaches. In relation to that, we have to take notice of the two sides of the compatibility approach. On the one hand, the derivation of $u(T)$ and $s(T)$ allows to do without explicit knowledge of the spatial dependence of the temperature or of the material properties, which is usually not known in a practical problem. On the other hand, $u(x)$ and $s(x)$ are more advantageous if we look for an optimum grading along a TE leg to achieve maximum performance from a fundamental and theoretical point of view. Doing so, Snyder's criterion has also been discussed in Ref. [5] as a local criterion in a sense that $u(x) = s(x)$ has to be simultaneously fulfilled, as the condition of "self-compatibility," for all infinitesimal segments of a TE element (within the interval $0 \leq x \leq L$). This is correct, because the variational problem can also be formulated with respect to the spatial coordinate if there is a steady and monotonic temperature profile $T(x)$. This applies for self-compatible elements (see also next subsection) and, similarly, if constant or temperature-dependent material properties are considered. By using the transformation $\int_{T_a}^{T_s} dT = \int_0^L T'(x) dx$ we find with a given $T'(x) \neq 0$

$$\delta \int_{T_a}^{T_s} K(u(T), T) dT = 0 \iff \delta \int_0^L K^\star(u(x), x) dx = 0 \tag{5.34}$$

with the kernel related to space

$$K^\star(u(x), x) = K[u(T(x)), T(x)] \, T'(x).$$

Generally, equivalent results are obtained from both formulations if $\alpha(x) = \alpha[T(x)]$, $\sigma(x) = \sigma[T(x)]$, and $\kappa(x) = \kappa[T(x)]$.

We expressively emphasize, however, that the temperature profile must be consistent if $u(x) = s(x)$ has to be fulfilled. In other words, Snyder's criterion can only be met exactly if the material gradients are locally compensating for the variation of the optimum temperature profile $T(x)$, see [5]. The appropriate compatibility factors can then be converted by means of the optimum $T(x)$

$$s(x) = s[T(x)] \implies s_{g,c}(x) = \frac{-1 \pm \sqrt{1 + z(x) \, T(x)}}{\alpha(x) \, T(x)}, \tag{5.35}$$

where the plus sign applies for s_g and the minus sign for s_c. From a mathematical point of view, the behavior of our variational problem due to a change in the independent coordinate is considered: The optimum u can be transferred from T to x if the (optimum) temperature profile is known.[14] The local implementation of the postulate $u = s$ results in the design of self-compatible elements, which are characterized by an optimal set of continuous profiles (temperature and materials). Fully self-compatible performance parameters η_{sc} and φ_{sc} can be calculated by the integrals

TEG: $\int_0^L K^\star(s_g(x), x) dx = \ln(1 - \eta_{sc}),$ \hfill (5.36)

TEC: $\int_0^L K^\star(s_c(x), x) dx = \ln\left(1 + \dfrac{1}{\varphi_{sc}}\right).$ \hfill (5.37)

With the abbreviation $\xi(x) = \dfrac{-1 + \sqrt{1 + z(x) \, T(x)}}{1 + \sqrt{1 + z(x) \, T(x)}}$ we find

$$K^\star(s_g(x), x) = \frac{T'(x)}{T(x)} \xi(x) \quad \text{resp.} \quad K^\star(s_c(x), x) = \frac{T'(x)}{T(x)} \frac{1}{\xi(x)}. \tag{5.38}$$

---

14) In principle, this transformation can also be performed within the framework of CPM (where $u \neq s$) based on analytical expressions for $u(T)$ and $T(x)$, respectively.

Clearly, the function $\xi(x) \equiv \eta_{r,opt}[T(x)]$ is nothing more than the optimal reduced efficiency transformed to the x-coordinate. Note that $z(x)$ contains all (optimal) material profiles.

5.7
Self-Compatibility and Optimum Material Grading

The previous considerations implicate that two strategies for optimizing the material can be established to achieve Snyder's criterion $u = s$ in a local sense, as the condition of self-compatibility, for all infinitesimal segments of a TE element (within the interval $0 \leq x \leq L$ or $T_a \leq T \leq T_s$):

A) optimization in the T-domain based on Eq. (5.2) and the criterion $u(T) = s(T)$ mainly used for temperature-dependent materials,
B) optimization in the x-domain based on Eq. (3.25) and the criterion $u(x) = s(x)$ for FGM containing an explicit dependence of the properties on x.

Consistent optimization results are obtained by both strategies if equivalent material profiles are used. However, in order to prevent the divergence of global performance in the optimization process, limits of the material properties have to conform to the process, be these upper limits of the Seebeck coefficient and the electrical conductivity and a lower limit of the thermal conductivity, or averages of the TE properties (or the figure of merit) or the power factor.

Results within optimization strategy A are discussed in the next section for the constraints $z =$const. and $zT =$const.

A central problem is that with $z = \alpha^2\sigma/\kappa$ and $u = s$, only two governing equations are available, constraining the local values of $\alpha, \sigma, \kappa, zT$ for the optimization strategies (see A and B) when referring to thermoelectricity from a phenomenological point of view. They are, in general, not sufficient for calculating all three optimal material profiles. In addition, the temperature profile $T(x)$ has to be calculated in a consistent manner when $u(x) = s(x)$ is used as a (thermodynamic) optimization criterion; this condition can be rewritten as a first-order differential equation for the optimum temperature profile based on the "coordinate" zT [5, 17]

$$\frac{dT}{dx} = \frac{j_o}{\sigma\alpha} f(zT) \quad \text{with} \quad f(zT) = \frac{zT}{1 \pm \sqrt{1+zT}}. \quad (5.39)$$

The positive sign applies to the TEC ($f = f^{(c)}$), but the negative one to the TEG ($f = f^{(g)}$).

An optimization strategy referring to item B has been proposed in Ref. [5]. It has become apparent that self-compatible elements can only be constructed based on an optimum combination of material profiles, whereas there is not only a single, uniquely defined set of $\alpha(x), \sigma(x)$, and $\kappa(x)$ but a manifold with two degrees of freedom. Only one profile out of the three properties can be calculated based on the optimization criterion found, while two material profiles can be specified arbitrarily to fix an optimum set. The remaining degrees of freedom

5.7 Self-Compatibility and Optimum Material Grading

can be used, for example, to involve interrelations between the thermoelectric properties due to the solid-state nature of the TE materials. This strategy has been tested in Ref. [5] with presumed constant gradients of α and σ having opposite directions, and the thermal conductivity κ has been optimized. From the first results [5], it can be concluded that there is only a little reserve for TEG efficiency improvement when using optimized material gradients, but much more potential for the performance improvement of a TEC. The reason is a stronger curved temperature profile in a TEC leading to a broader range of u. In any case, an ultimate performance limit has to be set, for example, by a $z_{\max}(T)$ curve or by a constraint $z = \text{const.}$ or $zT = \text{const.}$[15] As long as the same constraint $z = \text{const.}$ or $zT = \text{const.}$ is fulfilled, any combination of (optimal) material profiles obtains the same value of efficiency (TEG) or COP (TEC). In general, however, the choice of the predefined material profiles determines greatly the increase in performance from the self-compatibility effect.

Within optimization strategy B, an optimal $T(x)$ can be found from Eq. (5.39) together with one optimal set of spatial material profiles. Specifics are discussed here using the constraint $z(x) T(x) = k_o = \text{const.}$ where fully self-compatible performance values are given by the integrals (5.62) and (5.50), respectively.

From Section 1.10, we can conclude that the reduced "efficiencies" in a one-dimensional approach are given by

$$\eta_r(x) = \frac{1}{\varphi_r(x)} = \frac{\Phi'(x)/\Phi(x)}{T'(x)/T(x)} \quad \text{with} \quad \Phi(x) = \alpha(x)\, T(x) + \frac{1}{u(x)}. \tag{5.40}$$

Eq. (5.40) combines the compatibility concept with additional thermodynamic arguments. This opens up new opportunities for optimizing the material profiles as shown in Refs [17, 42]. We recall here the interrelation between the optimal temperature profile and the optimal Seebeck profile. By applying Eq. (5.40) for the case of optimal reduced efficiency where

$$\eta_{r0} \equiv \eta_{r,\text{opt}} = \varphi_{r,\text{opt}} = \frac{\sqrt{1+k_o}-1}{\sqrt{1+k_o}+1}, \tag{5.41}$$

we obtain with the optimal TE potential from Eq. (1.103)

$$\text{TEG:} \quad \eta_{r0} = \frac{T(x)\, \Phi'(x)}{T'(x)\, \Phi(x)} = \frac{\alpha'(x)\, T(x)}{T'(x)\, \alpha(x)} + 1, \quad \text{TEC:} \quad \frac{1}{\eta_{r0}} = \frac{\alpha'(x) T(x)}{T'(x)\alpha(x)} + 1,$$

leading to similar differential equations for both TEG and TEC

$$\text{TEG:} \quad \frac{\alpha'(x)}{\alpha(x)} = (\eta_{r0} - 1)\frac{T'(x)}{T(x)} \equiv k_g \frac{T'(x)}{T(x)} \tag{5.42a}$$

and

$$\text{TEC:} \quad \frac{\alpha'(x)}{\alpha(x)} = \left(\frac{1}{\eta_{r0}} - 1\right)\frac{T'(x)}{T(x)} \equiv k_c \frac{T'(x)}{T(x)}. \tag{5.42b}$$

---

15) For a direct comparison to a real material, the constant may be related to an average of the figure of merit z or zT.

A simple integration gives a correlation between the optimal temperature profile and the optimal spatial Seebeck coefficient for both TEG and TEC (again with $\alpha_{\text{ref}} = \alpha(T_{\text{ref}})$), which is equivalent to Eqs. (5.77) and (5.78) in Section 5.9.3:

$$\text{TEG:} \quad \alpha(x) = \alpha_{\text{ref}} \left[\frac{T(x)}{T_{\text{ref}}} \right]^{k_g}, \quad \text{TEC:} \quad \alpha(x) = \alpha_{\text{ref}} \left[\frac{T(x)}{T_{\text{ref}}} \right]^{k_c}. \quad (5.43)$$

Thus, only one material profile must be predefined when using the constraint $zT = \text{const}$. In fact, Eq. (5.43) represents a third optimization equation within variant B whereas the search for optimal spatial profiles is then based on only one given profile, for example, $\kappa(x)$ for maximum η (TEG) and, respectively, $\sigma(x)$ for maximum φ (TEC), or, as an alternative, constant conductivities (e. g., $\kappa = \text{const.}$ for maximum φ).

A novel method for optimizing inhomogeneous generators and coolers has been published by Gerstenmaier and Wachutka [38]. They used a generalized Euler–Lagrange multiplier method with the heat equation as constraint for any of the performance parameters to be optimized. This powerful method allows considering all relevant maximization/optimization problems (within both the T-domain and the x-domain) for functionally graded thermoelectric converters. We add here the following note:

In the mathematical formulation of constraint variational problems, one has to distinguish between constraints in the form of an integral (isoperimetric variational problems) and constraints that have to be fulfilled pointwise (holonomic and nonholonomic constraints, respectively). Constraints in an integral form in general occur if some average is fixed, for example, $\int_{T_c}^{T_h} Tz(T)\,dT = k_o\,(T_h - T_c)$. Then one has to use a constant Lagrange multiplier μ (as e. g., done in Ref. [32]). Gerstenmaier and Wachutka use the pointwise heat equation as a constraint; this is a nonholonomic constraint. In this case, they had to use a variable Lagrange multiplier $\mu(x)$.

While the proposed variational approach is an alternative to the compatibility approach, the optimization results in the T-domain coincide with those derived by Snyder and coworkers [1]. We emphasize here three points:

- Gerstenmaier's definition $u(T) = \frac{iS(T)}{v(T)}$ correlates with Snyder's definition of the relative current density, and the compatibility factors for TEG and TEC are reproduced (note that Seebeck is part of u due to the iS optimization!), see Eq. (21) in the original paper.[16]
- A solution to the optimization problem can only be found in the T-domain for the efficiency η of a TEG and the COP φ of a TEC. The equations for maximum η and maximum φ are reproduced by Gerstenmaier and Wachutka with another writing of the integrands, see Eqs. (22),(23) in the original paper.
- The optimized temperature dependencies of $iS(T)$, see Eqs. (27),(28) in Ref. [38], coincide with the optimal Seebeck profiles in the particular case of a constant electric current, see Eq. (17) in Ref. [19].

16) In Reference [38] a slightly different notation is used: $i\,S(T)$ represents the product of electric current i and Seebeck $S(T)$, and $v(T)$ denotes the Fourier heat flow.

Thus, the compatibility approach has been confirmed by an independent approach.

New results of optimizing the electrical power output of a TEG have been published in Ref. [39]. Based on a variational formulation found in the x-domain, this paper presents the application of the Gerstenmaier and Wachutka approach to a graded, prismatic thermogenerator element with fixed length. In evaluating an example calculation, a first quantitative comparison of the optimal Seebeck coefficients for maximum efficiency and maximum power has been shown. The results suggest that compromise gradients can be found, which nearly allow operation of both maximum efficiency and electrical output power simultaneously.

A detailed analysis of the Gerstenmaier/Wachutka approach reveals that maximization of the cooling power of a graded Peltier cooler of fixed length requires a separate consideration as an optimal control problem. By referring to an ideal single-element device, the mathematical analysis is presented in Ref. [43] together with the first results: a rapidly increasing Seebeck coefficient toward the cooler's hot side has also been found when optimizing the cooling power. Further results can be expected since the maximization of the cooling power is also a multidimensional optimization problem where the device geometry may also be considered as a design variable.

5.8
Thermodynamic Aspects of Compatibility

The compatibility approach focuses on an optimal adaptation of the thermal and electric impedance of the device. In particular, the ratio of electric and thermal fluxes is introduced as a function of temperature (or space), instead of considering both quantities separately.

As an alternative, we discuss here the ratio of dissipative to reversible heat fluxes. We begin with Clingman's "dimensionless heat flux c" introduced in Ref. [11] for device optimization, defined as the ratio of (total) heat flow $\dot{Q}$ to Peltier heat flow $I\alpha T$. At a local scale (with using relative current u and TE potential Φ), c translates to

$$c = \frac{q}{j\alpha T} = \frac{q_\kappa + q_\pi}{j\alpha T} = \frac{-\kappa \nabla T \cdot \mathbf{n} + j\alpha T}{j\alpha T} = \frac{j/u + j\alpha T}{j\alpha T} = 1 + \frac{1}{u\alpha T} = \frac{\Phi}{\alpha T}, \quad (5.44)$$

where $\mathbf{n}$ points to the flow direction of the electrical current density ($\mathbf{j} = j\mathbf{n}$), which is assumed to be parallel (or antiparallel) to the heat fluxes. For more details, see [42].

We introduce now r as the ratio of dissipative to reversible heat fluxes,

$$r = \frac{-\kappa \nabla T \cdot \mathbf{n}}{j\alpha T} = \frac{1}{u\alpha T}. \quad (5.45)$$

Hence, we have

$$c = 1 + r \quad \text{and} \quad \Phi = \alpha T(1 + r). \quad (5.46)$$

We also recall that $u\alpha = ez$, with e being the TE field factor introduced in Ref. [3] as the ratio of "Ohmic electric field" $V = IR$ to "Seebeck electric field" $V_\alpha = \alpha \Delta T$

$$e = -\frac{\mathbf{E}_\sigma \cdot \mathbf{n}}{\mathbf{E}_{TE} \cdot \mathbf{n}} = -\frac{\mathbf{j}/\sigma}{\alpha \nabla T \cdot \mathbf{n}} = \frac{u\kappa}{\sigma \alpha} = \frac{u\alpha}{z}, \quad (5.47)$$

with the (total) electric field $\mathbf{E} = \mathbf{E}_\sigma + \mathbf{E}_{TE} = \frac{1}{\sigma}\mathbf{j} + \alpha\nabla T$. Alternatively, e can also be interpreted as the ratio of the source terms of the heat fluxes (Joule heat density divided by thermoelectrically converted power)

$$e = -\frac{j^2/\sigma}{\alpha\nabla T \cdot \mathbf{j}} = \frac{u\alpha}{z}. \quad (5.48)$$

Note that u, e, and r are signed values when considering TEG and TEC; e ultimately is an *electrical loss ratio*, and r represents a *dissipation ratio*.

By using

$$u = \frac{1}{r\alpha T},$$

we finally find with Eq. (5.48)

$$\frac{1}{zT} = \frac{e}{u\alpha T} = re, \quad (5.49)$$

which gives a simple relation between the reciprocal figure of merit on the left side and the "degree of dissipation" r of the thermal fluxes and the "degree of dissipation" e of the source terms of the heat fluxes on the right side.

As examples, we list here some physical quantities written in terms of u and, alternatively and often more transparent, written as function of the ratio r:

$$\text{Fourier heat flux:} \quad \mathbf{q}_\kappa = -\kappa\nabla T = r\alpha T\mathbf{j} \quad (5.50)$$

$$\text{total heat flux:} \quad \mathbf{q} = \alpha T\mathbf{j} + \frac{\mathbf{j}}{u} = (1+r)\alpha T\mathbf{j} \quad (5.51)$$

$$\text{volumetric heat production:} \quad v_q = \mathbf{j}\cdot\nabla\left(T\alpha + \frac{1}{u}\right) = \mathbf{j}\cdot\nabla[(1+r)\alpha T] \quad (5.52)$$

$$\text{thermoelectric potential:} \quad \Phi = \alpha T + \frac{1}{u} = (1+r)\alpha T \quad (5.53)$$

$$\text{coupling of } \Phi \text{ and } T: \quad y(\Phi, T) = \frac{1}{u(\Phi, T)} = \Phi - \alpha T = r\alpha T. \quad (5.54)$$

Further examples can be found in Chapter 1, Sections 1.9.1 and 1.9.2.

The performance integrals for the efficiency η of a TEG and the COP of a cooler (Eqs. (5.4), (5.5)) can be formulated with one kernel $K(u, T)$ [17, 33]

$$K(u, T) = \frac{1}{T}\eta_r(u, T) = \frac{1}{T}\frac{1}{\varphi_r(u, T)} = \frac{1}{T}\frac{u\frac{\alpha}{z}(1 - u\frac{\alpha}{z})}{u\frac{\alpha}{z} + \frac{1}{zT}} = \frac{\alpha}{z}\frac{(z - u\alpha)}{(u^{-1} + \alpha T)}, \quad (5.55)$$

see Eq. (5.28). The target is to search for an extreme value (maximum) for the integral

$$\int_{T_a}^{T_s} K(u(T), T)\,dT \rightarrow \text{Max.} \quad (5.56)$$

As the integral kernel K does not depend on the derivative $u'(T)$, the Euler–Lagrange differential equation (see Section 5.6) reduces to the necessary condition

$$\frac{\partial K(u,T)}{\partial u} = 0. \tag{5.57}$$

The roots of the equation $\partial K(u,T)/\partial u = 0$ are the compatibility factors (positive for TEG and negative for TEC) given in Eq. (5.8). A similar result can be found for the electrical loss ratio e when we optimize the corresponding kernel

$$K(e,T) = \frac{z(1-e)}{(e^{-1}+zT)} = \frac{1}{T}\frac{1-e}{1+r}. \tag{5.58}$$

The compatibility factors (5.8) are then equivalent to

$$e_{opt} = \frac{-1 \pm \sqrt{1+zT}}{zT}, \tag{5.59}$$

and

$$r_{opt} = \frac{1}{zT\, e_{opt}} = \frac{1}{-1 \pm \sqrt{1+zT}}, \tag{5.60}$$

where the plus sign applies for TEG and the minus sign for TEC. To give an example also here, we mention the variants of the optimal TE potential (see Eq. (1.103))

$$\Phi_{opt}^{(g/c)} = \alpha T + \frac{1}{s^{(g/c)}} = \alpha T \left[\frac{\sqrt{1+zT}}{\sqrt{1+zT} \mp 1}\right] = (1+r_{opt})\,\alpha T. \tag{5.61}$$

Because of the relation $r = 1/(u\alpha T)$, any optimization procedure based on $u(T)$ corresponds to an optimization with respect to the dissipation ratio r (and is thus thermodynamically justified) as long as the product αT of the optimized material is a monotonic function of temperature. This is obviously true in the particular case when the constraint $z = $ const. is used and $\alpha_{opt}(T)$ is monotonic. This constraint has been used in a series of related publications [19, 39, 44].

5.9
Analytic Results for Self-Compatible TEG and TEC Elements

5.9.1
Performance of Self-Compatible TEG and TEC Elements

Let us consider the performance of TE devices in which the individual material properties are adjusted in suitable spatial profiles to maintain self-compatibility.

By applying $u = s$ locally, analytical expressions of the integrals (5.4) and (5.5) can be found for the following cases:

- Thermogenerator element (TEG) with $T_a > T_s$ (see also [9, 20])
  case A) $zT = k_o = $ const.

Fully self-compatible efficiency for constant reduced efficiency η_{r0} is given by

$$\eta \equiv \eta_{sc}^{(k_o)} = 1 - \left(\frac{T_s}{T_a}\right)^{\eta_{r0}}, \quad \eta_{r0} = \frac{\sqrt{1+k_o}-1}{\sqrt{1+k_o}+1}. \tag{5.62}$$

case B) $z =$ const.
Fully self-compatible efficiency (with optimal reduced efficiency $\eta_{r,opt} = \left(\sqrt{1+zT}-1\right)/\left(\sqrt{1+zT}+1\right)$, where $u = s_g$) is given by[17]

$$\eta = 1 - \left(\frac{1+M_s}{1+M_a}\right)^2 \exp\left[\frac{2(M_a - M_s)}{(1+M_a)(1+M_s)}\right] \tag{5.63}$$

with

$$M_s = \sqrt{1+zT_s} \quad \text{and} \quad M_a = \sqrt{1+zT_a}. \tag{5.64}$$

- Peltier cooler element (TEC) with $T_a < T_s$ (see also [9])
case A) $zT = k_o =$ const.
Fully self-compatible COP[18] for constant reduced efficiency η_{r0}

$$1 + \frac{1}{\varphi} = \left(\frac{T_s}{T_a}\right)^{1/\eta_{r0}} \quad \text{or} \quad \varphi \equiv \varphi_{sc}^{(k_o)} = \left[\left(\frac{T_s}{T_a}\right)^{1/\eta_{r0}} - 1\right]^{-1} \tag{5.65}$$

case B) $z =$ const.
Fully self-compatible COP (again with optimal reduced efficiency $\varphi_{r,opt} = \eta_{r,opt} = \frac{\sqrt{1+zT}-1}{\sqrt{1+zT}+1}$, where $u = s_c$) is given by[19]

$$1 + \frac{1}{\varphi} = \left(\frac{M_s - 1}{M_a - 1}\right)^2 \exp\left[\frac{2(M_s - M_a)}{(M_a - 1)(M_s - 1)}\right] \tag{5.66}$$

with M_s and M_a given above.

It need not be emphasized that k_o and z_o have to be specified in case A and case B, respectively, when a self-compatible element shall be constructed.

The TE material properties α, σ, and κ are in general temperature- and position-dependent quantities. Often the approach of decoupled dependencies is applied, that is, either temperature or spatial dependence of the material coefficients is assumed. An exact fulfillment of the aforementioned constraints requires that

$$z(T) = \alpha(T)^2 \sigma(T)/\kappa(T) = z_o = \text{const.}, \quad \text{or} \quad T\alpha(T)^2\sigma(T)/\kappa(T) = k_o = \text{const.}$$

which can be only approximately satisfied. (Similar relations hold for spatially dependent material.)

---

17) Another writing of formula (5.63) is
$\eta = 1 - \exp\left[2/(M_s + 1) - 2/(M_a + 1) + 2\ln(M_s + 1)/(M_a + 1)\right]$.
18) Eq. (4) in Ref. [9] is in error requiring the exponent e to be replaced by $1/e$.
19) Another writing of formula (5.66) is
$1/\varphi = -1 + \exp\left[2/(M_a - 1) - 2/(M_s - 1) + 2\ln(M_s - 1)/(M_a - 1)\right]$.

The usage of averaged material properties can be an alternative to meet the constraints; for an overview, see [17], Section 4.4.3. By using here the overbar for averages over temperature, for example, for the Seebeck coefficient

$$\bar{\alpha} = \frac{1}{T_h - T_c} \int_{T_c}^{T_h} \alpha(T) \, dT,$$

different averaging variants can be chosen to define a constant figure of merit:

i) locally averaged figure of merit: $z_{av} = \overline{\alpha^2 \sigma / \kappa}$,

ii) averaged z suggested by Ioffe and Borrego: $z_{av} = \dfrac{\overline{\alpha}^2}{\overline{\rho}\,\overline{\kappa}}$,

iii) z calculated from individual averages: $z_{av} = \dfrac{\overline{\alpha}^2 \overline{\sigma}}{\overline{\kappa}}$.

An equivalent averaging can be defined if spatial material profiles are used. Generally, the differences between the three averaging variants are small for weak gradients, but remarkable for strong gradients. An important argument to recommend variant i) as a uniform basis for a comparison of different TE materials is: if arbitrary transport parameters fulfill the condition $z(T) = \alpha(T)^2\, \sigma(T)/\kappa(T) = z_o =$ const., then we have $\bar{z} = z_o$ for variant i), but generally $\bar{z} \ne z_o$ for variants ii) and iii).

For investigations of FGM problems using spatial material profiles, especially for the construction of self-compatible elements, we recommend for case A the spatial average $[zT]_{av}$ and for case B the average z_{av} defined by

$$[zT]_{av} = \frac{1}{L}\int_0^L z(x)\, T(x)\, dx, \quad z_{av} = \frac{1}{L}\int_0^L z(x)\, dx.$$

However, when referring to CPM, then $z_{av}\, T_m$ with z_{av} calculated from averages of the individual spatial material profiles should be used. Doing so, the maximum performance values given in Eqs. (2.32) and (2.33) can immediately be considered as the reference for evaluating the self-compatibility effect.

5.9.2
Self-Compatible Elements and Optimal Figure of Merit

The question arises how the Eqs. (5.62), (5.65), both derived under the constraint $zT = k_o =$ const., can be useful to estimate an upper performance limit for self-compatible material with a nonconstant $z(T)$. A proof of the relations

$$\eta_{sc} < \eta_{sc}^{(k_o)} \quad \text{and} \quad \varphi_{sc} < \varphi_{sc}^{(k_o)} \tag{5.67}$$

is given in Ref. [31], if k_o is calculated as an average over the temperature of a *monotonically increasing function* $z(T)\, T$,

$$k_o = \frac{1}{T_s - T_a} \int_{T_a}^{T_s} z(T)\, T\, dT. \tag{5.68}$$

Note that this strict monotonicity is fulfilled in most cases when chemically homogeneous materials are used in practical applications. Then we get the following inequalities:

TEG: $\quad 1 - \exp\left(-\int_{T_s}^{T_a} \frac{1}{T} \frac{\sqrt{1+z(T)T}-1}{\sqrt{1+z(T)T}+1} dT\right) \leq 1 - \left(\frac{T_s}{T_a}\right)^{\frac{\sqrt{1+k_o}-1}{\sqrt{1+k_o}+1}}$, (5.69)

TEC: $\quad \left[\exp\left(\int_{T_a}^{T_s} \frac{1}{T} \frac{\sqrt{1+z(T)T}+1}{\sqrt{1+z(T)T}-1} dT\right) - 1\right]^{-1} \leq \left[\left(\frac{T_s}{T_a}\right)^{\frac{\sqrt{1+k_o}+1}{\sqrt{1+k_o}-1}} - 1\right]^{-1}$. (5.70)

Equality holds if $z(T)T = $ const. If $z(T)T$ is decreasing with T, however, the aforementioned inequalities do not hold in general.

If the restriction of a monotonically increasing product $z(T)T$ is dropped, then we can look for an optimal $z(T)T$ where $\eta_{sc} > \eta_{sc}^{(k_o)}$ and $\varphi_{sc} > \varphi_{sc}^{(k_o)}$ and η_{sc}, φ_{sc} will be maximal. Since the integrals cannot be optimized for arbitrary zT, we have considered in Ref. [32] a constraint optimization problem for the figure of merit including condition (5.68), that is, a fixed temperature average of the figure of merit zT was set as a limit. As a result, we obtain convex optimal functions $k(T) = z(T)T$, slightly decreasing with temperature, for both TEG and TEC. As is known, curves $k(T) = z(T)T$ decreasing with temperature does not often occur in practical applications. However, it has turned out that the optimal function $k(T)$ is almost a constant $k(T) = k_o$ for a TEC and close to this constant function for a TEG; for details, see [32]. This fact shows the practical relevance of the constraint $zT = k_o = $ const., which is naturally achieved in practice only approximately.

We still want to discuss the question whether it is possible to compare the fully self-compatible performance parameters of case A ($zT = k_o = $ const.) with case B ($z = $ const.) as given in the previous section for both TEG (η) and TEC (φ). As mentioned earlier, the inequalities (5.67) yield a comparison of both cases A and B provided that $z(T)T$ is monotonically increasing and the constant k_o from Eq. (5.68) is the same in both cases. The first condition is fulfilled: If $z = $ const. in case B then $z(T)T = zT$ is a monotonically increasing function. The second condition yields an assumption on the admissible z. We evaluate the integral in Eq. (5.68) for the case B: if $z = $ const., then

$$\frac{1}{T_s - T_a} \int_{T_a}^{T_s} z(T)T \, dT = \frac{z}{T_s - T_a} \int_{T_a}^{T_s} T \, dT = \frac{z}{T_s - T_a} \frac{T_s^2 - T_a^2}{2} = z \frac{T_s + T_a}{2}.$$
(5.71)

Let now the constant k_o in case A be fixed. Then, η and φ given by the Eqs. (5.63) and (5.66) fulfill the inequality (5.67) if the integral (5.71) is equal to that of k_o. This yields the condition

$$z = \frac{2 k_o}{T_s + T_a} \quad \text{resp.} \quad zT_m = k_o,$$
(5.72)

with mean temperature T_m. Then we obtain from Eq. (5.67) that the fully self-compatible performance parameters η and φ, respectively, in the case B are smaller than the corresponding quantities in case A. By writing down this relation in detail if $z = $ const. is given by Eq. (5.72), we obtain the following analogue to the inequalities (5.69) and (5.70)

$$\text{TEG: } 1 - \left(\frac{1+M_s}{1+M_a}\right)^2 \exp\left[\frac{2(M_a - M_s)}{(1+M_a)(1+M_s)}\right] \leq 1 - \left(\frac{T_s}{T_a}\right)^{\frac{\sqrt{1+k_o}-1}{\sqrt{1+k_o}+1}}, \tag{5.73}$$

$$\text{TEC: } \left[\left(\frac{M_s-1}{M_a-1}\right)^2 \exp\left(\frac{2(M_s-M_a)}{(M_a-1)(M_s-1)}\right) - 1\right]^{-1} \leq \left[\left(\frac{T_s}{T_a}\right)^{\frac{\sqrt{1+k_o}+1}{\sqrt{1+k_o}-1}} - 1\right]^{-1} \tag{5.74}$$

with M_s, M_a given by Eq. (5.64) and the constant k_o from case A.

We remark that if Eq. (5.72) is not fulfilled, then both cases cannot be compared. (This particularly applies for case A and a constant z calculated by averaging within CPM.) Then it is possible that for large $z > \frac{2k_o}{T_s+T_a}$, we may obtain better performance in case B. However, this comparison is not appropriate because we then compare both cases with different material properties. Indeed, the condition (5.72) means that case B arises from case A if the temperature T on the right-hand side of $z(T) = \frac{k_o}{T}$ is replaced by the mean temperature $T_m = (T_s + T_a)/2$.

Finally, a note on Bergman's theorem [45] is necessary. It states that the effective figure of merit Z_{eff} of a composite can never exceed the largest Z value of any of its components.

However, since our investigations are based on the average of the figure of merit, see Eq. (5.68), and not on its upper bound, this theorem is not affected or violated.

5.9.3
Optimal Seebeck Coefficients for Self-Compatible Material

The transformed heat equation for $u(T)$, Eq. (5.2), can be used to derive the temperature dependence of an optimal Seebeck coefficient. By using the material's figure of merit, $z = \alpha^2/(\varrho \kappa)$, Eq. (5.2) adopts the form

$$\frac{d}{dT}\left(-\frac{1}{u}\right) = T\frac{d\alpha}{dT} + \frac{\alpha^2}{z}u. \tag{5.75}$$

In a $u = s$ material, the relative current u is equal to the compatibility factor:

$$s_g = \frac{\sqrt{1+zT}-1}{\alpha T} \quad \text{for TEG, but} \quad s_c = -\frac{\sqrt{1+zT}+1}{\alpha T} \quad \text{for TEC.}$$

Then Eq. (5.75) gives

$$\frac{d}{dT}\left(\frac{\alpha T}{1 \pm \sqrt{1+zT}}\right) = T\frac{d\alpha}{dT} - \frac{\alpha}{z}\frac{1 \pm \sqrt{1+zT}}{T}, \tag{5.76}$$

where the plus sign applies for TEC, but the minus sign for TEG.

Eq. (5.76) can be solved analytically for both constraints $z = z_o = $ const. and $zT = k_o = $ const. We get the following expressions for the optimal Seebeck coefficient $\alpha(T)$:

case A) solution for $zT = k_o = $ const.

This solution has been discussed in Refs [17, 31]. We have found

$$\text{TEG:} \quad \alpha(T) = \alpha_{\text{ref}} \left(\frac{T}{T_{\text{ref}}}\right)^{\eta_{r0}-1} \quad \text{with} \quad \eta_{r0} = \frac{\sqrt{1+k_o}-1}{\sqrt{1+k_o}+1}, \tag{5.77}$$

$$\text{TEC:} \quad \alpha(T) = \alpha_{\text{ref}} \left(\frac{T}{T_{\text{ref}}}\right)^{1/\eta_{r0}-1}. \tag{5.78}$$

By expanding the square root in Eqs. (5.77), (5.78), we find the approximations for low zT: $\eta_{r,\text{opt}} - 1 \approx -4/(4+k_o)$, respectively, $1/\varphi_{r,\text{opt}} - 1 \approx 4/k_o$ giving

$$\text{TEG:} \quad \alpha(T) \approx \alpha_{\text{ref}} \left(\frac{T}{T_{\text{ref}}}\right)^{-4/(4+k_o)} \quad \text{with} \quad \alpha_{\text{ref}} = \alpha(T_{\text{ref}}); \tag{5.79}$$

$$\text{TEC:} \quad \alpha(T) \approx \alpha_{\text{ref}} \left(\frac{T}{T_{\text{ref}}}\right)^{4/k_o} \quad \text{with} \quad \alpha_{\text{ref}} = \alpha(T_{\text{ref}}). \tag{5.80}$$

case B) solution for $z = z_o = $ const.

From Eq. (5.76), we find for this particular case:

$$T \frac{d\ln\alpha}{dT} = \frac{\pm 1 \pm \frac{3}{2} z_o T + \sqrt{1+z_o T}}{(1+z_o T)(\sqrt{1+z_o T} - (\pm 1))} = \frac{1 + \frac{3}{2} z_o T \pm \sqrt{1+z_o T}}{(1+z_o T)(-1 \pm \sqrt{1+z_o T})}, \tag{5.81}$$

where the plus sign applies for TEC, but the minus sign for TEG.

A computer algebra software is a helpful tool to find the analytical solution for TEG and TEC (with a free constant α_o)

$$\alpha(T) = \alpha_o \frac{\sqrt{1+z_o T} \pm 1}{\sqrt{1+z_o T}} \exp\left(\frac{2}{1 \pm \sqrt{1+z_o T}}\right). \tag{5.82}$$

Note that the plus sign in Eq. (5.82) applies for TEG, but the minus sign for TEC. The solution for TEC has been discussed in Ref. [19].

Low $z_o T$ approximations can directly be derived from the differential equation (5.81):

$$\text{TEG:} \quad T \frac{d\ln\alpha}{dT} = -\frac{z_o T}{2} \Rightarrow \alpha(T) \approx \alpha_o \exp\left(-\frac{z_o T}{2}\right); \tag{5.83}$$

$$\text{TEC:} \quad T \frac{d\ln\alpha}{dT} = \frac{4}{z_o T} \Rightarrow \alpha(T) \approx \alpha_o \exp\left(-\frac{4}{z_o T}\right). \tag{5.84}$$

5.9.4
Temperature Profile for $u = s$ Material

The temperature profile for both $u = s$ cooler and generator can be evaluated from Snyder's criterion $u = s$. From the definition of the relative current density, we find

$$u(x) = -\frac{j}{\kappa T'(x)} \Rightarrow u(T) = -\frac{j}{\kappa} x'(T) \Rightarrow x'(T) = -\frac{\kappa}{j} u(T). \tag{5.85}$$

5.9 Analytic Results for Self-Compatible TEG and TEC Elements

In a $u = s$ material, the relative current u is equal to the compatibility factor. When $u = s_c$ for TEC and $u = s_g$ for TEG (and $j = j_{opt}$, $\alpha = \alpha_{opt}$), then we obtain the differential equations:

$$\text{TEC:} \quad x'(T) = \frac{\kappa \left(1 + \sqrt{1 + zT}\right)}{j_{opt} \, T \, \alpha_{opt}(T)}, \quad \text{TEG:} \quad x'(T) = \frac{\kappa \left(1 - \sqrt{1 + zT}\right)}{j_{opt} \, T \, \alpha_{opt}(T)}. \quad (5.86)$$

Optimal Seebeck profiles $\alpha_{opt}(T)$ can be derived from the governing equation for $u(T)$, see Eq. (5.75). Results for both constraints $z = z_o = \text{const.}$ and $zT = k_o = \text{const.}$ are listed in Section 5.9.3, see also [44].

case A) solutions for $zT = k_o = \text{const.}$ (see also [17, 31])

By inserting the optimal Seebeck profiles, see Eq. (5.77) and (5.78), and assuming a constant thermal conductivity κ_o, the differential equations for the inverse temperature profile read

$$\text{TEC:} \quad x'(T) = \frac{\kappa_o \left(1 + \sqrt{1 + k_o}\right)}{j_{opt}} \frac{1}{\alpha_o T^{k_c + 1}}, \quad \text{TEG:} \quad x'(T) = \frac{\kappa_o \left(1 - \sqrt{1 + k_o}\right)}{j_{opt}} \frac{1}{\alpha_o T^{k_g + 1}}, \quad (5.87)$$

with $k_g = \eta_{r0} - 1$ and $k_c = \frac{1}{\eta_{r0}} - 1$, respectively (see also Section 5.7). Both equations (5.87) can be solved analytically for $x(T)$ respectively $T(x)$.

By using $x'(T) = 1/T'(x)$, the results for the temperature profile are:

$$\text{TEC:} \quad T(x) = [\lambda_c k_c (x_o - x)]^{-1/k_c}, \quad \text{TEG:} \quad T(x) = [\lambda_g k_g (x_o - x)]^{-1/k_g} \quad (5.88)$$

with $\lambda_c = \frac{\alpha_o j_{opt}}{\kappa_o (1 + \sqrt{1 + k_o})}$, $\lambda_g = \frac{\alpha_o j_{opt}}{\kappa_o (1 - \sqrt{1 + k_o})}$.

The free constant x_o can be determined from a boundary condition, for example, at the cold side for TEC (index c):

$$T_c = T(x = 0) = [\lambda_c k_c x_o]^{-1/k_c} \implies x_o = (\lambda_c k_c T_a^{k_c})^{-1}. \quad (5.89)$$

case B) solution for $z = z_o = \text{const.}$

By inserting the optimal Seebeck profile (5.82) and assuming a constant thermal conductivity κ_o, the differential equations for the inverse temperature profile read

$$\text{TEC:} \quad x'(T) = \frac{\kappa_o}{\alpha_o j_{opt}} \frac{\sqrt{1 + z_o T} \left(1 + \sqrt{1 + z_o T}\right)}{T \left(-1 + \sqrt{1 + z_o T}\right)} \exp\left(\frac{-2}{1 - \sqrt{1 + z_o T}}\right) \quad (5.90)$$

$$\text{TEG:} \quad x'(T) = \frac{\kappa_o}{\alpha_o j_{opt}} \frac{\sqrt{1 + z_o T} \left(1 - \sqrt{1 + z_o T}\right)}{T \left(1 + \sqrt{1 + z_o T}\right)} \exp\left(\frac{-2}{1 + \sqrt{1 + z_o T}}\right). \quad (5.91)$$

By applying $T'(x) = 1/x'(T)$, both differential equations can be solved numerically for $T(x)$, but an analytical solution can also be calculated for $x(T)$.

We consider first the TEC, that is, Eq. (5.90). By using the abbreviation $k_o = \kappa_o/(\alpha_o j_{opt})$ and integrating from T_h to T with boundary condition $x(T_h) = L$, we obtain with the integration variable τ

$$x(T) = L + k_o \int_{T_h}^{T} \frac{\sqrt{1+z_o\tau}\,(1+\sqrt{1+z_o\tau})}{\tau(-1+\sqrt{1+z_o\tau})} \exp\left(\frac{-2}{1-\sqrt{1+z_o\tau}}\right) d\tau.$$

A substitution $v = \sqrt{1+z_o\tau}$ yields

$$x(T) = L + 2k_o \int_{v_h}^{v(T)} \frac{v^2}{(v-1)^2} \exp\left(\frac{2}{v-1}\right) dv \quad \text{with}$$

$$v_h = v(T_h) = \sqrt{1+z_o T_h},$$

and another substitution $w = 2/(v-1)$ followed by a partial integration then leads to

$$x(T) = L - k_o \int_{w_h}^{w(T)} \left(1 + \frac{4}{w} + \frac{4}{w^2}\right) e^w \, dw$$

$$= L - k_o \left[e^w - \frac{4}{w} e^w + 8\,\mathrm{Ei}(w)\right]_{w_h}^{w(T)}.$$

Here $\mathrm{Ei}(w) = \int_{-\infty}^{w} \frac{e^\xi}{\xi} d\xi$ denotes the exponential integral.[20] By resubstituting all variables, we finally obtain for a TEC

$$x(T) = L + \frac{\kappa_o}{\alpha_o j_{opt}} \left[(2\sqrt{1+z_o T} - 3)\exp\left(\frac{-2}{1-\sqrt{1+z_o T}}\right)\right.$$

$$- (2\sqrt{1+z_o T_h} - 3)\exp\left(\frac{-2}{1-\sqrt{1+z_o T_h}}\right)$$

$$\left. - 8\,\mathrm{Ei}\left(\frac{-2}{1-\sqrt{1+z_o T}}\right) + 8\,\mathrm{Ei}\left(\frac{-2}{1-\sqrt{1+z_o T_h}}\right)\right]. \quad (5.92)$$

By integrating Eq. (5.91) in the same way, we obtain the corresponding function $x(T)$ for a TEG,

$$x(T) = L - \frac{\kappa_o}{\alpha_o j_{opt}} \left[(2\sqrt{1+z_o T} + 3)\exp\left(\frac{-2}{1+\sqrt{1+z_o T}}\right)\right.$$

$$- (2\sqrt{1+z_o T_c} + 3)\exp\left(\frac{-2}{1+\sqrt{1+z_o T_c}}\right)$$

$$\left. + 8\,\mathrm{Ei}\left(\frac{-2}{1+\sqrt{1+z_o T}}\right) - 8\,\mathrm{Ei}\left(\frac{-2}{1+\sqrt{1+z_o T_c}}\right)\right]. \quad (5.93)$$

Since the aforementioned functions are monotone with respect to T [see e.g., Eqs. (5.90) and (5.91)] the Eqs in (5.92) and (5.93), respectively, may be inverted numerically in order to determine $T = T(x)$ and, especially, the temperature T_a at the heat absorbing side $x = 0$.

[20] This is not an elementary function, but implemented for example, in MATHEMATICA, see exponential integral function.

The analytical solution for $x(T)$ allows a direct comparison with the inverse temperature profile for constant material properties (CPM). By using the compatibility approach, $x(T)$ follows from the modified scaling integral, see Eq. (5.3),

$$x(T) = -\frac{1}{j}\int_{T_a}^{T} u\kappa \, dT, \qquad (5.94)$$

where $x(T_s) = L$. We remark that Eq. (5.94) can be applied not only for constant, but also for temperature dependent material properties.

For CPM, the inverse temperature profile can be derived by inverting the analytical $T(x)$:

$$T''(x) = -\frac{j^2}{\kappa\sigma} =: -c_o \;\rightarrow\; T(x) = -\frac{c_o}{2}x^2 + c_1 x + c_2,$$

where $c_1 = \frac{T_s - T_a}{L} + \frac{c_o}{2}L, c_2 = T_a$ if $T(0) = T_a$ and $T(L) = T_s$. Alternatively, the CPM differential equation

$$x''(T) = c_o \, (x'(T))^3 \qquad (5.95)$$

can be solved for $x(T)$.[21] By applying boundary conditions $x(T_a) = 0$ at the heat absorbing side and $x(T_s) = L$ at the heat sink side, the solution is given by

$$x(T) = \frac{T_s - T_a}{c_o L} + \frac{L}{2} \pm \frac{1}{c_o}\sqrt{\left(\frac{T_s - T_a}{L} + \frac{c_o}{2}L\right)^2 + 2c_o(T_a - T)}. \qquad (5.96)$$

Note that the plus sign in Eq. (5.96) applies for TEG, but the minus sign for TEC. Further note that Eq. (5.96) is only valid as long as $|T_a - T_s| \geq \frac{c_o}{2}L^2$; the TE heater with small $|T_a - T_s|$ and a maximum of $x(T)$ in the temperature range must be considered separately.

Figure 5.8 shows a TEC example calculation. Note that a convex $x(T)$ appears if $T(x)$ is concave (as for CPM indicating a zero Thomson coefficient) and vice versa. Interrelations between the curvature of the temperature profile and the Fourier heat divergence are discussed in the following section. Note that a similar figure for a TEG would show a smaller bowing (e.g., at the optimum current $j = j_{opt,\eta}^{cpm}$). So far, the authors have considered an increasing importance of the compatibility approach for TEC FGM design.

5.10
Thermoelectric Thomson Cooler

Peltier coolers are the most widely used solid-state cooling devices, enabling a wide range of applications from thermal management of optoelectronics and infrared detector arrays to some equipment in the semiconductor industry.

The traditional analysis of a Peltier cooler approximates the material properties as independent of temperature. Peltier coolers are characterized by the maximum

---

21) The differential equation $T''(x) = -c_o$ can be transformed with $x'(T) = 1/T'(x)$ and $x''(T) = -\frac{1}{(T'(x))^2}T''(x)\,x'(T) = -T''(x)/(T'(x))^3$.

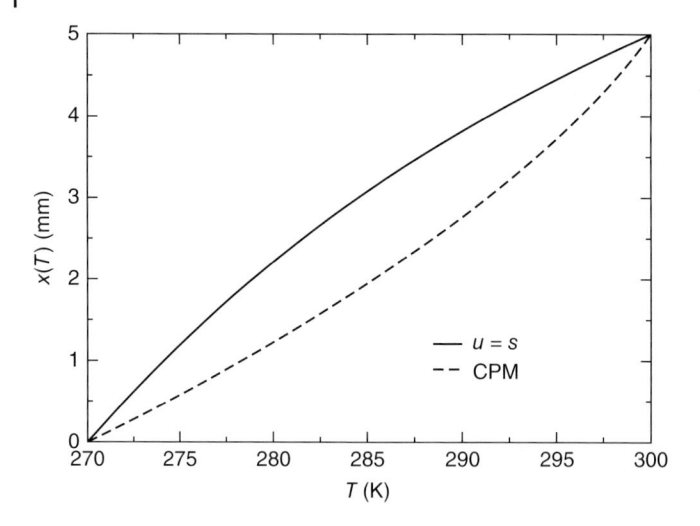

Figure 5.8 Inverse temperature profile $x(T)$: Comparison of CPM (dashed line) versus $u = s$ material (solid line) for a TEC ($T_c = 270$ K, $T_h = 300$ K, $L = 5$ mm, $z = 0.002$ K^{-1} at $j = j_{\text{opt, COP}}^{\text{cpm}} = 30.49$ A/cm^2. Note that $T(x)$ and $x(T)$ have opposite concavity: if $T(x)$ is convex, then $x(T)$ is concave and vice versa.

COP, which relates the rate of heat extraction at the cold end to the power consumption in the device, and by the temperature difference for maximum cooling $\Delta T_{\max}$ in the limit $\varphi \to 0$ [13, 46]:

$$\varphi_{\max} = \frac{T_c}{(T_h - T_c)} \frac{\sqrt{1 + ZT_m} - T_h/T_c}{1 + \sqrt{1 + ZT_m}}, \quad T_m = \frac{(T_h + T_c)}{2}; \quad (5.97)$$

$$\Delta T_{\max} = \frac{Z}{2} T_c^2. \quad (5.98)$$

From Eq. (5.98), the minimum cold side temperature $T_{c,\min}$ can be obtained when solving for $T_c = T_{c,\min}$ (with $\Delta T_{\max} = T_h - T_{c,\min}$). The result is

$$T_{c,\min} = \frac{1}{Z}(-1 + \sqrt{1 + 2ZT_h}). \quad (5.99)$$

A detailed listing of performance parameters for the TEC (and further applications) can also be found in Ref. [42] and in Section 2.3.2 in this book.

In the CPM, the device ZT is equal to the material zT, and the only way to increase $\Delta T_{\max}$ for a single stage is to increase zT. Even further cooling to lower temperatures can be achieved using multistage Peltier coolers [13, 46]. In principle, each stage can produce additional cooling to lower temperatures, regardless of the zT of the thermoelectric material in the stage. Since the current density can be chosen at each stage of the cooler independently, all stages can work close to a state of compatibility although made of homogeneous material, not much different from the CPM case, even more, since the temperature difference in each stage remains small (30 K and less).

The six-stage cooler of Marlow has a ΔT_{max} of 133 K, which translates to a device ZT at 300 K of 2.5 (even though $zT < 1$ thermoelectric materials are used similarly to the one-stage device) [47]. To achieve cryogenic cooling ($T_c \rightarrow 0$) within the CPM, zT must approach infinity [Eq. (5.99)]. For example, cooling to 10 K requires zT to be over 1000 if the hot side is 300 K.

In cascaded devices, the current density can be chosen at each stage of the cooler independently, all stages can work close to state of compatibility, although made of homogeneous material, not much different from the CPM case, even more, since the temperature difference in each stage remains small (30 K and less).

Along with continuing to improve the material's figure of merit zT, the concept of FGM [48–50] offers another way of gradual improvement of device performance. A central target of theoretical FGM studies is to elaborate recipes for optimum design of thermoelectric (TE) elements [5, 51], that is, to identify optimal (temperature dependent resp. spatial) profiles of the transport material properties. Such considerations can be seen as a complement to a more technological optimization of individual TE elements (concerning shape and size) integrated in Peltier cooling modules.

We emphasize at this point that the cooling limit (i.e., ΔT_{max}) is not required from classical thermodynamics but can be traced to problems of thermoelectric compatibility. In this context, it should be noted that maximum zT and $u = s$ are interrelated. In real materials, changing material composition also changes zT, so the effect of maximizing average zT is difficult to decouple from the effect of compatibility. As such, efforts that are focused on maximizing zT will generally fail to create a material with $u = s$ and may only marginally increase ΔT_{max}. Conversely, focusing on $u = s$ without consideration of zT could rapidly lead to unrealistic materials requirements. It should also be mentioned that thermal losses and complications of fabrication limit the performance of any cooler.

The concept of a "Thomson cooler", designed to maintain thermoelectric compatibility, has been introduced in 2012 by Snyder and Toberer [19] and shall be presented here again. In particular, we focus on discussing the cooling limit of such a device where the Thomson effect is a more significant thermoelectric effect than the Peltier effect. In a complementary paper [44], Snyder's concept (using the particular case of a $u = s$ cooler as a demonstration) has been compared with experimental data where the COP has been discussed for various temperature differences $\Delta T = T_h - T_c$. (In contrast, in Ref. [19], the focus is on maximum cooling.) A comparison of a conventional Peltier cooler made of $(Bi_{0.5}Sb_{0.5})_2Te_3$ material (as published in Ref. [52]) versus a $u = s$ cooler has shown that there is a similar self-compatibility effect for different ΔT. Furthermore, the extended concept of a $u = s$ cooler [53] allows for the discussion of further characteristics, in particular the temperature profile, as derived in Section 5.9.4, and the optimized spatial material profiles. The analysis presented in these papers opens a new strategy for solid-state cooling and creates new challenges for material optimization based on compatibility rather than only zT.

5.10.1
Cooling Performance

The overall COP of the entire device is related to the performance of its individual components. The summation metric for a continuous system in one dimension is attributed to Zener [29] as discussed in the appendix of [19]. At a local level, the reduced COP (relative to Carnot efficiency), φ_r, provides a measure of cooling performance (COP) at any point along the length of the device [31]. As discussed in Section 5.3.1, the relation between global φ and local φ_r is given by

$$\frac{1}{\varphi} = \exp\left(\int_{T_c}^{T_h} \frac{1}{T} \frac{1}{\varphi_r(T)} \, dT\right) - 1 \qquad (5.100)$$

with the reduced COP

$$\varphi_r(u, T) = \frac{u\frac{\alpha}{z} + \frac{1}{zT}}{u\frac{\alpha}{z}\left(1 - u\frac{\alpha}{z}\right)} = \frac{1 + \frac{1}{u\alpha T}}{1 - u\frac{\alpha}{z}} = \frac{u\alpha + 1/T}{u(\alpha - u\varrho\kappa)}. \qquad (5.101)$$

As φ approaches zero, no heat is extracted from the cold side, and the maximum temperature difference is reached.

In the typical CPM model used to analyze Peltier coolers, the Thomson effect is zero because α is constant along the leg ($d\alpha/dT = 0$). The exact performance (including the Thomson effect) of a thermoelectric device can be computed straightforwardly using the reduced variables of relative current density (u) and thermoelectric potential (Φ) for materials with arbitrary temperature dependence of $\alpha(T)$, $\varrho(T)$, and $\kappa(T)$ [1]. The relative current density $u = -j^2/(\kappa \nabla T \cdot j)$ is adjusted by tuning the electrical current density (j) relative to the resulting temperature gradient (∇T, which changes with differing j). The thermoelectric potential Φ is a state function, which simplifies Eq. (5.100) to

$$\varphi = \frac{\Phi(T_c)}{\Phi(T_h) - \Phi(T_c)} \quad \text{where} \quad \Phi = \alpha T + \frac{1}{u}, \qquad (5.102)$$

see also Eq. (5.31). Finally, the corresponding electrical current density can be found by means of the scaling integral [1]

$$j = \frac{1}{L}\int_{T_s}^{T_a} u(T)\,\kappa(T)\,dT = \frac{1}{L}\int_{T_s}^{T_a} \frac{\kappa(T)}{\alpha(T)\,r(T)\,T}\,dT, \qquad (5.103)$$

with $r(T)$ being the ratio of dissipative to reversible heat fluxes, see Section 5.8. By using this formalism, where the heat equation (2.5) simplifies to a one-dimensional first-order differential equation in $u(T)$, see Eq. (5.2), the reduced COP (φ_r) is simply defined for any point in the cooler and the overall COP (φ) can be calculated from this local value.

Figure 5.9a shows this relationship between u and φ_r. From Eq. (5.100), it can be shown that φ is largest when φ_r is maximized for every infinitesimal segment

5.10 Thermoelectric Thomson Cooler

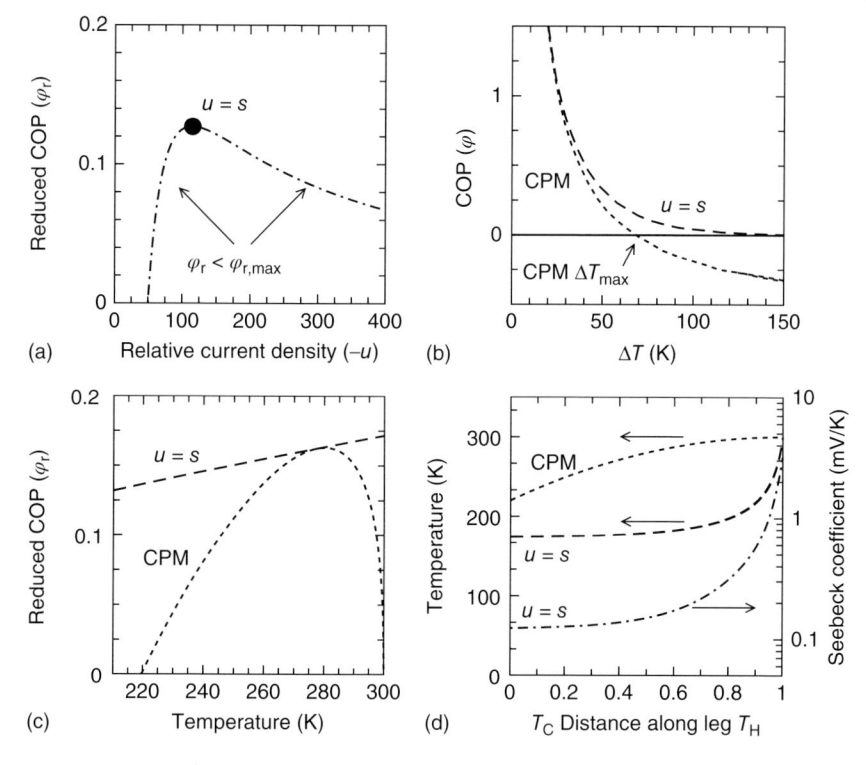

Figure 5.9 The traditional CPM Peltier cooler and a $u = s$ Thomson cooler are compared using the same constant $z = 1/300\text{K}$. (a) The local coefficient of performance φ_r is optimized only at the compatibility condition when the reduced current density (u) equals the local material compatibility factor (s). If $u \neq s$, the φ_r will be less than that predicted by the material zT. (b) The overall device φ of a CPM cooler crosses zero at a finite temperature, indicating that ΔT_{max} is reached, while the self-compatible cooler φ remains positive for all temperatures. (c) In CPM, $u = s$ holds at only one point along the leg, and φ_r is significantly compromised. In contrast, $\varphi_{r,max}$ is achieved at all temperatures when $u = s$. (d) The constant α CPM Peltier cooler has a distinctly different $T(x)$ temperature profile from the $u = s$ Thomson cooler where $\alpha(x)$ is strongly temperature dependent. (Arrows in this subfigure point to the respective ordinate of the curve.)

along the cooler. The maximum local φ_r, denoted as $\varphi_{r,max}$, occurs in a cooler when $u = s_c$:

$$s_c = \frac{-\sqrt{1+zT}-1}{\alpha T} \Rightarrow \varphi_{r,max} = \frac{\sqrt{1+zT}-1}{\sqrt{1+zT}+1}.$$

Then, $\varphi_{r,max}$ is an explicit function of the material zT and is independent of the individual material properties α, ϱ, and κ. This maximum allowable local efficiency provides a natural justification for the definition of zT as the material's figure of merit.

By evaluating φ [Eq. (5.100)] when $u = s_c$ with constant z (as also assumed in CPM), one obtains the maximum COP

$$\frac{1}{\varphi_{u=s}^{\max}} = \left(\frac{M_h - 1}{M_c - 1}\right)^2 \exp\left[\frac{2(M_h - M_c)}{(M_h - 1)(M_c - 1)}\right] - 1, \tag{5.104}$$

where $M_i = \sqrt{1 + zT_i}$ and $(T_i = T_h, T_c)$. For a $u = s$ cooler, inspection of Eq. (5.104), where $M_h > M_c > 1$, reveals that φ is always greater than zero. This difference can be observed in Figure 5.9b, with the φ of the $u = s$ cooler asymptotically approaching zero. Thus, in principle, if $u = s$ can be maintained, the $u = s$ cooler can achieve an arbitrarily low temperature at the cold side as long as the all of the materials have a finite zT. Because the material requirements to maintain $u = s$ become more difficult as the cooling temperature is reduced, the ultimate cooling will be finite, resulting in $T_c > 0$. Also, the cooling power itself will become very small for low T_c.

The remarkable difference in cooling performance can also be observed (Figure 5.9c) by comparing the φ_r of a traditional CPM Peltier cooler and that of a fully self-compatible TEC. Because compatibility is maintained at only one point in the CPM cooler, φ_r is less than $\varphi_{r,\max}$ for all but one point. The CPM cooler operates inefficiently (actually near $\varphi_r = 0$) at both the hot and the cold end. Once φ_r reaches below zero at low temperature, the thermoelectric device is no longer cooling the cold end and $\Delta T_{\max}$ is reached. Not only does $u = s$ lead to a greater $\Delta T_{\max}$, but also a fully self-compatible cooler achieves $\varphi_{r,\max}$ throughout the device, thus improving the overall cooling performance (φ) under a heat load.

To understand what is limiting the CPM cooler at $\Delta T_{\max}$, we finally derive the local reduced COP $\varphi_r^{\text{cpm}}(T)$. To obtain φ_r^{cpm}, we need u as a function of T. The solution to differential equation (5.2) for CPM is

$$\frac{1}{u(T)^2} = \frac{1}{u_h^2} + \frac{2\alpha^2}{z}(T_h - T), \tag{5.105}$$

where the value u_h of u at $T = T_h$ serves as an initial condition. This expression allows $u(T)$ to be determined for any CPM cooler, regardless of temperature drop ($\Delta T \leq \Delta T_{\max}$) and applied current density (j). The global maximum COP (φ) is obtained when the optimum u_h from Eq. (5.106) is employed,

$$\frac{1}{u_h} = \frac{-\alpha}{z} \frac{zT_c^2 - 2(T_h - T_c)}{T_h + T_c\sqrt{z\left(\frac{T_h + T_c}{2}\right) + 1}}. \tag{5.106}$$

Consideration of Eq. (5.106) reveals that the maximum T_c is obtained when $1/u_h$ approaches zero, that is, $|u|$ becomes infinite at T_h for the CPM cooler. In this limit, Eq. (5.106) can be simplified to give Eq. (5.98) with $Z = z$. Thus, a local approach to transport yields the classic CPM limit typically obtained through an evaluation of global transport behavior.

5.10.2
Thomson Cooler Phase Space

A detailed analysis of the optimized, functionally graded ($u = s$) cooler reveals that the dominant thermoelectric effect is the Thomson effect rather than the Peltier effect. The Peltier, Seebeck, and Thomson effects are all manifestations of the same thermoelectric property characterized by α. The Thomson coefficient ($\tau = T d\alpha/dT$) describes the Thomson heat absorbed or released when current flows in the direction of a temperature gradient. In a Peltier cooler, the production of heat is dominated by the Joule term ($\varrho j^2 > T j \nabla T \, d\alpha/dT$) in the heat divergence equation

$$\nabla \cdot \mathbf{q}_\kappa = \nabla \cdot (-\kappa \nabla T) = \varrho j^2 - \tau \, \mathbf{j} \cdot \nabla T. \tag{5.107}$$

In the CPM model where the Thomson effect does not occur ($T j \nabla T \, d\alpha/dT = 0$) this is obviously the case.

In the $u = s$ cooler, the Thomson effect dominates throughout the device $\varrho j^2 < T j \nabla T \, d\alpha/dT$. In terms of the relative current, this translates to $-\frac{a^2 u}{z} > T \, d\alpha/dT$ which with Eqs. (5.75) and (2.5) leads to a fundamental difference in the behavior of $u(T)$ and ∇T between the Peltier cooler and the Thomson cooler: In the Peltier cooler, $u(T)$ decreases while in the Thomson cooler $u(T)$ increases with temperature, and the two coolers have different concavity in the $T(x)$ profile (Figure 5.9d). This criterion can be particularly helpful to define the dominant cooling mechanism in experimental data. The constant relative current $u(T) = $ const. separates the Thomson-type and Peltier-type solutions; in this case, the Thomson effect just compensates Joule heat completely. This can be shown when we express the Fourier heat divergence in terms of reduced variables:

$$\nabla \cdot \mathbf{q}_\kappa = \mathbf{j} \cdot \nabla \left(\frac{1}{u}\right) = -\frac{1}{u^2} \frac{du}{dT} \mathbf{j} \cdot \nabla T = \frac{j^2}{\kappa u^3} u'(T) = \frac{j^2}{2\kappa u^4} \frac{d}{dT}(u^2). \tag{5.108}$$

In the last term, we have replaced du/dT by $\frac{1}{2u} d/dT(u^2)$. Thus, the sign of $\nabla \cdot \mathbf{q}_\kappa$ is determined by the sign of $d|u|/dT$, which is valid for both p- and n-type elements regardless of the sign of u. From the second last term of Eq. (5.108), we find

$$\nabla \cdot \mathbf{q}_\kappa \equiv \nabla \cdot (-\kappa \nabla T) = 0 \iff u'(T) = 0 \iff u = \text{const.}$$

For a more detailed discussion of the Thomson cooler phase space, we refer to Section V of [19].

For clarity, we suggest coolers that are predominately in the Thomson phase space ($\nabla \cdot \mathbf{q}_\kappa < 0$) but may not have $u = s$ be referred to as "Thomson coolers." Similarly, "Peltier coolers" should refer to coolers operating in the usual $\nabla \cdot \mathbf{q}_\kappa > 0$ Fourier heat divergence phase space where Joule heating dominates. These differences demonstrate that the study of Peltier cooling, particularly within the CPM framework, does not lend itself to finding solutions where the Thomson effect dominates. So, it is not surprising that the Thomson cooling side has not been explicitly examined before mathematically.

A Thomson cooler has two key advantages over state-of-the-art Peltier coolers: (i) For a given material zT, the performance (φ) of the cooler is optimized. (ii) In an ideal Thomson cooler without losses, the temperature minimum is not limited by zT explicitly as it is in a traditional Peltier cooler. This in principle leads to arbitrarily low cooling even for low zT, but in practice, the $u = s$ requirement of a Thomson cooler has stringent material requirements that become more demanding for small zT.

5.10.3
Performance Limits

The Thomson cooler requires elements with large Thomson coefficient ($\tau = T\,d\alpha/dT$) and therefore rapidly changing $\alpha(T)$ from the hot to the cold end. The optimal Seebeck coefficient $\alpha(T)$ for a $u = s$ cooler with constant z (as in a CPM cooler) has already been given in Section 5.9.3

$$\alpha(T) = \alpha_0 \frac{\sqrt{1+zT}-1}{\sqrt{1+zT}} \exp\left(\frac{-2}{\sqrt{1+zT}-1}\right). \tag{5.109}$$

For substantial cooling, $\alpha(T)$ should change by orders of magnitude, see Figure 5.9d. The greater the ratio of α_h/α_c, the greater the difference between T_h and T_c can be. However, there is a realistic range of Seebeck coefficients due to solid-state physics constraints. This will make sure that the ultimate cooling of a Thomson cooler will be finite.

We cite here Snyder's limiting model from Ref. [19], Section IV: Large values of α are found in lightly doped semiconductors and insulators with large band gaps (E_g) that effectively have only one carrier type, thereby preventing compensated thermopower from two oppositely charged conducting species. Using the relationship between peak α and E_g of Goldsmid and Sharp [54] allows an estimate for the highest $\alpha(T_h)$, which we might expect at the hot end (index "h"),

$$\alpha_h = E_g/(2eT_h). \tag{5.110}$$

Good thermoelectric materials with band gap up to 1 eV are common while 3 eV should be feasible. For a cooler with an ambient hot side temperature, this would suggest that α_h should be $\sim 1 - 5$ mV/K. Maintaining zT at such large α will require materials with both extremely high electronic mobility and low lattice thermal conductivity.

A lower bound to α_c also arises from the interconnected nature of the transport properties. We require zT to be finite; thus, the electrical conductivity σ must be large as α_c tends to zero. In this limit, the electronic component of the thermal conductivity (κ_e) is much larger than the lattice (κ_{lat}) contribution and $\kappa \sim \kappa_e$. To satisfy the Wiedemann–Franz law ($\kappa_e = L\sigma T$ where $L = \pi^2/3\ k_B^2/e^2$ is the Lorenz factor in the free electron limit), α_c has a lower bound given by

$$\alpha_c^2 = LzT_c = \frac{\pi^2}{3}\frac{k_B^2}{e^2}zT_c. \tag{5.111}$$

For example, a $z = 1/300$ K^{-1} and $T_c = 175$ K results in a lower bound to α_c of 119 µV/K.

The maximum cooling temperature T_c can be solved as a function of z, E_g and T_h from Eqs. (5.109)–(5.111). For small z, the approximate solution

$$\Delta T \approx \frac{z}{8} T_h^2 \ln\left(\frac{E_g^2}{\frac{4}{3}\pi^2 k_B^2 z T_h^3}\right) \tag{5.112}$$

gives an indication of the important parameters but quickly becomes inaccurate for zT above 0.1.

The solution of the maximum ΔT for the Thomson cooler compared to the CPM Peltier cooler with the same material assumption for z is shown in Figure 5.10 from $E_g = 0.5$ eV to 3 eV. The Thomson cooler provides significantly higher ΔT than the Peltier cooler with the same zT, nearly twice the ΔT for $E_g = 3$ eV.

These analytic results are possible because the compatibility approach does not require exact knowledge of the spatial profile for the material properties. In a real device, the spatial profile of thermoelectric properties will need to be engineered. Figure 5.9d shows an example of the Seebeck distribution $\alpha(x)$ along the leg that will provide the necessary $\alpha(T)$ where a constant $\kappa_{lat} = 0.5$ W/m K is assumed. If this rapidly changing α is achieved by segmenting different materials, low electrical contact resistance is required between the interfaces. We anticipate that such control of semiconductor materials may require thin film methods on active bulk thermoelectric substrates.

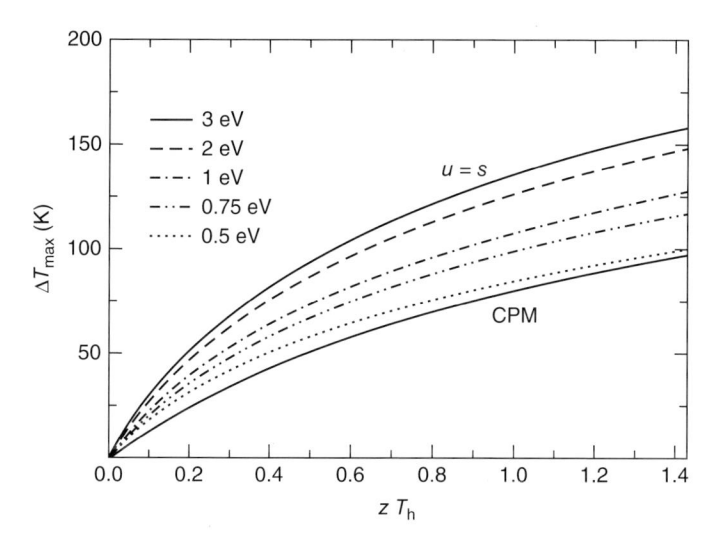

Figure 5.10 The maximum temperature drop ΔT_{max} of a Thomson cooler exceeds that of a Peltier cooler with the same z. Large band-gap E_g thermoelectric material at the hot junction improves the performance ($T_h = 300$ K, E_g plotted from 0.5 eV to 3 eV, model described in the text and in Ref. [19]).

There is improvement in compatibility and staging also for TEGs, but the improvement is small (< 10% compared to CPM). This is because the u does not typically vary by more than a factor of 2 across the device. However, in a cryogenic cooler, the compatibility requirement is much more critical. When operating a TEC to maximum temperature difference, the temperature gradient varies from zero to very high, which means u will range from a low value to infinity. Thus, unless compatibility is specifically considered, the poor compatibility will greatly reduce the performance of the TEC, resulting in the ΔT_{max} limit well known for Peltier coolers.

Minor improvements in thermoelectric cooling beyond increasing average zT by including the Thomson effect in a functionally graded material were predicted as early as 1960 [6]. More recently, Müller and Bian et al. describe modest gains in cooling from functionally grading [6, 50, 51, 55–57] where an average zT remains constant. The method of Bian et al. [56, 57], for instance, arrives at similar (but not equivalent) material requirements as the Thomson cooler – a rapidly increasing α at the hot side [56] – but focuses on redistributing the Joule heat. Such previous approaches to functionally grading have not, until now, focused on the compatibility criterion, $u = s$, nor identified the importance of the Thomson effect. In this analysis, we have focused on constant z (as opposed to zT [31]) to demonstrate the differences between a Thomson and a Peltier cooler typically analyzed with the CPM model; generally, any finite z, as long as $u = s$, will lead to lower temperature cooling.

A comparison of cooler models including the Bian–Shakouri type segmentation can be found in Ref. [58].

5.10.4
Further Characteristics of Self-Compatible Material for Cooling

Based on the new concept of a "Thomson cooler" (using the particular case of a $u = s$ cooler as a demonstration), we present here further characteristics of self-compatible cooler material.

For numerical calculations, $\alpha_{opt}(T)$ must be specified. We conclude from Eqs. (5.78), (5.82) that α is very large at the hot end (heat sink side $T_s \equiv T_h$) and decreases to a low value at the cold end (heat absorbing side $T_a \equiv T_c$). Due to the interconnected nature of the transport properties, there is a realistic range of α when $u = s$ is to be maintained for $z =$const. respectively $zT =$const. Again, we follow here Snyder's line of argument published in Ref. [19]: Using the relationship between peak α and E_g of Goldsmid and Sharp [54] $\alpha_h = E_g/(2eT_h)$ allows an estimate for the highest $\alpha_h = \alpha(T_h)$, which is typically limited by the band gap energy E_g. In this section, we use the peak $\alpha_h = 833$ µV/K ($E_g = 0.5$ eV)

5.10 Thermoelectric Thomson Cooler

as an example.[22] We stress at this point that peak α_h decides on the cooling performance of a $u = s$ cooler.

A lower bound to α_c can be estimated according to the Wiedemann–Franz law[23]: $\varrho\kappa = \alpha_c^2/z = L_N T_c \Rightarrow \alpha_c = \sqrt{z L_N T_c}$; for example, $z = 1/300$ K and $T_c = 220/200/180$ K results in a lower bound to α_c as $\alpha_c = 134/127/121$ µV/K. In order not to fall below the lower bound, we have used $T_c = 220$ K in our $u = s$ example calculation, which gives $\alpha_c = 154$ µV/K > 134 µV/K when $z = 1/300$ K, see Figure 5.13a.

Characteristics of the $u = s$ cooler are plotted in this section for $z = 1/300$ K and to some extent also for $zT = 1$, for a constant thermal conductivity $\kappa_o = 1.35$ W/(m K), and an element length $L = 5$ mm (with hot side temperature $T_h = 300$ K throughout).

We discuss first the electrical current density j as a function of the cooling temperature T_c when applying the constraint $zT = k_o =$const. In this case, the compatible Seebeck coefficient $\alpha_{opt}(T)$ is given by Eq. (5.78) and the compatibility factor s_c for TEC reads

$$s_c(T) = -\frac{1 + \sqrt{1 + k_o}}{T\,\alpha_{opt}(T)} = -\frac{1 + \sqrt{1 + k_o}}{\alpha_o} T^{-(k_c+1)} \qquad (5.113)$$

with $\alpha_o = \alpha_{ref}\, T_{ref}^{-k_c}$. For given boundary temperatures $T_c < T_h$, the optimal current density results from the scaling integral[24] [1]:

$$\begin{aligned}
j_{opt}^{u=s} &= \frac{1}{L} \int_{T_h}^{T_c} \kappa_o s_c(T)\,dT \\
&= \frac{\kappa_o}{\alpha_o L} \left(1 + \sqrt{1 + k_o}\right) \int_{T_c}^{T_h} T^{-(k_c+1)}\,dT \\
&= \frac{\kappa_o}{\alpha_o L} \frac{\left(1 + \sqrt{1 + k_o}\right)}{k_c} \left(T_c^{-k_c} - T_h^{-k_c}\right).
\end{aligned} \qquad (5.114)$$

Alternatively, for $z =$const., j_{opt} can be estimated in a similar way by numerical integration (with using Eq. (5.82) for cooling).

Figure 5.11 shows the optimal current density j_{opt} for both constraints in the range 180 K $< T_c < 300$ K (solid dashed curve). In the limit $T_c \to 0$, we find for both cases $j_{opt}^{u=s} \to \infty$; Obviously, zero Kelvin is not reachable.

Self-compatibility locally maximizes the cooler's COP for a given zT. Hence, a $u = s$ cooler always operates at maximum COP. For this reason, we can compare

---

22) Classical TE materials as Bi SbTe$_3$ and Pb Te have band gap energies below 0.5 eV; to manufacture $u = s$ material with large gap energies at room temperature (and below) is a challenge for future material design.

23) Classical theory and $T_c \to 0$ are incompatible. We expect that constraints on z violate Wiedemann–Franz law well before T_c gets close to zero.

24) Snyder has specified the scaling integral $j = \frac{1}{L}\int_{T_c}^{T_h} u\kappa\,dT$ for TEG where $T_h = T(x = 0)$, $T_c = T(x = L)$ and $u > 0$; it can easily be proved in a 1D approach using $dT = T'(x)\,dx = \frac{-j}{\kappa u}\,dx$.

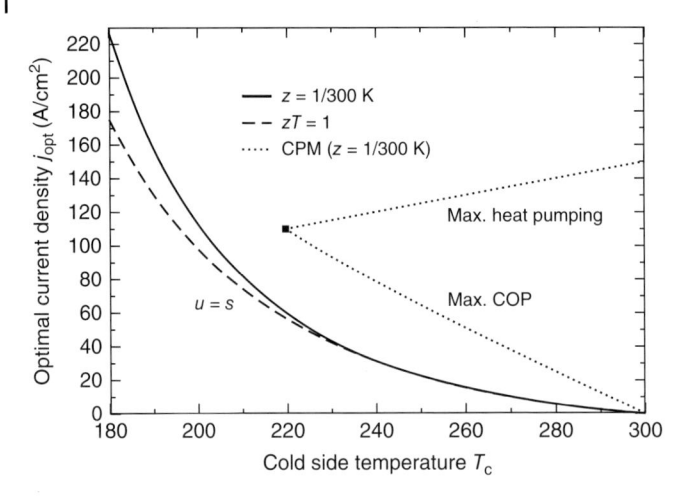

Figure 5.11 Optimal current density j_{opt}^{cpm} of a CPM Peltier cooler leg ($z = 1/300$ K, dotted; black square: maximum temperature difference) versus $j_{opt}^{u=s}$ of a $u = s$ cooler for varying cold side temperatures T_c and constraints $z = 1/300$ K (solid) with $zT = 1$ (dashed). Parameters: peak $\alpha_h = 833$ μV/K ($E_g = 0.5$ eV), element length $L = 5$ mm. For example, for $T_c = 220$ K, we obtain $j_{opt}^{u=s} = 59.4$ A/cm² ($z = \frac{1}{300\,K}$) and $j_{opt}^{u=s} = 56.2$ A/cm² ($zT = 1$). For a limiting model regarding T_c, see Figure 5.16 and [19].

$j_{opt}^{u=s}$ (Eq. (5.114)) to the optimal electrical current density of a CPM Peltier cooler [5]

$$j_{opt,COP}^{cpm} = \frac{2\kappa}{\alpha L}\frac{T_h - T_c}{T_h + T_c}(1 + \sqrt{1 + zT_m}), \quad T_m = \frac{T_c + T_h}{2}. \tag{5.115}$$

For the case of maximum temperature difference (CPM: $\varphi_{max}^{cpm} = 0$ at $T_c = T_{c,min}$), we obtain from Eq. (5.115)

$$j_{opt,\varphi=0}^{cpm} = \frac{2\kappa}{\alpha L}\frac{T_h - T_{c,min}}{T_h + T_{c,min}}\left(1 + \sqrt{1 + \frac{z}{2}(T_{c,min} + T_h)}\right) \tag{5.116}$$

with the lowest attainable cooling temperature [59] (see also Eq. (5.99))

$$T_{c,min}^{cpm} = \frac{1}{z}\left(-1 + \sqrt{1 + 2zT_h}\right). \tag{5.117}$$

Alternatively, but also within CPM, the optimal current density for maximum ΔT is given by Seifert et al. [5]

$$j_{opt,\Delta T_{max}} = \frac{\kappa}{\alpha L}\left(-1 + \sqrt{1 + 2zT_h}\right). \tag{5.118}$$

From Eqs. (5.117), (5.118), we obtain $j_{opt,\Delta T_{max}} = \frac{\alpha \sigma}{L}T_{c,min}^{cpm}$. Hence, the fundamental relation for maximum heat pumping that the optimum current is always proportional to the temperature of the cold side is also valid in this case.[25] For

---

[25] The equivalence of both equations (5.116), (5.118) can be shown with Eq. (5.117), see the appendix of [39].

5.10 Thermoelectric Thomson Cooler

example, for constant values $z = 1/300$ K, $\kappa = 1.35$ W/(m K) and $\alpha = 180$ μV/K, we get within CPM: $T_{c,min}^{cpm} = 219.6$ K at $j_{opt,\varphi=0}^{cpm} = 109.8$ A/cm², see the black square in Figure 5.11. The lower dotted curve in Figure 5.11 shows $j_{opt, COP}$ for $T_h > T_c > T_{c,min}$ according to Eq. (5.115). Figure 5.11 further shows that

A) maximum heat pumping within CPM becomes less efficient when the temperature difference $\Delta T = T_h - T_c$ decreases, and
B) a $u = s$ cooler needs less electrical current to reach the same cooling temperature T_c.

The temperature profile [as derived in Section 5.9.4, see Eq. (5.88)] can now be evaluated for a given cold side temperature T_c. Figure 5.12a shows the optimal $T(x)$ for $T_c = 220$ K and both constraints (with marginal differences < 1.5 K). We point out that the curvature of $T(x)$ of a $u = s$ cooler is opposite to that of a conventional Peltier cooler because of the different sign of the Fourier heat divergence [19]: $T(x)$ is a convex function with a low gradient at the cold side. When operating the $u = s$ cooler to maximum temperature difference $\Delta T \to T_h$, the temperature gradient varies from zero (when $T_c \to 0$) to very high values indicating that u will have a broader range in a TEC than in a TEG. Figure 5.12b shows the relative current for our calculation example. Since we have optimal conditions, Snyder's criterion $u(x) = s(x)$ is fulfilled across the device for both constraints.

The compatible material profiles $\alpha_{opt}(x)$ (with peak $\alpha_h = 833$ μV/K) and $\sigma_{opt}(x)$ are plotted in Figure 5.13, where the optimal electrical conductivity profiles have been calculated using the constraints: $\sigma_{opt}(x) = \dfrac{\kappa_o k_o}{\alpha_{opt}^2(x) T(x)}$ when $zT = k_o = 1$ and $\sigma_{opt}(x) = \dfrac{\kappa_o z_o}{\alpha_{opt}^2(x)}$ when $z = z_o = \dfrac{1}{300 \text{ K}}$.

The overall heat flux and its components are shown in Figure 5.14. We want to point out the qualitative differences to a conventional Peltier cooler. This particularly concerns the Fourier heat flow, which goes to zero at the cold side when $T_c = T(x = 0) \to 0$. To sum up, we can state that a $u = s$ cooler leads to an

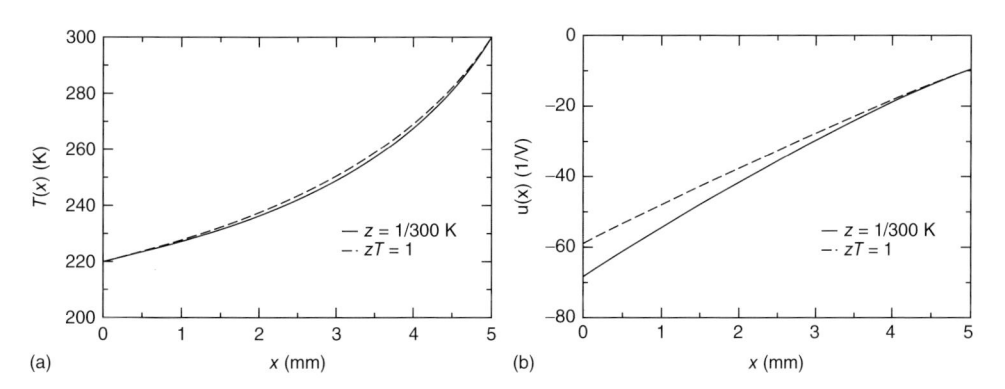

Figure 5.12 (a) Optimal, convex temperature profile $T(x)$, and (b) relative current density $u(x) = s_c(x)$ at optimal electrical current (parameters as given in the legend of Figure 5.11) for constraints $z = 1/300$ K (solid) and $zT = 1$ (dashed) and fixed boundary temperatures $T_c = 220$ K, $T_h = 300$ K.

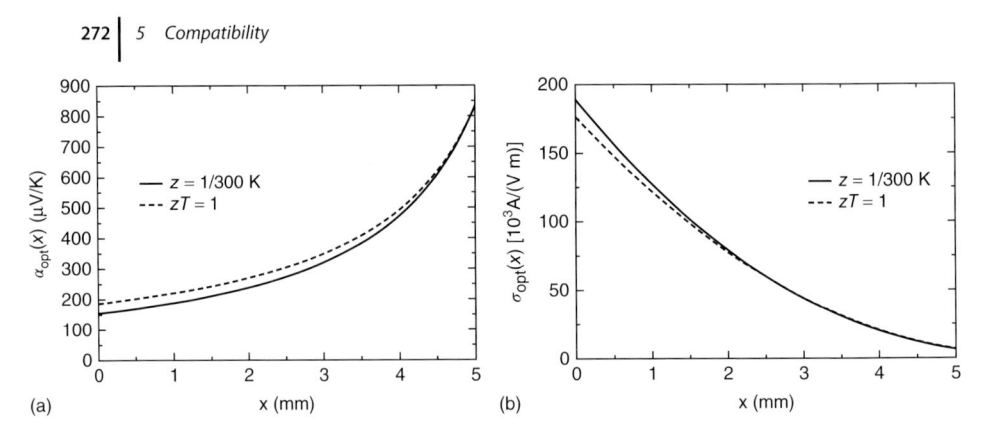

Figure 5.13 Optimal, spatial material profiles at optimal electrical current (j_{opt} values given in the legend of Figure 5.11) for fixed boundary temperatures $T_c = 220$ K, $T_h = 300$ K. (a) Optimal Seebeck profile $\alpha_{opt}(x)$ [with boundary values $\alpha_h = 833$ μV/K, $\alpha_c = 154$ μV/K ($z = \frac{1}{300\,K}$), $\alpha_c = 186$ μV/K ($zT = 1$)] and (b) optimal electrical conductivity $\sigma_{opt}(x)$.

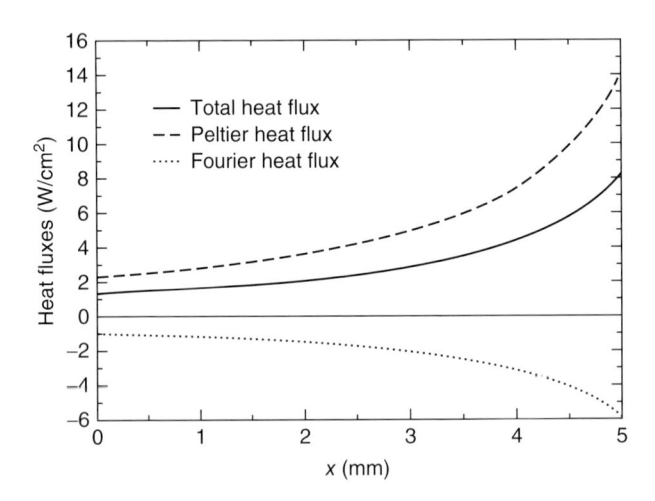

Figure 5.14 Total heat flux $q(x)$ (solid line), and its components Peltier heat flux $q_\pi(x) = j\,\alpha(x)T(x)$ (dashed) and Fourier heat flux $q_\kappa(x) = -\kappa\, T'(x)$ (dotted) for a $u = s$ cooler with constraint $z = 1/300$ K, fixed boundary temperatures $T_c = 220$ K, $T_h = 300$ K, and $j^{u=s}_{opt} = 59.4$ A/cm^2. (An appropriate figure using the alternative constraint $zT = 1$ shows only marginal differences.)

improved performance (ΔT and COP, for the latter, see also the next section), and that it may contribute to realizing solid-state cooling to cryogenic temperatures. However, as already outlined in Ref. [19], the material requirements to maintain $u = s$ become exceedingly difficult to achieve when the cooling temperature is reduced. One possible strategy could be an approximation of $u = s$ material by segmentation schemes based on controlled charge carrier concentration. There is no doubt that the related technological problems pose another challenge.

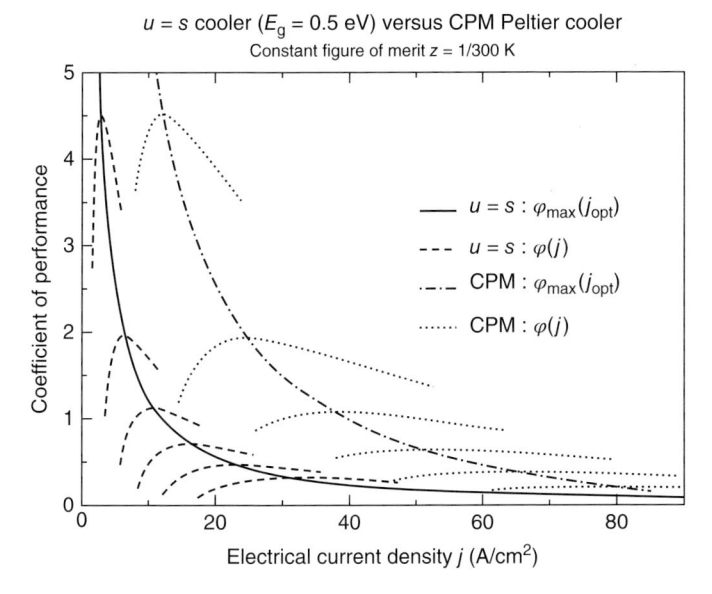

Figure 5.15 Comparison of cooler models: COP as a function of the electrical current density for different cold side temperatures $T_c = 290$ K, $\cdots$, 240 K (in 10 K steps, from top to bottom).

We complete the set of performance solutions with plots of the COP and the cooling temperature as function of the electrical current.

For a CPM Peltier cooler (with boundary temperatures T_c, T_h), this relation is given by Seifert *et al.* [5]

$$\varphi(j) = \frac{2\sigma L\alpha T_c j - 2\sigma\kappa(T_h - T_c) - L^2 j^2}{2\sigma\alpha(T_h - T_c)Lj + 2L^2 j^2}. \tag{5.119}$$

The dependence $\varphi^{u=s}(j)$ of a self-compatible element can be calculated as described in Ref. [44]: Numerically solve the differential equation for $u(T)$, Eq. (5.2), using optimal material properties. Evaluate the scaling integral to get the corresponding electrical current density j, and integrate $\varphi^{u=s}(j)$ within the compatibility approach. (All this is done in a loop with a varying boundary value of u.) Then, the maximum value $\varphi_{max}^{u=s}$ can be evaluated. The $\varphi(j)$ curves for a $u = s$ cooler (dashed) with $z = 1/300$ K ($E_g = 0.5$ eV) are also plotted in Figure 5.15 and compared to that of a CPM Peltier cooler (dotted) with the same z.[26] An equivalent result has been found in Ref. [44] for a lower figure of merit (see Figure 8).

We have already pointed out that the peak α_h (which is typically limited by the band gap energy) significantly affects the cooling performance of a $u = s$ cooler (Tables 5.1 and 5.2). Though there is no limited temperature drop, even in a $u = s$ cooler solid-state physics constraints will make sure that the ultimate cooling will be finite resulting in $T_{c,min} > 0$.

---

26) Figure 6 with results based on the alternative constraint $zT = 1$ shows only marginal differences.

Table 5.1 Optimal performance data of a $u = s$ cooler ($z = 1/300$ K).

T_c	290	280	270	260	250	240
$\varphi_{max}^{u=s}$	4.517	1.961	1.121	0.711	0.473	0.323
$j_{opt}^{u=s}$ (A/cm^2)	2.88	6.40	10.74	16.13	22.87	31.38

Table 5.2 Optimal performance data of a CPM Peltier cooler ($z = 1/300$ K).

T_c	290	280	270	260	250	240
φ_{max}^{cpm}	4.501	1.929	1.072	0.643	0.387	0.216
$j_{opt,COP}^{cpm}$ (A/cm^2)	12.25	24.85	37.84	51.22	65.03	79.28

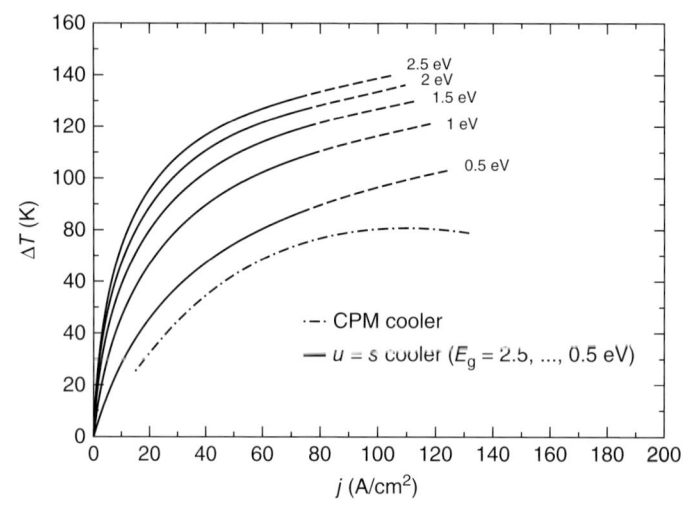

Figure 5.16 Temperature difference $\Delta T = T_h - T_c$ in dependence on the electrical current density j for a $u = s$ cooler ($z = 1/300$ K) with different gap energies E_g. Solid curves ($u = s$) up to the ΔT_{max} estimation, see Table 5.3. Since ΔT_{max} (end of the solid curves) was theoretically estimated, the ΔT_{max} curves are extended as dashed. CPM cooler (dot–dot–dashed curve: $z = 1/300$ K, $\Delta T_{max}^{cpm} = 80.4$ K at $j_{opt,\varphi=0}^{cpm} = 109.8$ A/cm^2) plotted for comparison.

The results for $z = 1/300$ K are plotted in Figure 5.16 where Snyder's limiting model for T_c (see [19] and the beginning of Section 3) has been applied. The estimated ΔT_{max} for varying gap energies are listed in Table 5.3; the results coincide with those published in Figure 4 of [19].

Table 5.3 $\Delta T_{max}^{u=s}$ estimation for varying gap energies ($z = 1/300$ K).

E_g	0.5	1.0	1.5	2.0	2.5
$\Delta T_{max}^{u=s}$	85.3	108.1	119.3	126.5	131.3
$\Delta T_{max}^{u=s}/\Delta T_{max}^{cpm}$	1.06	1.34	1.48	1.57	1.64

We sum up the results of this section as follows: By using a peak $\alpha_h = 833$ µV/K ($E_g = 0.5$ eV) as an example, the characteristics of the $u = s$ cooler have been discussed, in particular the temperature profile, the optimized spatial material profiles in a self-compatible element, and the cooling performance. Although our calculation example is still far from a "true" FGM optimization strategy (where only possible or available materials are considered), the new cooler concept demonstrates an increasing importance of the compatibility approach to TEC FGM design. However, a rapidly rising, optimal Seebeck coefficient places bounds on the maximum cooling obtainable. This happens in particular when z is low because then the exponential $\alpha(T)$, see Eq. (5.82) for cooling, decreases faster.

In a classical Peltier cooler, there is a competition between a reversible effect ($\propto j$) and an irreversible effect ($\propto j^2$). The maximal temperature difference is obtained in the limit $\varphi \to 0$. In a $u = s$ cooler where an optimal adaptation of the thermal and electric fluxes is realized at a local level, the classical definition of maximum cooling as the ΔT_{max} case at adiabatic cold side condition is no longer valid; $u = s$ and $\varphi = 0$ are incompatible since a $u = s$ cooler always operates at maximum COP. The consequence is a strictly monotonic function between the optimal electrical current and the cooling temperature (while $\varphi_{max}^{u=s} > 0$ for any current). This holds as long as the range of α is a realistic one.

While increasing zT is important for the improvement of Peltier coolers, engineering the compatibility of thermoelectric materials through functional grading can potentially lead to greater gains in the temperature difference. All previously published results clearly highlight the benefits when using $u = s$ material for cooling: the use of self-compatible elements is the most efficient way to accomplish direct energy conversion in thermoelectrics. Even though the CPM Peltier cooler and the $u = s$ Thomson cooler with constant z are both idealizations that can only be realized approximately in practice because of the constraints of real materials, this analysis demonstrates the fundamental difference between the two mechanisms for cooling and gives a general strategy as well as a new challenge for material optimization and for realizing solid-state cooling to cryogenic temperatures.[27]

---

27) We remember here a quotation from E. Altenkirch, who stated already in 1911 (see [60], p. 922):
    Die Erzeugung von Kälte wird um so schwieriger, je tiefer die absoluten Temperaturen sind. (in Engl.: The lower the absolute temperature, the more difficult thermoelectric cooling becomes.)

5.11
Compatibility Approach versus Device Optimization

The performance of a thermoelectric device is dependent on many variables, which could be optimized globally to find the optimum design.[28] However, by using a reduced variable approach to the design problem, interdependencies of the design variables can be eliminated, which allows a better understanding of the effect of each variable. In this context, the compatibility approach is certainly an alternative to Ioffe's global description, which is very often used for technological applications, but is surely not suitable for locally characterizing TE processes or even for local optimization purposes. Nevertheless, there must be interrelations between both approaches based on different quantities such as local j and κ, but appropriate global quantities electric current $I = j\, A_c$ and the thermal conductance $K = \kappa\, A_c/L$.

The philosophy of the compatibility approach is ultimately a consideration of the ratio of dissipative and reversible heat fluxes as a function of temperature (or space), instead of considering thermal and electrical quantities separately (see Section 5.8). The definition of the relative current density u reflects at a local scale the definition of the efficiency of a TEG as the ratio of the net electrical power output to the thermal power supplied to the system at the hot end. Obviously, a Peltier cooler can also be described using relative current; this is simply the consequence of the reciprocal definitions of the global performance parameters efficiency η and COP φ, see Section 5.3.1. In this context, we recall Sherman et al. [6], who used the inverse function $y(T) = 1/u(T)$ for the cooler (where y is nothing else than a relative Fourier heat flux, see Section 2.2):

$$u(T) = \frac{j}{\mathbf{q}_\kappa \cdot \mathbf{n}} \quad \Longleftrightarrow \quad y(T) = \frac{\mathbf{q}_\kappa \cdot \mathbf{n}}{j}. \tag{5.120}$$

The usage of y may simplify equations including the thermoelectric potential Φ but is less suitable for an open-circuit generator when $j = u = 0$.

Let us assume now a steady state, that is, a constant electric current $(\mathbf{j} = j\,\mathbf{n})$ flows through a TE leg. We start at the local scale with the scaling integral [1],

$$\int_{T_s}^{T_a} u(T)\kappa(T)\, dT = \int_{T_a}^{T_s} \frac{j}{dT/dx}\, dT = j\int_0^L dx = j\, L$$

$$\Rightarrow\ j = \frac{1}{L}\int_{T_s}^{T_a} u(T)\,\kappa(T)\, dT. \tag{5.121}$$

This integral can be transformed into a global expression along a TE leg with boundary temperatures T_a and T_s:

$$I = \frac{A_c}{L}\int_{T_s}^{T_a} u(T)\kappa(T)\, dT = \int_{T_s}^{T_a} u(T)K(T)\, dT \tag{5.122}$$

with $K(T) = \frac{A_c}{L}\kappa(T)$ and $I = j\, A_c$.

28) For optimization of thermoelectric conversion efficiency on a global scale see, for example, [61].

The mean value theorem of integral calculus finally gives (with $\Delta T = T_a - T_s$)

$$I = [u(T)\ K(T)\ ]_{T=T_x} \Delta T \quad \text{where} \quad T_x \in [T_a, T_s]. \tag{5.123}$$

We expect that an optimal product $u(T_x)K(T_x)$ ensures an optimal adaptation of the thermal and electric impedance of the leg resulting in an optimal current I when the temperature difference ΔT is given for a TEG. A reversed relation holds for the cooler.

Note that Eq. (5.123) refers to an ideal TE system. For real systems, an ansatz $I = K_{exp} \Delta T$ can be considered as equivalent to (5.123) whereby realistic thermal coupling may be included into K_{exp}, see [62]. In this case, the application of the compatibility method must be critically examined.

For system optimization, we should recall that thermal impedance system design [62, 63], size or cost constraints [64], electrical and mechanical interface issues can overwhelm the gains achieved by segmentation.

References

1. Snyder, G.J. and Ursell, T.S. (2003) Thermoelectric efficiency and compatibility. *Phys. Rev. Lett.*, **91** (14), 148301.
2. Ursell, T.S. and Snyder, G.J. (2002) Compatibility of segmented thermoelectric generators. 21st International Conference on Thermoelectrics, Piscataway, NJ, September 25–29, 2002, IEEE (Institute of Electrical and Electronics Engineers), pp. 412–417.
3. Seifert, W., Müller, E., and Walczak, S. (2006) Generalized analytic description of one-dimensional non-homogeneous TE cooler and generator elements based on the compatibility approach, in *25th International Conference on Thermoelectrics, Vienna, Austria*, August 06–10, 2006 (ed. P. Rogl), IEEE, Piscataway, NJ, pp. 714–719.
4. Seifert, W., Müller, E., Snyder, G.J., and Walczak, S. (2007) Compatibility factor for the power output of a thermogenerator. *Phys. Status Solidi RRL*, **1** (6), 250–252.
5. Seifert, W., Müller, E., and Walczak, S. (2008) Local optimization strategy based on first principles of thermoelectrics. *Phys. Status Solidi A*, **205** (12), 2908–2918.
6. Sherman, B., Heikes, R.R., and Ure, R.W. Jr. (1960) Calculation of efficiency of thermoelectric devices. *J. Appl. Phys.*, **31** (1), 1–16.
7. Egli, P.H. (1960) *Thermoelectricity*, John Wiley & Sons, Inc., New York.
8. Harman, T.C. and Honig, J.M. (1967) *Thermoelectric and Thermomagnetic Effects and Applications*, McGraw-Hill Book Company, New York.
9. Ybarrondo, L.J. (1967) Improved expressions for the efficiency of an infinite stage thermoelectric heat pump and generator. *Solid State Electron.*, **10**, 620–622.
10. Clingman, W.H. (1961) Entropy production and optimum device design. *Adv. Energy Convers.*, **1**, 61–79.
11. Clingman, W.H. (1961) New concepts in thermoelectric device design. *Proc. IRE*, **49** (7), 1155–1160.
12. Moizhes, B.Ya., Petrov, A.V., Shishkin, Yu.P., and Kolomoets, L.A. (1962) On the choice of the optimal mode of operation of a cascade thermoelectric element. *Sov. Phys.-Tech. Phys.*, 7, 336.
13. Heikes, R.R. and R.W. Ure Jr. (1961) *Thermoelectricity: Science and Engineering*, Interscience Publishers, Inc., New York.
14. Power, M. and Handelsman, R.A. (1961) Generalized calculation of thermoelectric efficiency. *Adv. Energy Convers.*, **1**, 45–60.
15. Vining, C.B. (1997) The thermoelectric process, in *Materials Research*

Society Symposium Proceedings: Thermoelectric Materials - New Directions and Approaches (eds T.M. Tritt, M.G. Kanatzidis Jr., H.B. Lyon, and G.D. Mahan), Materials Research Society, pp. 3–13.
16. Liu, L. (2012) A continuum theory of thermoelectric bodies and effective properties of thermoelectric composites. *Int. J. Eng. Sci.*, **55**, 35–53.
17. Müller, E., Zabrocki, K., Goupil, C., Snyder, G.J., and Seifert, W. (2012) Functionally graded thermoelectric generator and cooler elements, in *CRC Handbook of Thermoelectrics: Thermoelectrics and its Energy Harvesting*, Chapter 4 (ed. D.M. Rowe), CRC Press, Boca Raton, FL.
18. Anatychuk, L.I. and Vikhor, L.N. (2012) *Functionally Graded Thermoelectric Materials*, Thermoelectricity, vol. **4**, Institute of Thermoelectricity Bukrek Publishers, Chernivtsi.
19. Snyder, G.J. and Toberer, E.S. (2012) Raghav Khanna, and Wolfgang Seifert. Improved thermoelectric cooling based on the Thomson effect. *Phys. Rev. B*, **86**, 045202.
20. Snyder, G.J. (2006) Thermoelectric power generation: efficiency and compatibility, in *CRC Handbook of Thermoelectrics: Macro to Nano*, Chapter 9 (ed. D.M. Rowe), Taylor and Francis, Boca Raton, FL.
21. Caillat, T., Fleurial, J.-P., Snyder, G.J., and Borshchevsky, A. (2001) Development of high efficiency segmented thermoelectric unicouples. Proceedings ICT 2001 – 20th International Conference on Thermoelectrics, pp. 282–285.
22. Kelley, C.M. and Szego, G.C. (1964) Colloque on Energy Sources and Energy Conversion, Cannes, p. 651.
23. Skrabek, E.A. and Trimmer, D.S. (1995) *CRC Handbook of Thermoelectrics - Properties of the General TAGS System*, Chapter 22, CRC Press LLC, pp. 267–275.
24. Snyder, G.J. (2004) Application of the compatibility factor to the design of segmented and cascaded thermoelectric generators. *Appl. Phys. Lett.*, **84**, 2436.
25. Danielson, L.R., Raag, V., and Wood, C. (1985) Thermoelectric properties of rare earth chalcogenides, in *20th Intersociety Energy Conversion Engineering Conference*, vol. **3**, Society of Automotive Engineers, Miami Beach, FL, pp. 3.531–3.535.
26. Cox, C.A., Toberer, E.S., Levchenko, A.A., Brown, S.R., Snyder, G.J., Navrotsky, A., and Kauzlarich, S.M. (2009) Structure, heat capacity, and high-temperature thermal properties of $Yb_{14}Mn_{1-x}Al_xSb_{11}$. *Chem. Mater.*, **21**, 1354–1360.
27. Snyder, G.J. and Caillat, T. (2003) Using the compatibility factor to design high efficiency segmented thermoelectric generators. Materials Research Society Proceedings, vol. 793, p. 37.
28. Ngan, P.H., Christensen, D.V., Snyder, G.J., Hung, L.T., Linderoth, S., Nong, N.V., and Pryds, N. (2014) Towards high efficiency segmented thermoelectric unicouples. *Phys. Status Solidi A*, **211** (1), 9–17.
29. Zener, C. (1960) The impact of thermoelectricity upon science and technology, in *Thermoelectricity*, Chapter 1 (ed. P.H. Egli), John Wiley & Sons, Inc., New York, pp. 3–22.
30. Freedman, S.I. (1966) Thermoelectric power generation, in *Direct Energy Conversion*, Inter-University Electronic Series, vol. **3**, Chapter 3 (ed. G.W. Sutton), McGraw-Hill Book Company, pp. 105–180.
31. Seifert, W., Pluschke, V., Goupil, C., Zabrocki, K., Müller, E., and Snyder, G.J. (2011) Maximum performance in self-compatible thermoelectric elements. *J. Mater. Res.*, **26** – Focus Issue – Advances in Thermoelectric Material (15), 1933–1939.
32. Seifert, W. and Pluschke, V. (2012) Exact solution of a constraint optimization problem for the thermoelectric figure of merit. *Materials*, **5** (3), 528–539.
33. Seifert, W., Zabrocki, K., Snyder, G.J., and Müller, E. (2010) The compatibility approach in the classical theory of thermoelectricity seen from the perspective of variational calculus. *Phys. Status Solidi A*, **207** (3), 760–765.
34. Schilz, J., Helmers, L., Müller, Eckhard., and Niino, M. (1998) A local selection criterion for the composition of graded

thermoelectric generators. *J. Appl. Phys.*, **83** (2), 1150–1152.
35. Seifert, W., Zabrocki, K., Müller, E., and Snyder, G.J. (2010) Power-related compatibility and maximum electrical power output of a thermogenerator. *Phys. Status Solidi A*, **207** (10), 2399–2406.
36. Zabrocki, K., Müller, E., and Seifert, W. (2010) One-dimensional modeling of thermogenerator elements with linear material profiles. *J. Electron. Mater.*, **39**, 1724–1729, doi: 10.1007/s11664-010-1179-3.
37. Zabrocki, K., Müller, E., Seifert, W., and Trimper, S. (2011) Performance optimization of a thermoelectric generator element with linear material profiles in a 1d setup. *J. Mater. Res.*, **26** – Focus Issue – Advances in Thermoelectric Material (15), 1963–1974.
38. Gerstenmaier, Y.C. and Wachutka, G. (2012) Unified theory for inhomogeneous thermoelectric generators and coolers including multistage devices. *Phys. Rev. E*, **86**, 056703.
39. Seifert, W. and Pluschke, V. (2014) Optimizing the electrical power output of a thermogenerator with the Gerstenmaier/Wachutka approach. *Phys. Status Solidi A*, **211**, 685–695. WILEY online library, doi: 10.1002/pssa.201330176.
40. Apertet, Y., Ouerdane, H., Goupil, C., and Lecoeur, P. (2014) Comment on "'effective thermal conductivity in thermoelectric material". *J. Appl. Phys.*, **115** (12), 126101.
41. Baranowski, L.L., Snyder, G.J., and Toberer, E.S. (2014) Response to "comment on 'effective thermal conductivity in thermoelectric material'". *J. Appl. Phys.*, **115** (12), 126102.
42. Goupil, C., Seifert, W., Zabrocki, K., Müller, E., and Snyder, G.J. (2011) Thermodynamics of thermoelectric phenomena and applications. *Entropy*, **13** (8), 1481–1517.
43. Seifert, W. and Pluschke, V. (2014) Maximum cooling power of a graded thermoelectric cooler. *Phys. Stat. Sol. B*, **251**, 1416–1425. WILEY online library, doi: 10.1002/pssb.201451038.
44. Seifert, W., Snyder, G.J., Toberer, E.S., Goupil, C., Zabrocki, K., and Mueller, E. (2013) The self-compatibility effect in graded thermoelectric cooler elements. *Phys. Status Solidi A*, **210** (7), 1407–1417.
45. Bergman, D.J. and Levy, O. (1991) Composite thermoelectrics - exact results and calculational methods, in *Modern Perspectives on Thermoelectrics and Related Materials*, vol. **234** (eds G.A. Slack, D.D. Allred, and C.B. Vining), Pittsburgh, PA, pp. 39–45.
46. Goldsmid, H.J. (2010) *Introduction to Thermoelectricity*, Springer-Verlag.
47. Inc. Marlow Industries (2011) Six state - thermoelectric modules.
48. Shiota, I. and Miyamoto, Y. (eds) (1996) *Functionally Graded Material 1996*, AIST Tsukuba Research Center, Tsukuba. Proceedings of the 4th International Symposium on Functionally Graded Materials, Elsevier.
49. Helmers, L., Müller, E., Schilz, J., and Kaysser, W.A. (1998) Graded and stacked thermoelectric generators - numerical description and maximisation of output power. *Mater. Sci. Eng., B*, **56** (1), 60–68.
50. Müller, E., Drašar, v.C., Schilz, J., and Kaysser, W.A. (2003) Functionally graded materials for sensor and energy applications. *Mater. Sci. Eng., A*, **362** (1–2), 17–39. Papers from the German Priority Programme (Functionally Graded Materials).
51. Müller, E., Walczak, S., and Seifert, W. (2006) Optimization strategies for segmented Peltier coolers. *Phys. Status Solidi A*, **203** (8), 2128–2141.
52. Seifert, W., Ueltzen, M., and Müller, E. (2002) One-dimensional modelling of thermoelectric cooling. *Phys. Status Solidi A*, **1** (194), 277–290.
53. Seifert, W. and Pluschke, V. (2014) The extended concept of a self-compatible thermoelectric cooler. *Phys. Status Solidi A*, **211**, 917–923. WILEY online library, doi: 10.1002/pssa.201330392.
54. Goldsmid, H.J. and Sharp, J.W. (1999) Estimation of the thermal band gap of a semiconductor from Seebeck measurements. *J. Electron. Mater.*, **28**, 869–872.
55. Müller, E., Karpinski, G., Wu, L.M., Walczak, S., and Seifert, W. (2006) Separated effect of 1D thermoelectric

material gradients, in *25th International Conference on Thermoelectrics* (ed. P. Rogl), IEEE, Piscataway, NJ, pp. 204–209.
56. Bian, Z. and Shakouri, A. (2006) Beating the maximum cooling limit with graded thermoelectric materials. *Appl. Phys. Lett.*, **89** (21), 212101.
57. Bian, Z., Wang, H., Zhou, Q., and Shakouri, A. (2007) Maximum cooling temperature and uniform efficiency criterion for inhomogeneous thermoelectric materials. *Phys. Rev. B: Condens. Matter Mater. Phys.*, **75** (24), 245208.
58. Seifert, W., Pluschke, V., and Hinsche, N.F. (2014) Thermoelectric cooler models and the limit for maximum cooling. *J. Phys.: Condens. Matter*, **26** (25), 255803; Corrigendum in *J. Phys. Condens. Matter*, **26** (29), 299501.
59. Goldsmid, H.J. (1986) *Electronic Refrigeration*, Pion, London.
60. Altenkirch, E. (1911) Elektrothermische Kälteerzeugung und reversible elektrische Heizung. *Phys. Z.*, **12**, 920–924.
61. Dashevsky, Z., Gelbstein, Y., Edry, I., Drabkin, I., and Dariel, M.P. (2003) Optimization of thermoelectric efficiency in graded materials. Proceedings of the 22nd International Conference on Thermoelectrics, pp. 421–424.
62. Apertet, Y., Ouerdane, H., Glavatskaya, O., Goupil, C., and Lecoeur, P. (2012) Optimal working conditions for thermoelectric generators with realistic thermal coupling. *Europhys. Lett.*, **97** (2), 28001.
63. Baranowski, L.L., Snyder, G.J., and Toberer, E.S. (2013) Effective thermal conductivity in thermoelectric materials. *J. Appl. Phys.*, **113**, 204904.
64. Kristiansen, N.R., Snyder, G.J., Nielsen, H.K., and Rosendahl, L. (2012) Waste heat recovery from a marine waste incinerator using a thermoelectric generator. *J. Electron. Mater.*, **41** (6), 1024–1029.

6
Numerical Simulation
Knud Zabrocki and Wolfgang Seifert†

In the previous chapters, analytical methods for characterizing the behavior of TE elements and devices have been discussed. In this last chapter, we want to analyze some problems with numerical methods.

Domenicali's heat equation, which gives a generalized description of thermoelectric phenomena on continuum level combining thermal and electrical transport equations, is in general a nonlinear partial differential equation (PDE) with state-dependent coefficients due to the temperature dependence of the material properties [1–3]. As the temperature T is the target function in the solution of the PDE, the coefficients are said to be *state-dependent*. Furthermore, complex geometries and complicated boundary conditions can lead to the situation that an analytical solution is not possible or is too cumbersome. In such a case, computational methods are an opportunity to study physical properties and processes in thermoelectric materials and systems, and it is more or less convenient to address such problems with numerical methods. A review on numerical methods in thermoelectricity is given by Strutynsky [4]. In this review, modern numerical methods are compared and analyzed to show their capability in the framework of thermoelectric systems and processes. Depending on the length and timescales, several methods are established in thermoelectricity today. The general principles of physical modeling are introduced in Ref. [5].

In Figure 6.1, an overview of different methods used in thermoelectricity is given. Here we focus on the macroscopic or system level, where classical approaches are used. In that case, the numerical methods are very well known from pure heat transfer and electrical transport problems. Their mathematical description is based on PDEs or systems of PDEs (or eventually integro-partial differential equations). Problems of this kind can be solved by structural-functional simulations.

From Figure 6.2, it can be seen that the numerical methods are just a part of this kind of simulations. The numerical methods are subgrouped into three major approaches:

- *finite difference method* (FDM),
- *finite volume method* (FVM), and
- *finite element method* (FEM)

Continuum Theory and Modelling of Thermoelectric Elements, First Edition. Edited by Christophe Goupil.
© 2016 Wiley-VCH Verlag GmbH & Co. KGaA. Published 2016 by Wiley-VCH Verlag GmbH & Co. KGaA.

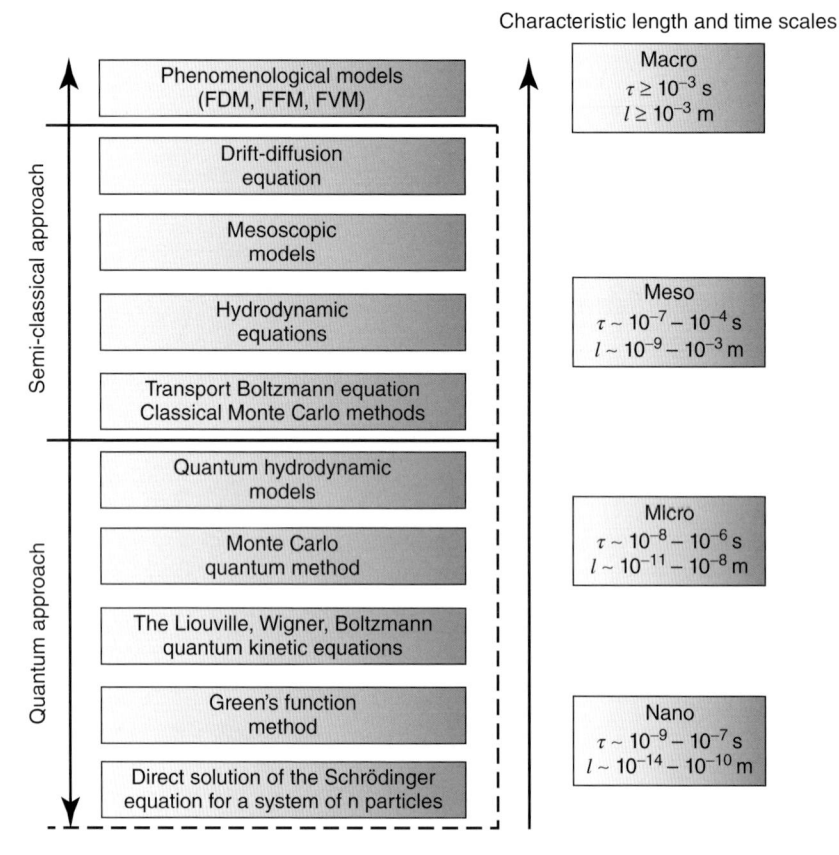

Figure 6.1 A hierarchy of simulation methods (adapted from Figure 2 in Ref. [4]).

As a matter of fact, each of these methods has its advantages depending on the nature of the underlying physical problem. Several factors and questions come into play to decide which choice is best for a specific problem, for example:

- Which type is the PDE?
- What are the relevant physical dimensions?
- Are there any relevant symmetries that allow for a special kind of coordinate system?
- Are the governing equations and the corresponding boundary conditions non-linear or linear?
- Is the problem a steady-state or transient problem?

Going beyond the ideal 1D CPM thermoelectric element, there is no other choice than using a numerical method. Therefore, we highlight these methods exemplarily, which can be found as either implemented in commercial programs or coded in common programming languages. In Reference [6], three major steps

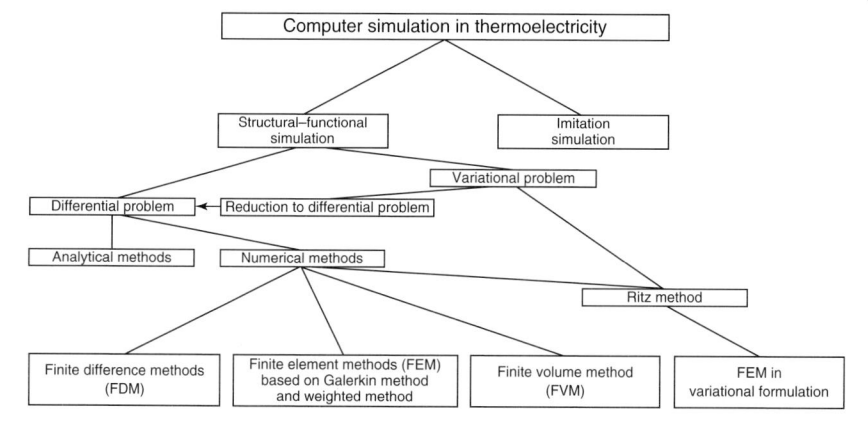

Figure 6.2 Classification of computer simulations especially of structural-functional simulation methods (adapted from Figures 3 and 4 in Ref. [4]).

are emphasized to be involved with finite volume methods, which are valid for all three methods (FDM, FVM, FEM):

1) Discretize the domain,
2) Replace the PDEs by algebraic equations,
3) Specify the solution algorithm.

For the choice of the numerical method and particular schemes, the type of the PDE plays a significant role. It is known that the PDEs are classified into three different categories denoted as *parabolic*, *elliptic*, and *hyperbolic*, depending on the derivatives present in the PDEs. For example, the classical 1D Fourier heat equation is a *parabolic* PDE, whereas non-Fourier heat conduction, which can exhibit a wave-like propagation of the temperature field with a finite propagation speed, is generally described by a *hyperbolic* PDE, see [7–9]. For every category, there are particular methods, which have been developed [10]:

Parabolic PDEs For the parabolic PDEs, an approximation can be done with a system of ordinary differential equations (ODEs), which are characterized as to be large, sparse, and stiff. For these properties, implicit methods are needed and care has to be taken concerning the sparsity. Special methods concerning parabolic PDEs are introduced in Ref. [11].

Elliptic PDEs The approximations in the case of elliptic PDEs result in large systems of algebraic equations, which are, in general, sparse. Special numerical methods for such a kind of property have to be considered for their solution. The numerical treatment of elliptic PDEs is shown in Ref. [12].

Hyperbolic PDEs For this type of PDEs, a special property is observed, the propagating discontinuities, for which special attention is needed. The methods concerning hyperbolic PDEs and its special properties are covered in Ref. [13].

Clearly, numerical solutions of PDE are connected with errors that originate from the discretization process. Other errors are due to the computational solution of the resulting (system of) algebraic equations. A threefold classification of errors can be performed in *truncation*, *discretization*, and *round-off* errors. For a detailed explanation and definition of these errors, the dedicated reader is advised to refer to [7, 8].

In the next sections, the different standard numerical methods used in classical thermoelectricity are briefly described. Nevertheless, we want to draw the reader's attention to an essential fraction out of the uncountable number of books containing the (numerical) solution of PDEs as a topic [14–32].

6.1
Finite Difference Methods

The *finite difference methods* (FDM) are known as the simplest and most straightforward interpolation methods. From all numerical methods treating differential equations, it is the best studied and established one. There is a large number of articles and textbooks where this method has been introduced and described in detail [33–38], some of them particularly concentrating on the heat transfer, for example, [7, 8, 39–41].

The main idea of the FDM is to approximate the differential quotient by the difference quotient. This approximation comes directly from the definition of the derivative, as the partial derivative of an arbitrary function $F(x,y)$ at $x = x_0$ and $y = y_0$ with respect to x is given by:

$$\frac{\partial F}{\partial x} = \lim_{\Delta x \to 0} \frac{F(x_0 + \Delta x, y_0) - F(x_0, y_0)}{\Delta x}, \tag{6.1}$$

see [7]. If the function $F(x,y)$ is supposed to be a continuous one, then the partial derivative can be approximated for a *sufficiently small* but finite Δx. With Figure 6.3 the nomenclature for a Taylor series expansion of a function $f = f(x)$ at $\bar{x}$ is illustrated. The parameter h denotes the (fixed) Δx such that for the Taylor series expansion for the functions $f(\bar{x} + h)$ and $f(\bar{x} - h)$, respectively,

$$f(\bar{x} + h) = f(\bar{x}) + hf'(\bar{x}) + \frac{h^2}{2!}f''(\bar{x}) + \frac{h^3}{3!}f'''(\bar{x}) - \ldots , \tag{6.2a}$$

$$f(\bar{x} - h) = f(\bar{x}) - hf'(\bar{x}) + \frac{h^2}{2!}f''(\bar{x}) - \frac{h^3}{3!}f'''(\bar{x}) + \ldots , \tag{6.2b}$$

where the primes denote as usual the derivatives with respect to x. Both relations can be used to express the first- and second-order derivatives in the point $x = \bar{x}$ in the form of finite differences. These expressions can be derived in different ways, which has already been shown for the first-order derivatives by just solving either

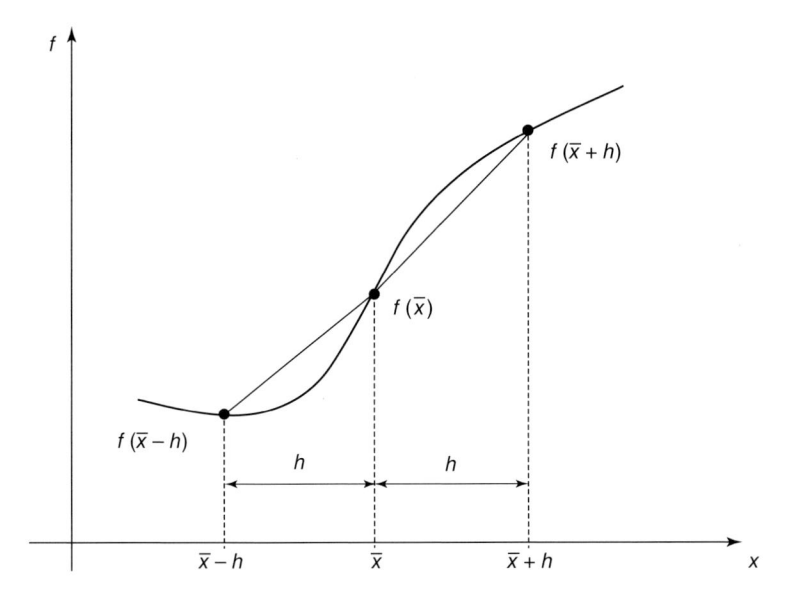

Figure 6.3 Nomenclature for a Taylor series expansion of a function $f = f(x)$ at $\bar{x}$ (adapted from Figure 12.1 in Ref. [7]).

Eqs. (6.2a) or (6.2b) for $f'(x = \bar{x})$, that is,

$$f'(\bar{x}) = \frac{f(\bar{x}+h) - f(\bar{x})}{h} - \frac{h}{2}f''(\bar{x}) - \frac{h^2}{6}f'''(\bar{x}) - \ldots \quad \text{(forward)}, \quad (6.3a)$$

$$f'(\bar{x}) = \frac{f(\bar{x}) - f(\bar{x}-h)}{h} + \frac{h}{2}f''(\bar{x}) - \frac{h^2}{6}f'''(\bar{x}) + \ldots \quad \text{(backward)}. \quad (6.3b)$$

A central version is gained if Eq. (6.2a) is subtracted from Eq. (6.2b)

$$f'(\bar{x}) = \frac{f(\bar{x}+h) - f(\bar{x}-h)}{2h} - \frac{h^2}{6}f'''(\bar{x}) - \ldots \quad \text{(central)}. \quad (6.3c)$$

In an analogous manner, higher and mixed order derivatives can be expressed.

In Figure 6.4, a standard notation in the FDMs is illustrated by showing a 1D temperature profile. At equidistant grid points $x_i = ih$, where $i \in \mathbb{N}$ and $h = \Delta x$ is the step size, the temperature profile $T(x = x_i) = T_i$ is calculated by FDM. Straightforwardly, the abbreviations $x_{i+1} = x_i + h = (i+1)h$ and $x_{i-1} = x_i - h = (i-1)h$, as well as $T(x = x_{i+1}) = T_{i+1}$ and $T(x = x_{i-1}) = T_{i-1}$, can be deduced. In this notation, the first derivatives for the temperatures at the grid points are given by

$$T'_i = \frac{T_{i+1} - T_i}{h} + \mathcal{O}(h) \quad \text{(forward)}, \quad (6.4a)$$

$$T'_i = \frac{T_i - T_{i-1}}{h} + \mathcal{O}(h) \quad \text{(backward)}, \quad (6.4b)$$

$$T'_i = \frac{T_{i+1} - T_i}{2h} + \mathcal{O}(h^2) \quad \text{(central)}, \quad (6.4c)$$

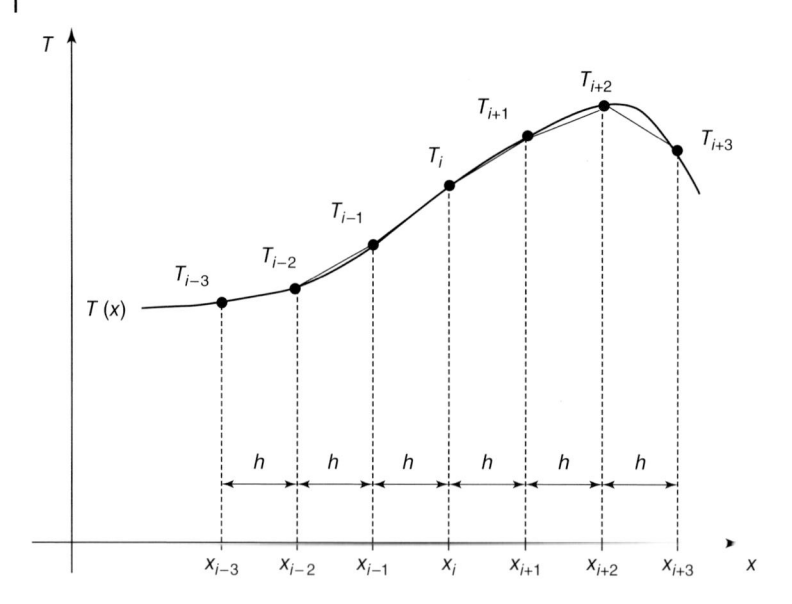

Figure 6.4 Nomenclature for the finite difference representation of a 1D temperature distribution}.

where the notation $\mathcal{O}(h)$ shows that the error involved is of the order of the step size h. In the case of the central representation, the error is of the order of h^2.

PDEs contain functions of two variables at least such that the discretization introduced has to be generalized. This is demonstrated with a classical example, the 1D time-dependent heat conduction problem for a finite region $0 \leq x \leq L$, which is given as

$$\frac{\partial T}{\partial t} = \lambda \frac{\partial^2 T}{\partial x^2} \quad \text{for} \quad 0 < x < L, t > 0 \tag{6.5}$$

This is as usual for PDEs subject to additional initial and boundary conditions

$$T(x,t) = f(x) \quad \text{at} \quad t = 0, \tag{6.6a}$$
$$T(x,t) = T_0(t) \quad \text{at} \quad x = 0, \tag{6.6b}$$
$$T(x,t) = T_L(t) \quad \text{at} \quad x = L, \tag{6.6c}$$

where $f(x)$, $T_0(t)$, and $T_L(t)$ are known functions of the spatial coordinate and the time, respectively. Instead of first-type boundary conditions (Dirichlet boundary conditions), other types can also be used.

The solution of the PDE in Eq. (6.5) can be obtained using the finite differences just defined. If for the partial derivative $\partial T/\partial t$ the forward difference and for $\partial^2 T/\partial x^2$ the central difference are used to discretize the PDE, this results in

$$\frac{T_{i,j+1} - T_{i,j}}{\Delta t} = \lambda \frac{T_{k-1,j} - 2T_{k,j} + T_{k+1,j}}{(\Delta x)^2} + \mathcal{O}[\Delta t, (\Delta x)^2], \tag{6.7}$$

where $T(x,t) \Rightarrow T(x_i, t_j) = T(i\Delta x, j\Delta t) \equiv T_{i,j}$. This kind of discretization is called explicit Euler method and a finite difference molecule is shown in Figure 6.5, which

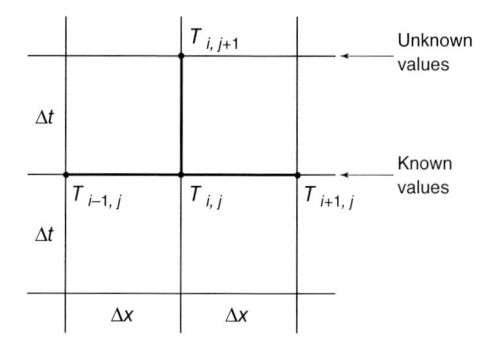

Figure 6.5 The finite difference molecules for the explicit Euler method (adopted from Ref. [7] Figure 12-12).

schematically illustrates the method. Equation (6.5) can be rearranged as

$$T_{i,j+1} = \xi T_{i-1,j} + (1-2\xi) T_{i,j} + \xi T_{i+1,j} \tag{6.8a}$$

with

$$\xi = \frac{\lambda \Delta t}{(\Delta x)^2} \tag{6.8b}$$

and the indices

$$i = 1, 2, \ldots, N-1 \quad \text{and} \quad j = 0, 1, 2, \ldots \tag{6.8c}$$

This method is called *explicit* because it is possible to determine explicitly the unknown temperature $T_{i,j+1}$ at the time t_{j+1} from the known temperatures $T_{i-1,j}$, $T_{i,j}$, and $T_{i+1,j}$ at the previous time step t_j such that a direct recursive determination can be performed. In the case of implicit methods, a step in between the calculations is a matrix inversion to obtain the resulting temperatures in the recursion. They are more involved computationally than the explicit methods, but they have the advantage that they are unconditionally stable, which is not the case for the explicit ones.

The deciding parameter for the stability of the explicit Euler method for the problem discussed here is ξ, see Eq. (6.8b). The numerical scheme becomes unstable if the following criterion

$$0 < \xi \equiv \frac{\lambda \Delta t}{(\Delta x)^2} \leq \frac{1}{2} \tag{6.9}$$

is violated. Practically, this means that, as the thermal diffusivity λ and the step size in the spatial domain Δx are fixed, there exists a limiting maximum for the temporal step size permitted.

The computational procedure of the explicit Euler method can be summarized in three main steps [7]:

1) Start of the calculation with $j = 0$. In this first step, the computation of $T_{i,1}$ is performed for $i = 1, 2, \ldots, N-1$. Here the right-hand side of Eq. (6.8a) is known from the initial condition, that is, Eq. (6.6a), meaning $f(x_i) \equiv f_i$ for $i = 1, 2, \ldots, N-1$.

2) Now set $j = 1$ and compute $T_{i,2}, i = 1, 2, \ldots, N-1$, that is, the values from the previous step and the right-hand side of Eq. (6.8a) are used to determine the second time step.
3) The computation is repeated for each subsequent time step. A termination of this continued procedure is performed until, on the one hand, a specific time or, on the other hand, a specified value of the temperature is reached.

The procedure is illustrated later on in this chapter by an example.

6.2
Finite Volume Method

Another} method that can be used for the modeling of thermoelectric elements and devices is the *finite volume method* (FVM). Although this method is not yet extensively used in thermoelectricity as some specialized algorithms are needed, it is another promising method for solving problems connected with thermoelectric materials and devices. The FVM is a method that has more often been used for finding the solution of hydrodynamical and fluid flow problems until now, see, for example [15].

In Reference [42], the FVM is explained by treating a thermoelectric couple within a 2D approach. In this calculation, a system of two PDEs is solved to gain information about the temperature field T and the electrical potential field V. The two PDEs require two boundary conditions each and some initial conditions as usual for PDEs. The material properties are supposed to be temperature dependent, which causes the PDEs to be nonlinear, and hence a numerical treatment is needed. An important feature of the FVM is that the physical domain is divided into nonoverlapping control volumes in which the physical conservation laws (mass, momentum, energy) are supposed to be fulfilled. An example for a 2D representation of a control volume is shown in Figure 6.6.

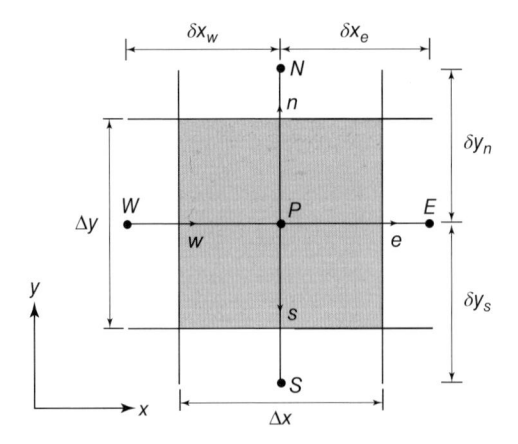

Figure 6.6 Typical control volume in 2D (adapted from Figure 3 (a) in Ref. [42]).

An integration cell of the discretized domain is represented by the center of the control volume. The field properties (T, V) are calculated at theses centers of the control volumes, whereas the fluxes (**q**, **j**) are determined at the surface of the control volume.

In Reference [6], the FVM is explained for the 3D case of a TE element. After the discretization of the domain and the transformation of the governing equations from differential to algebraic equations, the general procedure of the FVM can be given as follows:

1) Input of the leg geometry (by specifying the dimensions),
2) Input of the TE material properties in dependence on the temperature,
3) Specification of the boundary conditions,
4) Calculation of the cell properties: cell volume, cell face areas, spacing between the centers of the control volumes,
5) Specification of the time steps, convergence, and stability properties to ensure a stable and rapid convergence toward the steady-state solution,
6) Set the time level index to zero, starting the calculation,
7) Specification of the initial conditions or initial guess for the two fields T and V,
8) Calculation of the TE material properties for the given temperature (taken from the previous step) at every cell center,
9) Calculation of the temperature at the next time level,
10) Calculation of the TE material properties according to the predetermined temperature,
11) Calculation of the electrical potential V, and
12) Checking the convergence: either go on with the iteration from step 9 or terminate it if a certain criterion is reached.

For a detailed explanation of the procedure with examples, see [6].

To conclude this section, it is shown how Fourier's heat equation, as a PDE that involves flux terms, can be handled in the framework of FVM (without loss of generality in 1D), see for example, [43–45]. This can be generalized easily for the thermal energy balance containing the thermoelectric contribution as well as for the 2D or 3D case. With the heat equation

$$\rho_{md} c_p \frac{\partial T}{\partial t} = \frac{\partial}{\partial x}\left(\kappa \frac{\partial T}{\partial x}\right) \tag{6.10}$$

the time-dependent temperature profile $T = T(x, t)$ is the target function. The 1D element, which is investigated, is divided into subintervals or meshes, see Figure 6.7.

In Figure 6.7, an equidistant mesh is shown. In general, it is possible to use different mesh sizes for different locations or nonuniform meshes within regions. In most cases, this would lead to minor changes. For the calculation of the fields, here just the temperature field is observed, they are averaged in every subinterval such

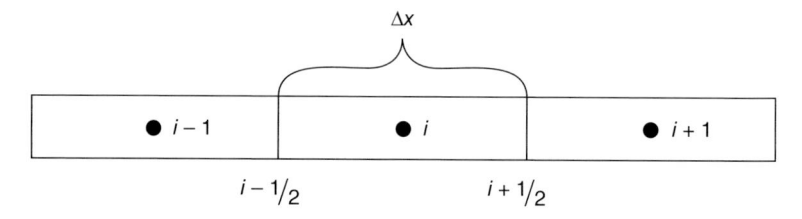

Figure 6.7 A section of the 1D mesh. The center node is x_i, whereas the boundaries are located at $x_{i-1/2}$ and $x_{i+1/2}$. (Figure adapted from Figure 3 in Ref. [45].)

that $T(x, t)$ is substituted by a vector of time-dependent values, for example, the ith entry of $\overline{T}(t)$ is

$$\overline{T}_i(t) := \frac{1}{\Delta x} \int_{x_{i-1/2}}^{x_{i+1/2}} T(x, t)\,dx. \tag{6.11}$$

The PDEs have to be averaged over the subintervals such that Eq. (6.10) results in

$$\varrho_{md} c_p \frac{d\overline{T}_i}{dt} = \frac{1}{\Delta x}\left(\kappa \frac{\partial T}{\partial x}\bigg|_{x_{i+1/2}} - \kappa \frac{\partial T}{\partial x}\bigg|_{x_{i-1/2}}\right). \tag{6.12}$$

If the material properties are supposed to be constant within the subvolume, then $\overline{\varrho_{md} c_p\, dT/dt} = \varrho_{md} c_p\, d\overline{T}_i/dt$ without any approximation. If they are functions of the position or temperature, they also have to be averaged. It is not possible to determine the right-hand side in terms of the averaged $\overline{T}$; therefore, the approximation

$$\kappa \frac{\partial T}{\partial x}\bigg|_{x_{i+1/2}} \approx \frac{1}{\Delta x}(\kappa \overline{T}_{i+1} - \kappa \overline{T}_{i-1}) \tag{6.13}$$

is used to describe the heat flux. The PDE is then approximated by a system of ODEs

$$\varrho_{md} c_p \frac{d\overline{T}_i}{dt} = \frac{\kappa}{\Delta x}\left[\left(\frac{\overline{T}_{i+1} - \overline{T}_i}{\Delta x}\right) - \left(\frac{\overline{T}_i - \overline{T}_{i-1}}{\Delta x}\right)\right] \tag{6.14}$$

Special care has to be taken concerning the discretization at the boundary. For details on that, see [43–45].

6.3
Finite Element Method

The finite element analysis is a well-established method in many areas of natural and technical sciences. In the mathematical modeling of many biological, chemical, physical phenomena, and engineering problems, PDEs are used. Very often, a closed or an analytical solution cannot be found or is impracticable. Therefore, numerical approximations are used to find the unknown solution. A particular class of numerical techniques for such an approximate solution of PDE are the *finite element methods* (FEM). These methods have been proposed first in Richard Courant's seminal work in 1943 [46]. However, the relevance of this work was not properly recognized at that time and the idea was forgotten. In the early 1950s, the FEM was rediscovered by engineers. Its mathematical analysis began much later in the 1960s, see, for example, [47]. There is a great number of introductory books on FEM in general and in particular with focus on fluid dynamics and heat transfer, see for example, [48–55]. The advantages of FDM have already been demonstrated in Section 6.1. It is conceptually simple and easy to implement. A crucial limitation of FDM is that the domain of interest has to be divided into rectilinear cells [49]. This means that especially in the vicinity of inclined boundaries, a step-like pattern is found as geometrically improper. FEM is not subject to such a limitation. The boundaries of the domain of interest can be matched accurately. The nonintersecting elements in FEM can have variable size. The field variable, for example, the temperature, is allowed to vary inside an element according to a chosen interpolation function such that with FEM, a more accurate approximation can be found. Therefore, this method is often used in heat transfer problems.

In thermoelectricity, this method is used because it allows modeling of arbitrary geometrical shapes of the components of a TE device, temperature-dependent material properties can be used, and various forms of boundary and load conditions can be applied. Clearly, an advantage of this method is that it can be easily adapted to different sets of constitutive equations [56].

In the late 1990s, this method was introduced in the framework of thermoelectricity in several works by Buist and Lau [57–59]. Nowadays, more and more commercial software solutions allow the treatment of thermoelectric problems, see for example, [4, 56, 60–66].

In Figure 6.8, the general algorithm scheme of the FEM is illustrated and the following steps are used in a finite element analysis (taken from Ref. [4, 48]):

- Discretization of the solution domain,
- Selection of a solution interpolation on the finite elements,
- Formation of the basis functions,
- Calculation of the differential problem residual with the use of approximate solutions in the form of series,
- Formation of the stiffness matrices for elements,
- Residual orthogonalization,
- Assembly of the stiffness matrices according to the elements,
- Accounting the boundary conditions, and
- Solution of algebraic equations system.

6.4
Performance Calculation of a TEG - A Case Study

To calculate the performance of a thermoelectric module (TEM) in a simplifying 1D approach, a basic unit is chosen consisting of two semiconductor pellets, that is, one p-type and one n-type pellet, which are electrically connected in series by

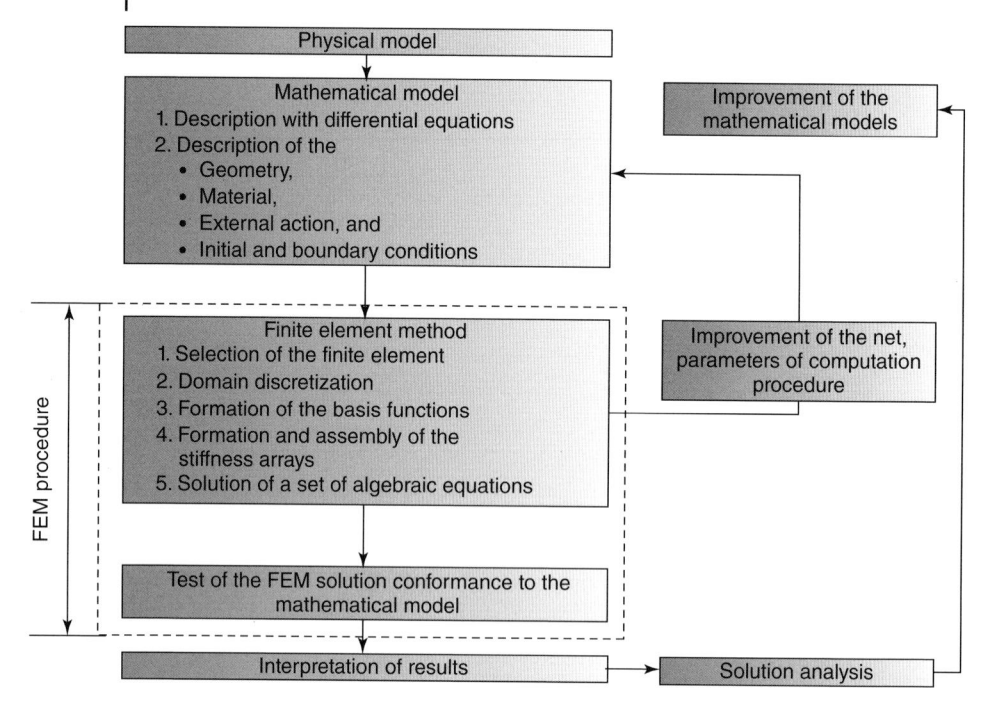

Figure 6.8 General scheme of problem solution by FEM (adapted from Figure 4 in Ref. [4]).

a conducting (metal) bridge. Usually, an additional electrically insulating layer, for example, made of ceramics, is positioned on top of the electrical connection such that a coupling of the module to the heat source and sink via heat exchangers is possible without causing short circuits between neighboring bridges.

Such a basic unit, forming a Π shape thermocouple, is shown in Figure 6.9. An interconnection of several basic units electrically in series and thermally in parallel

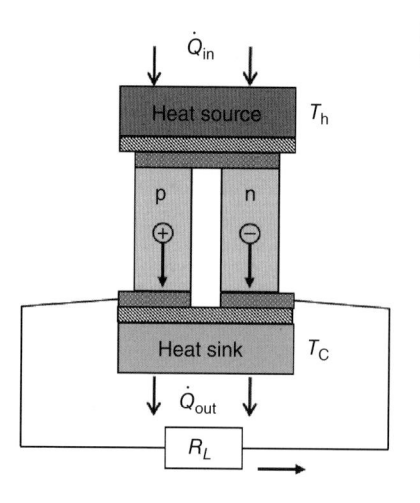

Figure 6.9 Classical design of a thermocouple (Π-shaped).

Table 6.1 Analogy between thermal and electrical quantities.

Thermal quantities		Electrical quantities	
Parameter	Unit	Parameter	Unit
Thermal conductivity κ	W/mK	Electrical conductivity σ	S/m
Temperature difference $\Delta T = T_h - T_c$	K	Electric voltage $V = \varphi_1 - \varphi_2$	V
Thermal resistance $R_{th} = \frac{1}{\kappa} \cdot \frac{L}{A_c}$	K/W	Electrical resistance $R = \frac{1}{\sigma} \cdot \frac{L}{A_c}$	Ω
Heat flow $\dot{Q} = \frac{\Delta T}{R_{th}}$	W	Electrical current $I = \frac{U}{R}$	A

is the principle to end up with a TEM. By neglecting the boundary effects, the investigation of a basic unit is sufficient to estimate the performance of a TEM.

For the calculation of the performance of a TEM, several parameters and quantities have to be known. There is an analogy between the thermal and electrical system parameters of a TEM, which is highlighted in Table 6.1.

The material properties are among the most important parameters. Along with the thermal and electrical conductivity, κ and σ, the Seebeck coefficient α of the p-type and n-type thermoelectric material, each, has to be known. If the transient behavior has to be determined, the mass density ϱ_{md} and the specific heat c_p have to be known. The geometry and topology of the elements, bridges, and modules represent the next group of parameters. With this knowledge, the boundary conditions of the problem can be applied. The initial state of the system is needed in the case of a transient calculation, and this is compiled in the initial conditions. Additional and external parameters that complete the problem formulation are, for example, the load resistance and the electrical and thermal contact resistances, with the latter being neglected in an idealized model.

In Table 6.2, the relations between the system properties, operational and performance parameters of the single pellets, thermocouples, and finally, a TEM consisting of N thermocouples are compiled. Ideal conditions are assumed here and the boundary effects are neglected as a thermocouple at the margin of the module is normally exposed to other thermal conditions than inside the TEM because of a different radiation background. A homogeneous current density distribution due to ideal contacts and ideally conducting bridges leads to quasi-1D conditions.

Table 6.3 shows the dependence of the performance parameters of a TEM on its thickness L. Again an ideal TEM free of losses is assumed and the temperature difference ΔT is kept constant.

As can be seen, some of the quantities such as the efficiency, the Seebeck coefficient, and the terminal voltage are independent of the thickness L. Most of the quantities show an inverse dependence related to the thickness, for example, the smaller the thickness, the larger is the heat flux for a fixed temperature difference.

Table 6.2 Relations between the parameters and quantities for the single pellets (p-type/n-type), the thermocouple, and the TEM (idealized conditions).

Quantity	Pellet	Thermocouple	TEM
Seebeck coefficient	α_p, α_n	$\alpha_{pn} = \alpha_p + |\alpha_n|$	$N \cdot \alpha_{pn}$
Electrical resistance	R_p, R_n	$R_{pn} = R_p + R_n$	$N \cdot R_{pn}$
Thermal conductance $K = R_{th}^{-1}$	K_p, K_n	$K_{pn} = K_p + K_n$	$N \cdot K_{pn}$
Voltage	V_p, V_n	$V_{pn} = V_p + V_n$	$N \cdot V_{pn}$
Current	$I_p = I_n$	I_{pn}	$I_m = I_{pn}$
Power	P_p, P_n	$P_{pn} = P_p + P_n$	$N \cdot P_{pn}$
Heat flow	$\dot{Q}_p, \dot{Q}_n$	$\dot{Q}_{pn} = \dot{Q}_p + \dot{Q}_n$	$N \cdot \dot{Q}_{pn}$
Efficiency	η_p, η_n	$\eta_{pn} = P_{pn}/\dot{Q}_{pn}$	$\eta_m = \eta_{pn}$

Table 6.3 Thickness dependence of the properties of an ideal TEM for a fixed temperature difference.

Quantity	L = 1	L
Seebeck coefficient	α_m	α_m
Electrical resistance	R_m	$R_m \cdot L$
Thermal conductance	K_m	K_m/L
Terminal voltage	V_α	V_α
Optimal current	I_{opt}	I_{opt}/L
Electrical power	P_{el}	P_{el}/L
Heat flow	$\dot{Q}$	$\dot{Q}/L$
Efficiency	η	η

6.4.1
Averages of the Material Properties

In Chapter 2, the classical CPM calculation as developed by Ioffe was introduced. If measured temperature-dependent material properties are given, then this method can also be used just by inserting a suitable average of the material properties into the CPM formulae (Table 6.4).

If the boundary conditions are fixed values, T_h and T_c, then a straightforward choice is the average over temperature. The WAV was introduced by Ioffe for averaging the temperature-dependent resistivity [67, p. 62]. For the Seebeck coefficient, Ioffe suggested the VMT, whereas the thermal conductivity was averaged by

$$\overline{\kappa} = \frac{2}{\frac{1}{\kappa_h} + \frac{1}{\kappa_c}}, \tag{6.15}$$

6.4 Performance Calculation of a TEG - A Case Study

Table 6.4 Different averages of the material properties (here shown for the Seebeck coefficient).

Notation	Short	Formula
Temperature average	TAv	$\bar{\alpha} = \frac{1}{\Delta T} \int_{T_c}^{T_h} \alpha(T) dT$
Value at the mean temperature	VMT	$\alpha_{T_m} = \alpha(T_m)$
mean of the boundary values	MBV	$\alpha_m = \frac{\alpha(T_h)+\alpha(T_c)}{2} = \frac{\alpha_h+\alpha_c}{2}$
Weighted average by Ioffe	WAv	$\alpha_{av,w} = \frac{1}{\ln(T_h/T_c)} \int_{T_c}^{T_h} \frac{\alpha(T)}{T} dT$

which is related to a good approximation of the thermal resistance of the element as $1/\kappa$ increases nearly linearly with temperature in the extrinsic range for many TE materials, see [67, p. 61]. The value at the mean temperature is often chosen because it is easy to determine. In the following, it is shown which average resembles the fully temperature-dependent case best. For this, no general rules can be given but only specific ones for a material class in a certain temperature range. This is because the curvature of the TE properties over temperature is different from one material group to another, depending in a complex way on band structure and scattering of carriers and phonons.

6.4.2 Processing Measured Material Properties

Material properties are measured at certain temperatures. As a result, a list of value pairs consisting of the temperature and the corresponding value of the measured material property is recorded. To enter numerical calculations, continuous functions of the material properties are needed since interpolated values might be needed at any temperature. To get this functional relation, the measurement data can be suitably fitted by a polynomial using commercial software. Such a commercial software could be a computer algebra system. Fitting by a polynomial is mathematically straightforward, but if a physical law of the temperature dependence of a material property is known, for example, $\kappa \propto T^{-1}$, fitting can also be performed with respect to such a law.

Here a TEM built of skutterudite material is chosen as an example. The first step is importing and manipulating the data such that fitting and displaying are possible.

In the MATHEMATICA snippet shown after the first step is demonstrated schematically. This passage can be understood as an intuitive pseudocode, which can also be used in other software with special languages.

```
SetDirectory["PATH"];
(* setting a special path *)
kappa-p-data=Import["kappa-p.dat","Table"];
(* importing data of the thermal conductivity  (p-type ele-
ment) *)
kappa-p-fit=Fit[kappa-p-data,{1,T,T^2,T^3},T]
(* fitting with respect to a third-degree polynomial *)
kappa-p-Tav=Integrate[kappa-p-fit,{T,Tc,Th}]/(Th-Tc)
(* determination of the temperature average *)
```

Having a continuous function of $\kappa_p(T)$ an integration is possible and the averages can be determined. The coefficients of the polynomial, that is, $\kappa(T) = A_\kappa + B_\kappa T + C_\kappa T^2 + D_\kappa T^3$, are given in Table 6.5.

The coefficients of the polynomial fits of the other material parameters are listed in the appendix in Table A.1.

There is another way to fit the measured data where more statistic features can be gained. Besides the coefficients, different statistical measures can be found, such as the standard error, the t-statistic, the p-value, and R^2 (R squared).

Here just a short definition of each of these values is given. For more detailed information, a look into [68, 69] is suggested. The *standard error* is the standard deviation of the sampling distribution of a statistic. The *t-statistic* is a ratio of the departure of an estimated parameter from its notional value and its standard error. The *p-value* is a function of the observed sample results (a statistic) that is used for testing a statistical hypothesis. The *coefficient of determination*, denoted as R^2 or r^2 (R squared), is a number that indicates how well the data fit to a statistical model. This coefficient ranges from 0 to 1, where $R^2 = 1$ indicates that the regression line perfectly fits the data.

```
kappa-p-fit=LinearModelFit[kappa-p-data,{1, T, T^2, T^3},T];
kappa-p-fit["ParameterTable"]
```

Then following result is found

	Estimate	Standard Error	t-Statistic	P-Value
1	1.62114	0.464967	3.48657	0.0101756
T	0.00224319	0.00286433	0.783144	0.459214
T^2	-0.0000100511	$5.593920711989988 \cdot 10^{-6}$	-1.79678	0.115427
T^3	$9.627993536462141 \cdot 10^{-9}$	$3.488715363292524 \cdot 10^{-9}$	2.75975	0.028106

```
kappa-p-fit["RSquared"]
R^=0.957983
```

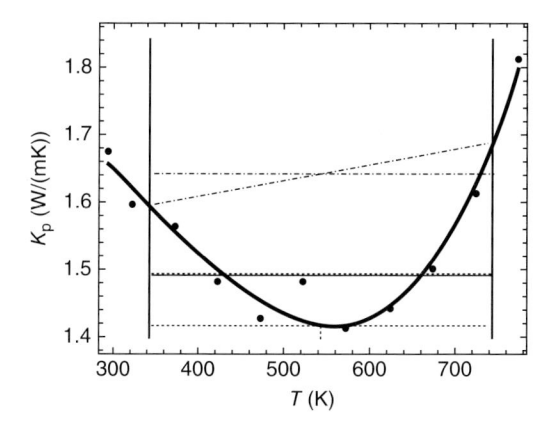

6.4 Performance Calculation of a TEG - A Case Study

Figure 6.10 Thermal conductivity and corresponding averages. (Solid line - TAv, dotted line - VMT, dash-dotted line - MBV, and dashed line - WAv, for explanation see Table 6.4, for values see Table 6.7)

Table 6.5 Coefficients of the polynomial fit of the thermal conductivity of a p-type skutterudite.

$\kappa_p(T)$ (W/m K)	A_κ	B_κ	C_κ	D_κ
Coefficients	1.62114	0.00224319	−0.0000100511	$9.62799 \cdot 10^{-9}$

Table 6.6 Temperature intervals of the single measurements.

κ_p	σ_p	α_p
[294.85 K, 773.45 K]	[342.1 K, 753.32 K]	[342.18 K, 758.08 K]
κ_n	σ_n	α_n
[294.35 K, 773.35 K]	[333.85 K, 748.39 K]	[334.07 K, 750.59 K]

For further calculations involving more than one material property, the maximum overlapping temperature interval has to be determined. This is done to prevent extrapolation of the material data.

By considering all TE properties of a pn couple, from Table 6.6, an intersectional temperature interval is fixed by setting $T_c = 343.15K(= 70°C)$ and $T_h = 743.15K(= 470°C)$. This choice of boundary temperatures results in a temperature difference of $\Delta T = 400K$ and a mean temperature $T_m = 543.15K(= 270°C)$. Knowing these values and having continuous functions of the material properties, all averages as introduced in Table 6.4 can be calculated.

6.4.3
Different Averages and the Corresponding Performance Values

As an example, the thermal conductivity of the p-type skutterudite is shown in Figure 6.10. The dots are the measured data, which are fitted to a third-order

Table 6.7 Values of the average of the thermal conductivity.

Average	κ (W/m K)	Linestyle
TAv	1.49226	Solid
VMT	1.41710	Dotted
MBV	1.64259	Dash-dotted
WAv	1.49313	Dashed

polynomial illustrated as thick solid line. The values of the different averages are calculated for the given temperature interval limited by the two vertical lines and summarized in Table 6.7. The temperature average and the weighted average are almost the same here, whereas the MBV is much larger and the VMT smaller.

The same procedure as for the thermal conductivity has to be performed analogously with the other material properties. In Figure A.1, the material properties of the p-type skutterudite and the corresponding averages are shown, whereas in Figure A.2, the same are shown for the n-type skutterudite, see appendix. The values of the different averages can be found in Tables A.2 and A.3.

6.4.4
Power Factor and Figure of Merit

As for the material properties, continuous functions have been determined, it is possible to obtain the temperature-dependent curve for the power factor f and the figure of merit zT for both the p-type and the n-type materials in the chosen temperature interval.

$z_p T$	T (K)	Value
$zT(T_c)$	343.15 K	0.199716
$zT(T_h)$	743.15 K	0.689838
$zT(T_m)$	543.15 K	0.573409
$(zT)_{max}$	680.29 K	0.728812

$f_p(T)$	T (K)	10^{-3} W/m K^2
$f(T_c)$	343.15 K	0.929113
$f(T_h)$	743.15 K	1.56763
$f(T_m)$	543.15 K	1.49604
f_{max}	665.38 K	1.63816

The figure of merit of the p-type material has a maximum in the chosen temperature interval at 680K with a value of about 0.73. The smallest value of the figure of merit is found at T_c with a value of about 0.2. A monotonous increase from this value to the maximum can be seen in Figure 6.11(a). Qualitatively, the same behavior as for the figure of merit is observed for the power factor. The maximum peak of the power factor is found at a smaller temperature at 665K in comparison to the maximum found for the figure of merit. In Appendix A.4, the figures and values for the n-type material are illustrated. In general, the values for the p-type and n-type materials are different. This can also be seen in the example here. For example, the maximum of the figure of merit of the n-type material is with $(zT)_{max} = 0.91$ larger than the value found for the p-type material.

6.4.5
Optimal Performance Based on Averaged Material Properties

Calculations of approximate values of optimal performance can be carried out based on averaged material properties by means of the CPM theory, that is, the maximum power output Eq. (2.55) and the maximum efficiency Eq. (2.57) as well as the corresponding current densities from Eqs. (2.54) and (2.56). For this calculation, some geometrical data (the length L and cross-sectional area A_c of the TE element) have to be given besides the material properties. Dirichlet boundary conditions are assumed for the calculation, that is, fixed hot and cold side temperatures. In Table 6.8, the calculated values utilizing different averages are summarized for a TE element that is built of a p-type skutterudite. The length of the element is chosen to be 5mm. Besides the calculation of the performance values from the averages, the same calculation is performed using the maximum values of the figure of merit and the power factor to give an estimation of the upper limit, which would be valid if these values would be present over the whole temperature interval. Often, but erroneously, the maximum ZT values from the measured temperature curves are inserted into the well-known Ioffe formula for maximum efficiency (2.32) in crude estimations, resulting in misleading overestimations of expected performance of a TE material. Our comparison illustrates the deviation involved.

An analogous calculation is performed for an element consisting of n-type skutterudite material. The results can be found in the appendix in Table A.4.

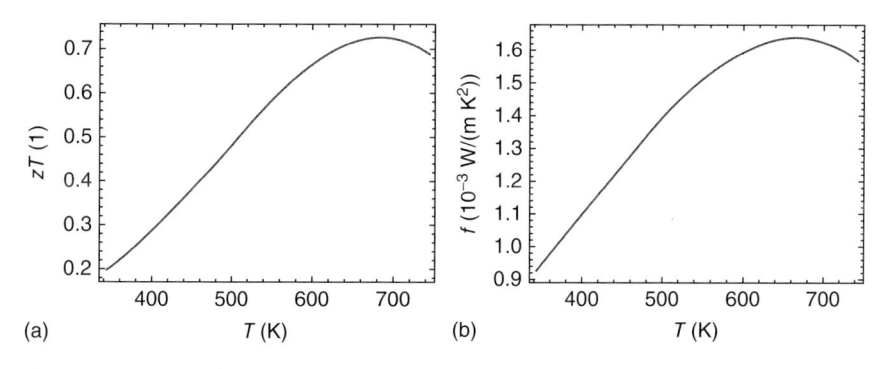

Figure 6.11 Figure of merit (a) $z_p(T)T$ and power factor (b) $f_p(T)$ of the p-type element.

6 Numerical Simulation

Table 6.8 Calculation results of optimal performance values obtained from different averages using the formulae of CPM theory.

	Unit	TAv	VMT	MBV	WAv	Max.
zT_m	1	0.51	0.57	0.41	0.49	0.73
f	10^{-3} W/m K^2	1.41	1.50	1.24	1.36	1.64
p_{max}	W/m^2	11260.9	11968.3	9891.36	10859.4	13105.3
$j_{opt,p}$	A/m^2	319731	328272	302098	315131	338552
η_{max}	%	7.31	7.98	6.10	7.10	9.54
$j_{opt,\eta}$	A/m^2	286784	291234	276273	283620	366005

a) For the p-type element, a length of 5mm and a fixed temperature difference of 400K are assumed.

As the power factors and figure of merit for both materials (p-type and n-type skutterudites) are different, so do the maximum performance and the corresponding current density. The electrical current through a thermocouple (pair of p-type and n-type TE elements) is the same as the elements are connected in series. The current densities can be changed by changing the cross-sectional area. In this sense, the elements can be adjusted just that the performance is maximized. The procedure for this is explained later in this section.

6.4.6
Comparison between CPM, FDM, and FEM Simulations

The next step is to calculate the performance of a p-type TE element by means of both FDM and FEM simulation. The temperature dependence of the material is taken into account with both numerical methods. A comparison between these exact results and the results gained from the CPM calculations with regard to the different choices of averages allows us to decide which average is best for this material and the temperature interval. It shall be noted here that the usage of averages in the CPM formulae as described earlier is a purely empirical ansatz driven by the desire for a simple formula. Actually, it is not possible to make a decision on a physically correct average from theoretical consideration in the framework of the continuum theory, see also the introductory remarks of Chapter 3.

6.4.6.1 Finite Difference Scheme for a TE Element

In commercial computer algebra systems, the numerical solution of differential equations is often already implemented. In MATHEMATICA it is the `NDSolve[...]` command that can be used to numerically solve the generalized heat equation for thermoelectric material with temperature-dependent material properties introduced in Chapter 2 as Eq. (2.6), that is,

$$\kappa(T)\frac{d^2 T}{dx^2} + \frac{d\kappa}{dT}\left(\frac{dT}{dx}\right)^2 - jT(x)\frac{d\alpha}{dT}\frac{dT}{dx} = -\frac{j^2}{\sigma(T)} \quad (6.16)$$

```
numsol=NDSolve[{kapfit[T[x]]*T"[x]+kapfit'[T[x]]*(T'[x])^2
-jc*T[x]*alpfit'[T[x]]*10^(-6)*T'[x]==-jc^2/(sipfit[T[x]]
*100),
T[0]==743.15,T[0.005]==343.15},T[x],{x,0,0.005}]
```

For the calculation of the temperature profile, the current density j has to be fixed in advance.[1] Here $j = 319731 \text{A/m}^2$ is used, that is, the value for $j_{\text{opt},p}$ found for TAv. The material properties are implemented by the fitted functions denoted here as `kapfit[T[x]]` for $\kappa(T)$, `alpfit[T[x]]` for $\alpha(T)$, and `sipfit[T[x]]` for $\sigma(T)$. Additional factors are introduced to guarantee that the units are handled in the right way, as, for example, the electrical conductivity is measured in S/cm, it has to be transformed to S m. Furthermore, the Dirichlet BCs are supplied, that is, $T(0) = T_h$ and $T(L = 0.005 \text{ m}) = T_c$. Several possibilities in the choice of the numerical method are given for the `NDSolve` command. If no computer algebra software is available, the temperature profile can be calculated with the (forward) Euler method directly, which can be programmed in an arbitrary language. As Eq. (6.16) is a differential equation of second order, the first step is to build a system of two differential equations of first order, that is,

$$Y(x) = \frac{dT}{dx} \tag{6.17a}$$

$$-\frac{j^2}{\sigma(T)} = \kappa(T)\frac{dY}{dx} + \frac{d\alpha}{dT}[Y(x)]^2 - jT\frac{d\alpha}{dT}Y(x) \tag{6.17b}$$

The material properties are given as a polynomial function, that is,

$$\kappa_p(T) = A_\kappa + B_\kappa T + C_\kappa T^2 + D_\kappa T^3, \tag{6.18a}$$

$$\sigma_p(T) = A_\sigma + B_\sigma T + C_\sigma T^2 + D_\sigma T^3, \quad \text{and} \tag{6.18b}$$

$$\alpha_p(T) = A_\alpha + B_\alpha T + C_\alpha T^2 + D_\alpha T^3, \tag{6.18c}$$

where the coefficients can be found in Table A.1. The material data as well as T and Y are calculated at the grid point $x_n = n\Delta x = nh$ with $n = 0, 1, \ldots, N$. The grid points are determined by the subdivision of the interval $0 < x < L$. In the Euler method, the differential quotient is substituted by a difference quotient, that is,

$$Y(x_n) = Y_n = \frac{T(x_{n+1}) - T(x_n)}{\Delta x} = \frac{T_{n+1} - T_n}{h} \Rightarrow T_{n+1} = T_n + hY_n. \tag{6.19}$$

For Y, which is a substitute for the derivative $T'(x)$, the following recursion formula can be determined from Eq. (6.17b)

$$Y_{n+1} = Y_n + h\left[j\frac{T_n\alpha'(T_n)}{\kappa(T_n)} - \frac{\kappa'(T_n)}{\kappa(T_n)}Y_n^2 - \frac{j^2}{\sigma(T_n)\kappa(T_n)}\right]. \tag{6.20}$$

With the formulae in Eqs. (6.19) and (6.20), the temperature distribution $T(x)$ can be recursively calculated. Note that for both the temperature at $x = 0$, $T(0) = T_0$

---
1) here denoted as `jc` to prevent confusion with the imaginary unit.

and the derivative $Y(0) = Y_0$ have to be known. The latter is not known a priori. Only T_0 is known and equivalent to T_h. For the determination of Y_0, the so-called *shooting method* is applied where a certain Y_0 is chosen, the temperature profile is calculated with the recursion algorithm, and the resulting temperature distribution is checked at the end point $T(x_N)$. This should be the cold side temperature. A bisection method is chosen to determine the right starting derivative Y_0. The main idea of this method is to first find one Y_0 that overestimates the temperature at the end point, that is, $T(x_N) > T_c$ and another one that underestimates it, that is, $T(x_N) < T_c$. The resulting interval of starting derivatives is reduced until the absolute of the difference is less than a small number ϵ, that is, $|T(x_N) - T_c| < \epsilon$.

The following program shows the recursion algorithm of the Euler method. It has to be performed in two steps. In the first step, the starting derivative Y_0, which is denoted as s0 in the program, has to be determined by means of the bisection method. In the second step, the resulting temperature profile is calculated with regard to the predetermined s0.

```
NumEulerTE[Th_, Tc_, jc_, s0_, Aka_, Bka_, Cka_, Dka_,
Asi_, Bsi_, Csi_, Dsi_, Bal_, Cal_, Dal_, L_, N_]:=
Module[{x=0, t=Th, n, m=N, s=s0}, (* def. of local vari-
ables *)
(* definition of the step size *)
h=L/N;
(* definition grid of positions *)
X = Table[x + n*h, {n, 0, N}];
(* definition of temperature derivative distribution *)
Y = Table[s, {n, 0, N}];
(* definition of temperature distribution *)
T = Table[t, {n, 0, N}];
(* Recursion loop *)
For[n = 1, n <= N, n++,
    T[[n + 1]] = T[[n]] + h*Y[[n]];
    Y[[n + 1]] = Y[[n]] - h*((Bka + 2*Cka*T[[n]] +
    3*Dka*T[[n]]^2)/
    (Aka + Bka*T[[n]] + Cka*T[[n]]^2 + Dka*T[[n]]^3)*
    Y[[n]]*Y[[n]]+
        jc^2/((Asi + Bsi*T[[n]] + Csi*T[[n]]^2 +
        Dsi*T[[n]]^3)*
        (Aka + Bka*T[[n]] + Cka*T[[n]]^2 + Dka*T[[n]]^3)) -
        jc*T[[n]]*Y[[n]]*(Bal + 2*Cal*T[[n]] + 3*Dal*
        T[[n]]^2)/
        (Aka + Bka*T[[n]] + Cka*T[[n]]^2 + Dka*T[[n]]^3));
    ];
(* Choose one of the following commands *)
```

```
(* First for the determination of the starting deriva-
tive *)
Return[T[[N + 1]] - Tc]
(* Second command is meant to write out the temperature
profile *)
Return[Transpose[{X, T}]]
]
```

In the next program example, the bisection method is illustrated. To do so, the first command in the latter program is chosen to get the difference between T_N and the aimed T_c. An auxiliary function $f[s0]$ is defined, where all the parameters are already defined, such as the coefficients from the material data, the number of steps N, the length L, the current density j, and the boundary temperatures. The only free parameter in this function is $s0$. On the one hand, the starting temperature derivative is supposed to be smaller than zero as it should decrease from $x = 0$ to $x = L$. On the other hand, the starting slope should be larger than $-\Delta T/L$. Both values fix a starting interval for the bisection algorithm.

```
(* Definition of an auxiliary function *)
f[s0_] =
  NumEulerTE[743.15, 343.15, 319731, s0, 1.62114177,
    0.00224319, -0.00001005,9.62799354*10^(-9),
375.27238112*100,
    0.88983346*100,-0.00221557*100,1.53044273*10^(-6)*100,
    -0.20817440*10^(-6),0.00114915*10^(-6), -9.90329205*
10^(-7)*10^(-6),
    0.005,10]

Bisection[a0_, b0_, lambda_] :=
  Module[{},
    a = N[a0];
    b = N[b0];
    c = (a + b)/2;
    i = 0;
    output = {{i, a, c, b, f[c]}};
    While[i < lambda,
      If[Sign[f[b]] == Sign[f[c]], b = c, a = c;];
      c = (a + b)/2;
      i = i + 1;
      output = Append[output, {i, a, c, b, f[c]}];];
    Print[NumberForm[TableForm[output, TableHeadings ->
        {None, {"i", "ak","ck","bk","f[ck], 16]];
    Print["   c = ", NumberForm[c, 16]];
```

```
            Print["   Delta c   = ±", (b - a)/2];
            Print["f[c] = ", NumberForm[f[c], 16]];
         ]
```

As the interval is fixed, the bisection algorithm can be started. With the command `Bisection[-400/0.005, 0, 20]`, the interval switching is repeated 20 times. The result of this recursive interval halving is summarized in Table 6.9.

The resulting value for starting slope is $s0 = -65832.48138427734$. Now it is possible to evaluate the temperature distribution for the proposed value of $j = j_{opt,p} = 319731 \text{ A/m}^2$.

In Figure 6.12, three different numerical methods for the calculation of the temperature profile are compared with an example. The p-type semiconductor material is chosen for a TE element with a length of $L = 5$ mm. The calculation is performed for a current density of $j = 319731 \text{A/m}^2$. With the solid line, the numerical solution with respect to MATHEMATICA's NDSolve is illustrated. The solid points are obtained from a FEM calculation, which is introduced in more detail in the next subsection. Both solutions are congruent. The solution evaluated by the Euler method is plotted with the open circles. There is a small deviation in the middle of the element, but this can be prevented if the number

Table 6.9 Example bisection algorithm.

i	a_k	c_k	b_k	$f[c_k]$
0	−80000.	−40000.	0.	157.153
1	−80000.	−60000.	−40000.	34.7865
2	−80000.	−70000.	−60000.	−24.5
3	−70000.	−65000.	−60000.	4.93024
4	−70000.	−67500.	−65000.	−9.8396
5	−67500.	−66250.	−65000.	−2.46824
6	−66250.	−65625.	−65000.	1.22763
7	−66250.	−65937.5	−65625.	−0.621147
8	−65937.5	−65781.3	−65625.	0.303032
9	−65937.5	−65859.4	−65781.3	−0.15911
10	−65859.4	−65820.3	−65781.3	0.071948
11	−65859.4	−65839.8	−65820.3	−0.0435843
12	−65839.8	−65830.1	−65820.3	0.014181
13	−65839.8	−65835.	−65830.1	−0.0147018
14	−65835.	−65832.5	−65830.1	−0.000260441
15	−65832.5	−65831.3	−65830.1	0.00696029
16	−65832.5	−65831.9	−65831.3	0.00334992
17	−65832.5	−65832.2	−65831.9	0.00154474
18	−65832.5	−65832.4	−65832.2	0.000642149
19	−65832.5	−65832.4	−65832.4	0.000190854
20	−65832.5	−65832.5	−65832.4	−0.0000347931

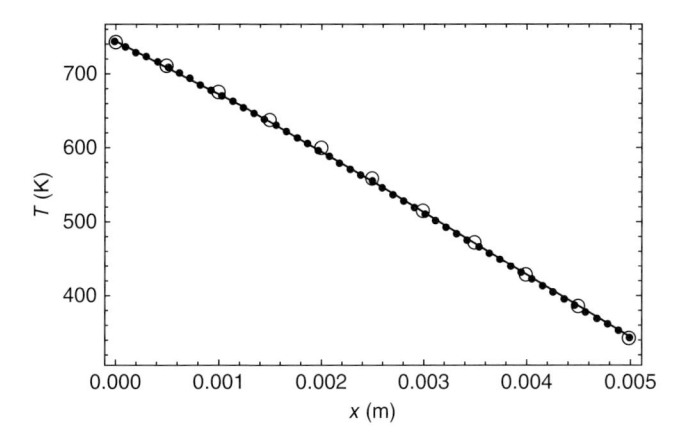

Figure 6.12 Numerical calculation of the temperature profile $T(x)$ in the interval $0 < x < L = $ 5mm for $j = 319731 A/m^2$. Solid line: NDSolve solution (FDM, MATHEMATICA), solid points: FEM (ANSYS), and open circles: forward Euler method (FDM)

of divisions of the interval N is increased. Here it is has been chosen as small as 10 for demonstration purposes.

6.4.6.2 Performance Parameters Dependent on the Current Density

The performance parameters depend on the current density j. The calculation performed so far in this subsection is for a particular fixed current density. The current density has to be varied to find the dependence of the performance parameters of a TEG element on this parameter. The program paragraph of this loop is shifted to the appendix, see Section A.5.1. The numerical results (FDM) in numbers for the chosen example are given in Table A.5 in Section A.5.2.

In Figures 6.13, the performance parameters of a TE element are illustrated graphically. Two numerical methods FDM (MATHEMATICA's NDSolve) and FEM (ANSYS) are used and compared. No significant deviation was observed for either of the parameters. The comparison of these "exact" (numerical) values can be used to value the different introduced averages as already explained. A cuboid with a length $L = 5$ mm and a cross-sectional area $A_c = 1$ mm^2 was taken for the FEM calculations. In Figure 6.13(a), the V-I characteristics are plotted. From the fit of the numerical data, two important quantities can be determined: on the one hand, the Seebeck voltage (open-circuit voltage), that is, $V_\alpha = V(I = 0)$, and on the other hand, the internal resistance R_i. Note that the internal resistance is, in general, not constant as σ depends on T. The temperature distribution $T(x)$ changes with changing electric current I and therefore R_i depends on I. The assumption of a constant R_i delivers a very good approximation in most cases. From V_α and R_i, it is possible to gain two effective values for the material properties, which can be compared with the introduced averages, because $V(I) = V_\alpha - R_i I \Rightarrow V = V(I) = \alpha_{\text{eff}} \Delta T - L/A_c \sigma_{\text{eff}}$ (Table 6.10).

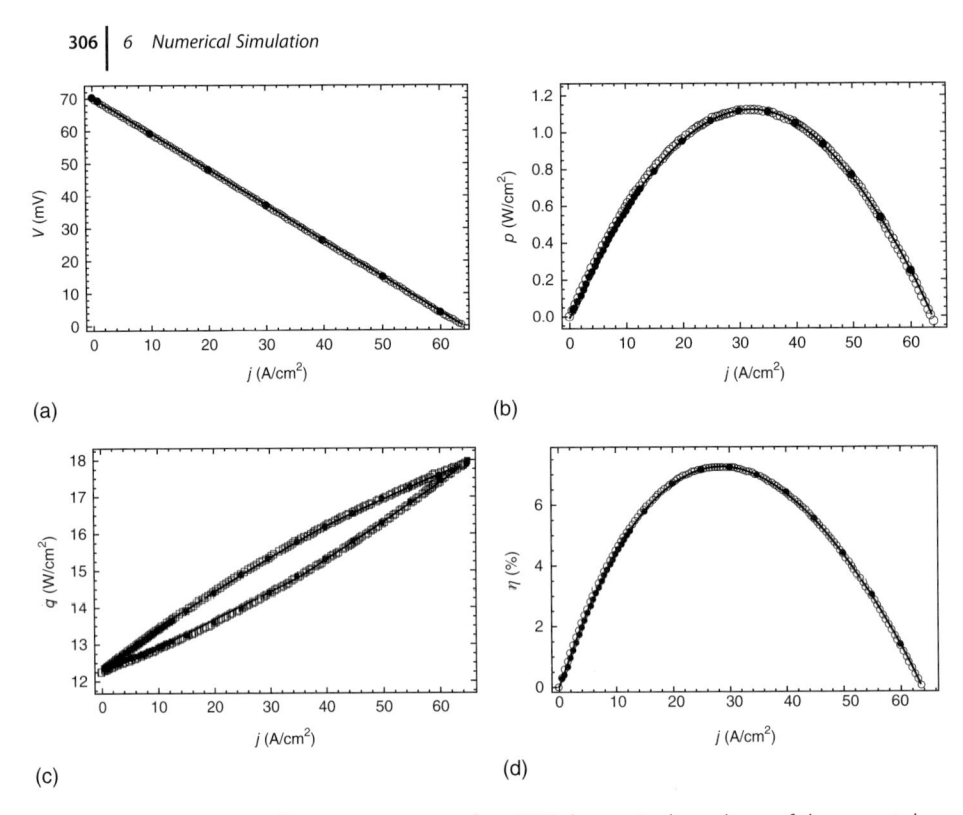

Figure 6.13 Performance parameters for a TEG element in dependence of the current density; Open circle: FDM (NDSolve), Points: FEM (ANSYS), Solid line: Fit of the FDM data (a) V–I characteristics (b) Power output density p (c) Heat flux $\dot{q}$ (d) Efficiency η

Table 6.10 Linear fit values for the $V(I)$ characteristics in case of FDM and FEM, respectively.

Unit	V_α (V)	R_i (Ω)	R^2 1	α_{eff} (μV/K)	σ_{eff} (S/cm)
FDM	0.0705051	0.110759	0.999997	176.263	451.431
FEM	0.07048	0.11066	1	176.2	451.834

An effective value for the thermopower α_{eff} can be determined from the Seebeck voltage as the temperature difference ΔT is constant and $U_\alpha = \alpha_{\text{eff}} \Delta T$. This effective value of the Seebeck coefficient of a TE element with temperature-dependent material data, that is, temperature-dependent Seebeck coefficient, is the same value that an equivalent TE element with constant Seebeck coefficient (CPM) would have under the same boundary conditions (boundary temperatures) to develop the same Seebeck voltage. An effective value for the electrical conductivity σ_{eff} or the resistivity $\varrho_{\text{eff}} = \sigma_{\text{eff}}^{-1}$ can be observed from the internal resistance R_i in an analogous manner. The classical relation $R_i = \varrho_{\text{eff}} L / A_c = L/(A_c \varrho_{\text{eff}})$ is used for this.

In Table 6.11, the effective values determined from FDM and FEM are compared to the averaged values. It is found that the temperature average (TAv) is the best choice for this example.

With the knowledge of V_α and R_i as parameters of the linear fit function, the short-circuit current (density) can be determined by their ratio $I_{sc} = V_\alpha/R_i$ ($j_{sc} = I_{sc}/A_c$), see Table 6.12. Half of the short-circuit current gives the current where the maximum electrical power is found, that is, $I_{opt,p} = I_{sc}/2 \Rightarrow P_{el,max} = V(I_{opt,p})I_{opt,p}$.

In Figure 6.13(c), the hot and cold side heat fluxes are plotted. The hot side heat flux is the upper concave curve, whereas the lower convex curve is the cold side heat flux. Here an effective value of the thermal conductivity can be observed by observing the heat flux for $j = 0$, that is, $\dot{q}_h(I = 0) = \dot{q}_c(I = 0) = \frac{\Delta T}{L}\kappa_{eff} \cdot \dot{q}_h(I = 0)$ and $\dot{q}_c(I = 0)$ obtained from the fit functions are slightly different due to numerical inaccuracy; therefore, a mean value is calculated. Finally, an effective thermal conductivity of $\kappa_{eff} = 1.49267$ W/m K is obtained. This effective value is almost the same as the temperature average (TAv).

The heat fluxes can be used to calculate the power output from thermal quantities alone. The difference in both heat fluxes is the power output density. In Figure 6.13(b), the electrical power output is plotted as a function of the current density. From the fit function, a maximum value of $p_{max} = 11233.2$ W/m² at a current density of $j_{opt,p} = 31.82$ A/cm² is found. An effective power factor can be defined from Eq. (2.55) in accordance to the CPM formula, that is, $p_{max} = f_{eff}(\Delta T)^2/4L \Rightarrow f_{eff} = p_{max} 4L/(\Delta T)^2$.

Table 6.11 Comparison between the effective values for α and σ and the averages.

Average	FDM		FEM		FDM		FEM	
Units	α_{eff} (μV/K)	Dev. (%)	α_{eff} (μV/K)	Dev. (%)	σ_{eff} (S/cm)	Dev. (%)	σ_{eff} (S/cm)	Dev. (%)
Effective	176.263	0	176.2	0	451.431	0	451.834	0
TAv	176.099	−0.09	176.099	−0.06	453.908	0.55	453.908	0.46
VMT	182.293	3.42	182.293	3.46	450.198	−0.27	450.193	−0.36
MBV	163.711	−7.12	163.711	−7.09	461.327	2.19	461.327	2.10
WAv	172.3	−2.25	172.3	−2.21	457.243	1.29	457.243	1.20

a) The remaining deviation between the obtained values of α_{eff} and α_{TAv} originates completely from numerical inaccuracy.

Table 6.12 Results from the V–I characteristics.

V_α (V)	R_i (Ω)	I_{sc} (A)	$I_{opt,p}$ (A)	$P_{el,max}$ (W)
0.07048	0.11066	0.6340	0.3170	0.01132

The same procedure can be performed for the treatment of the efficiency, which is shown in Figure 6.13(d). Here again, the maximum is determined from a fit function. Besides the polynomial fit (of third order), another fit is tested, based on the physical dependence of η on j, see Eq. (2.53), where the fit function

$$\eta = \eta(j) = \frac{aj + bj^2}{c + dj + ej^2} \tag{6.21}$$

is used in a nonlinear fitting procedure.

```
(* Nonlinear fit function *)
nlm = NonlinearModelFit[Table[{i*0.5, eta[i][[1]]},
{i, 0, 127}],
     (a j + b j^2)/(c + d j + e j^2), {a, b, c, d, e}, j]
(* Parameters of the statistics *)
nlm[{"ParameterTable"}]
```

	Estimate	Standard Error	t-Statistic	P-Value
a	31904.1	0.293771	108602.	$9.4895 \cdot 10^{-493}$
b	-501.724	0.00499953	$-100354.$	$1.5716 \cdot 10^{-488}$
c	54080.8	0.173956	310889.	$6.2358 \cdot 10^{-549}$
d	603.929	0.0566361	10663.3	$9.0050 \cdot 10^{-369}$
e	-2.54539	0.000994811	-2558.66	$1.5819 \cdot 10^{-292}$

```
(* Value of R^2 *)
nlm[{"RSquared"}]
R^2 = 1
```

As $R^2 = 1$, this nonlinear model fit matches the data perfectly.

In Table 6.13, the results based on the polynomial fit of third order and the nonlinear fit with regard to the function defined in Eq. (6.21) are compared. An effective figure of merit is already defined for a TE device in Section 2.3.3 in Eq. (2.40), which can be adopted to define an effective figure of merit for a temperature-dependent material under a certain temperature gradient

$$(zT_m)_{\text{eff}} = \frac{\left(1 + \frac{\eta_{\max}}{\eta_C}\frac{T_c}{T_h}\right)^2}{1 - \frac{\eta_{\max}}{\eta_C}} - 1 = \left(\frac{\Delta T + \eta_{\max} T_c}{\Delta T - \eta_{\max} T_h}\right)^2 - 1 \tag{6.22}$$

Table 6.13 Comparison between polynomial and nonlinear fits.

	Polynomial third order	Nonlinear fit
$\eta_{\max}$ %	7.297	7.248
$j_{\text{opt},\eta}$ (A/cm^2)	28.64	28.48

By using this equation, a value of $(zT_m)_{\text{eff}} = 0.506673$ is calculated. For a consistency check, the effective figure of merit can be determined from the single value α_{eff}, σ_{eff}, and κ_{eff}, i.e. $(zT_m)_{\text{eff}} = (\alpha_{\text{eff}}^2 \sigma_{\text{eff}}/\kappa_{\text{eff}}) T_m = 0.510442$, which is only slightly different.

6.4.7
Calculation for a p-n Thermocouple

In Table 6.14, the global parameters needed for a performance calculation as the internal resistance R_i and thermal conductance K of a TE element ($L = 5$ mm and $A_c = 1$ mm) and the performance parameters are summarized for both materials (p-type/n-type). These values are based on the TAv, which was found to deviate only slightly from the effective values.

It can clearly be seen from Table 6.14 that the performance values, especially the optimum currents, are different for p-type and n-type materials. However, in a thermocouple where the p-type and n-type elements are connected in series, the current is unique.

In Table 6.2, the relations between single element values and values with regard to a thermocouple built of one p-type and one n-type semiconductor element are summarized. For the Seebeck coefficient, it is $\alpha_{pn} = \alpha_p + |\alpha_n|$. The internal resistance and the thermal conductance are additive quantities in the configuration chosen in a thermocouple, that is, $R_i = R_{i,p} + R_{i,n}$ and $K = K_p + K_n$. By considering the values for the single elements from Table 6.14 for the pair where both elements have the same A_c, the following results can be found, see Table 6.15.

Table 6.14 Global parameters for the performance calculation based on TAv for a TE element of $L = 5$ mm and $A_c = 1$ mm.

Unit	α (μV/K)	R_i (mΩ)	K (mW/K)	ZT_m	η_{max} (%)	$I_{opt,\eta}$ (A)	f (μV/K²)	P_{max} (mW)	$I_{opt,p}$ (A)
p-type	176.099	110.154	0.2985	0.5124	7.31	0.29	0.2815	11.2609	0.32
n-type	−225.001	72.2349	0.5371	0.7088	9.35	0.54	0.7008	28.0338	0.62

Table 6.15 Combined global parameters and parameters for maximum performance of a TE thermocouple (both TE elements $L = 5$ mm and $A_c = 1$ mm²) based on the TAv of temperature-dependent material properties.

α_{pn} (μV/K)	R_i (mΩ)	K (mW/K)	ZT_m	η_{max} (%)	$I_{opt,\eta}$ (A)	f (μV/K²)	P_{max} (mW)	$I_{opt,p}$ (A)
401.1	182.389	0.8355	0.5734	7.98	0.39	0.8821	35.2831	0.44

It is noteworthy that the maximum power output of the thermocouple is less than the sum of the maximum power output of the p-type and n-type as both do not function with the optimal current density but with a compromise. In the next section, it is shown how this can be overcome.

6.4.8
Adjustment of Cross-Sectional Areas

As already mentioned, the p-type and n-type material properties are, in general, asymmetric, that is, $\alpha_p \neq |\alpha_n|$, $\sigma_p \neq \sigma_n$ and $\kappa_p \neq \kappa_n$.[2] This asymmetry in the material properties leads to the fact that the maximum performance of the single elements is reached at different current densities. Therefore, an adjustment of the aspect ratio, which is the ratio between the length and the cross-sectional area will result in an increase in the overall performance of a thermocouple. The length is fixed in most cases such that the adjustment is limited to an adjustment in the cross-sectional areas, $A_{c,p}$ and $A_{c,n}$. To do this, the ratio of the aspect ratios is defined as follows

$$g = \frac{(L/A_c)_n}{(L/A_c)_p}, \tag{6.23}$$

see, for example, [70, p. 141 ff] or [71, p. 464 ff]. If L of both elements is fixed, then Eq. (6.23) simplifies, that is, $g = A_{c,p}/A_{c,n}$. For an increase in the efficiency, the figure of merit ZT_m has to be increased while both boundary temperatures are kept constant. This can be done by minimizing the product

$$R_i K = \left[\left(\varrho \frac{L}{A_c}\right)_p + \left(\varrho \frac{L}{A_c}\right)_n\right]\left[\left(\kappa \frac{A_c}{L}\right)_p + \left(\kappa \frac{A_c}{L}\right)_n\right]$$

and the optimum is reached for

$$g_{\text{opt},\eta} = \sqrt{\frac{\kappa_n \varrho_p}{\kappa_p \varrho_n}} = \sqrt{\frac{\kappa_n \sigma_n}{\kappa_p \sigma_p}}. \tag{6.24}$$

For this ratio, the product is minimized for

$$(R_i K)_{\text{opt}} = (\sqrt{(\varrho\kappa)_p} + \sqrt{(\varrho\kappa)_n})^2$$

such that the optimum of the figure of merit for this special area ratio is

$$Z_{\text{opt}} = \frac{(\alpha_p - \alpha_n)^2}{(\sqrt{(\varrho\kappa)_p} + \sqrt{(\varrho\kappa)_n})^2}. \tag{6.25}$$

---

2) This is valid for the functional form of the temperature dependence of the material properties as well as for the averages.

For the adjustment of the electrical power output, which is different from the adjustment of the efficiency, only the electrical parameters are relevant for determining the ratio, which is

$$g_{opt,P} = \left(\frac{A_{c,p}}{A_{c,n}}\right)_{P_{max}} = \sqrt{\frac{\varrho_p}{\varrho_n}} = \sqrt{\frac{\sigma_n}{\sigma_p}}. \tag{6.26}$$

This results in a maximum value of the power output

$$P_{max} = \left[\frac{\alpha_p - \alpha_n}{\sqrt{\varrho_p} + \sqrt{\varrho_n}}\right]^2 \frac{(\Delta T)^2}{4L} A_{c,n}(1 + g_{opt,P}). \tag{6.27}$$

For the adjustment of the efficiency, the ratio $g_{opt,\eta} = 1.66$ is obtained, that is, the cross-sectional area of the p-type element should be about 66% larger than that of the n-type element. An increase from $\eta_{max} = 7.98\%$ to 8.36% can be observed. For the electrical power output, $A_{c,p}$ should be about 23% larger than $A_{c,n}$, that is, $g_{opt,P} = 1.23$. For this adjustment of the cross-sectional areas, an increase from $P_{max} = 35.28$ to 39.86 mW is achieved.

6.4.9
FEM Simulation

As already mentioned, the FEM calculations are performed with commercial software. ANSYS WORKBENCH has a module implemented, covering the thermoelectric material behavior. The modular character is illustrated in Figure 6.14. The classical steps in a FEM procedure, which were introduced in Section 6.3, are reflected in the module. The handling is very intuitive as no detailed mathematical operation steps are needed to do it because, for example, the field equations are already implemented. It is a kind of black box model. Nevertheless, it is very helpful in questions of designing a TE device.

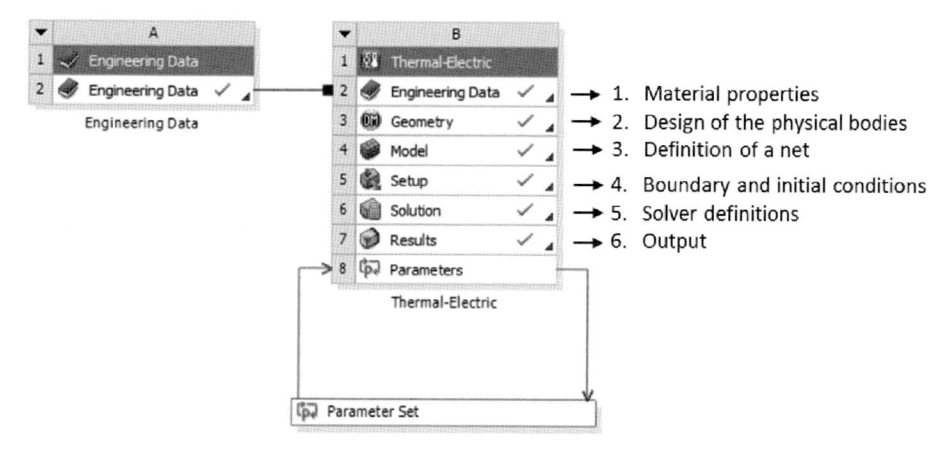

Figure 6.14 The thermoelectric module in ANSYS WORKBENCH.

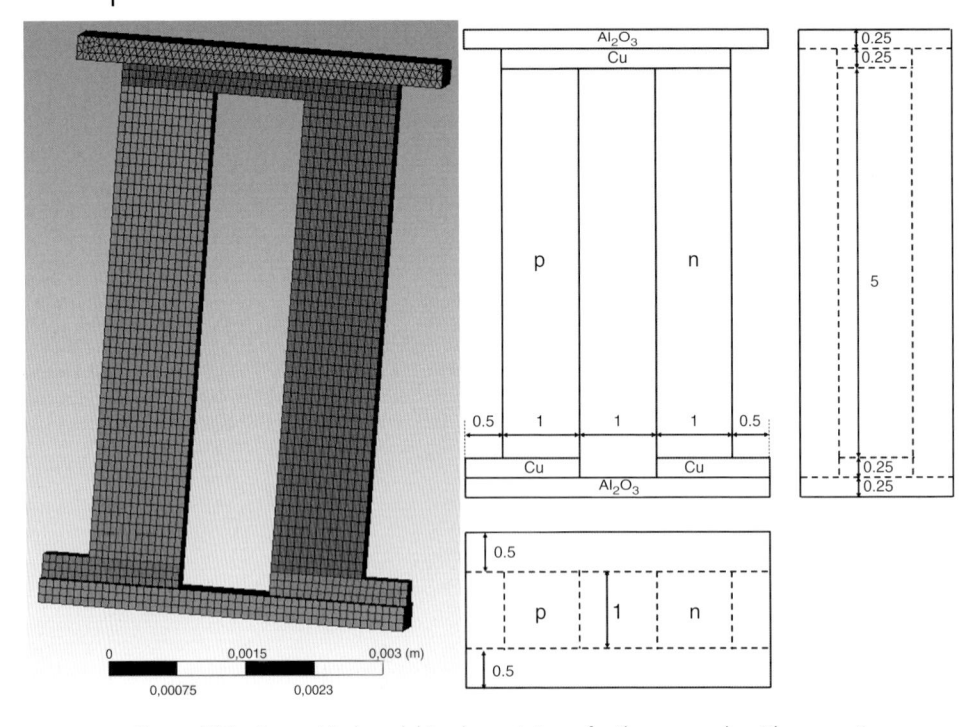

Figure 6.15 Geometrical model implementation of a thermocouple with connecting bridges and an additional insulation layer.

For a FEM calculation, a real geometry is considered. In Figure 6.15, the geometrical model of a thermocouple is illustrated. Besides the two elements, the connecting bridges as well as insulating layers are included. These additional parts can shift the results a bit due to their thermal and electrical resistance. A larger contribution in this sense could come from thermal and electrical contact resistances. Here these contact resistances are neglected but an ideal device is studied. As a real geometry is implemented, the problem is in general 3D and this could also have an effect on the performance. The boundary temperatures are not applied directly to the TE elements, but now they are set constant at the outer insulation layer.

The electrical properties can be determined easily. The electrical current is varied and the voltage drop is recorded for this variation. As a result, the characteristics and the electrical power output are found. The determination of the thermal properties such as the heat flux or the efficiency is important, as the heat flux at the hot and cold side is not constant anymore, but a local distribution over the outer interfaces is found for both (Figure 6.16).

The heat flux is a function of positions $\dot{q} = \dot{q}(x, y)$. The overall heat flow over the outer interfaces $\dot{Q} = \iint \dot{q}(x, y) \, dxdy$ can be determined numerically. The distribution of the heat flux can be exported from ANSYS to a text file, and

(a)

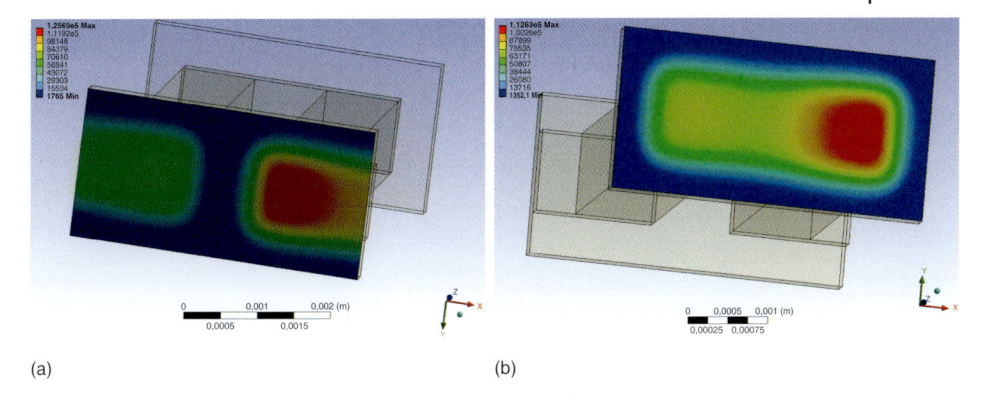

(b)

Figure 6.16 Heat flux at the hot and cold side (outer insulation layer) for a current of $I = 0.1$ A. Geometrical details can be found in Figure 6.15. The p- and n-type pellets can easily be distinguished since the thermal conductance of the n-type pellet is almost double that of the p-type (see Table 6.14). (a) Hot side heat flux $\dot{q}_h$ (b) Cold side heat flux $\dot{q}_c$

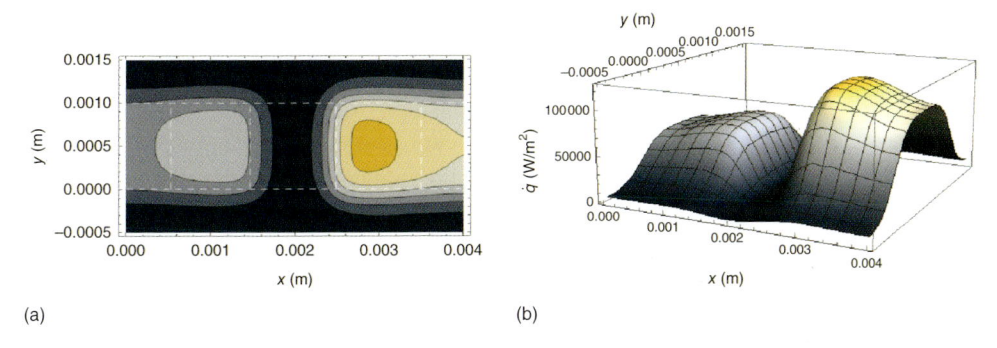

(a)

(b)

Figure 6.17 Plots of the data and their interpolation for $I = 0.1$ A. (a) ListContourPlot (b) Plot3D

after importing to MATHEMATICA, it can be processed further to execute the numerical integration (Figure 6.17).

```
(* Importing a text file *)
QhzIc01a = Import["SOURCE PATH\\Qhz-I0-1.txt", "Table"];
(* List Manipulation *)
QhzIc01b = Drop[QhzIc01a, 1];
QhzIc01c = Drop[QhzIc01b, None, {3}]
(* ListPlot of the data *)
ListPlot3D[QhzIc01c]
(* ContourPlot *)
```

```
ListContourPlot[QhzIc01c, BoxRatios -> Automatic, AspectRa-
tio -> 1]
(* Interpolation of the data *)
fQhzIc01 = Interpolation[QhzIc01c, InterpolationOrder -> 1]
(* Plot of the interpolated data *)
Plot3D[fQhzIc01[x, y], {x, 0, 0.004}, {y, -0.0005, 0.0015}]
(* Numerical integration of the heat flux *)
QhzIc01 =
 NIntegrate[
  NIntegrate[fQhzIc01[x, y], {x, 0, 0.004}, PrecisionGoal -
> 10,
    WorkingPrecision -> 11], {y, -0.0005, 0.0015}, Preci-
sionGoal -> 10,
   WorkingPrecision -> 11]
```

As a result, the value $\dot{Q}_h = 0.3614757$ W can be evaluated numerically for a current $I = 0.1$ A. The area where the flux passes through is $A = 4$ mm $\times$ 2 mm. With respect to this area, an averaged hot side heat flux of $\dot{q}_{av,h} = 45184.5$ W/m^2 can be determined. For a full variation of the current, the procedure has to be repeated at every point for the hot and the cold side.

The V-I characteristics led to a value of $\alpha_{pn} = 399.326$ µV/K and an internal resistance of $R_i = 185.665$ mΩ. These values are almost the same as found in a 1D CPM calculation where no contacting and insulating layers are present, compared with the values in Table 6.15. For the electrical power output, a value of $P_{max} = 34.5574$ mW is found at a current of $I_{opt,P} = 0.43$A. For these parameters, only a slight deviation from the values presented in Table 6.15 based on 1D CPM calculations is observed.

In further calculations, the adjustment of the cross-sectional areas as discussed in Section 6.4.8 has been done in two ways besides the already investigated case, see Figure 6.18: on the one hand, the adjustment is done in one direction, and on the other hand, a squared cross-sectional area is chosen. For the case b) $P_{max} = 38.8631$ mW at $I_{opt,P} = 0.484$A is obtained whereas for case c) $P_{max} = 38.9389$mW at $I_{opt,P} = 0.485$A is found. It is clear that for both almost identical values are found

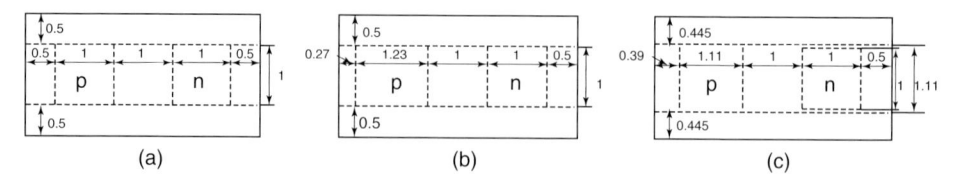

Figure 6.18 Different choices of cross-sectional areas: (a) Reference (both equal), (b) increase in the area according to maximization of power, keeping the width constant, and (c) increase in the area, keeping the square shape of the cross-sectional area.

and an increase in comparison to the case where both elements have the same cross-sectional area can be confirmed.

For the example introduced in this section, the maximum performance can be described with a certain precision within the CPM. FEM simulations are, on the one hand, more precise, but on the other hand, they are more complicated and time-consuming. Hence it can be deduced that a performance estimation should be performed first in the framework of CPM. The remaining uncertainties due to the approximations related to the applied model have to be rated against the uncertainties of the measurements of the material properties, which could bear a much larger uncertainty of the obtained performance.

In the next section, another example of temperature-dependent material is chosen to demonstrate how to solve the performance determination in the framework of the compatibility approach.

6.5
Nonlinear Material Parameters

When the material properties depend on further parameters, numerical methods are indispensable because nonlinear problems result typically. We consider here another example with temperature-dependent material properties

6.5.1
Temperature-Dependent Material Properties

For exact calculation, the temperature dependence of thermoelectric transport parameters has to be taken into account when solving the isotropic thermal energy balance

$$\nabla \cdot (-\kappa \nabla T) = \varrho j^2 - \tau \mathbf{j} \cdot \nabla T, \tag{6.28}$$

where ϱj^2 is the Joule heat per volume, $\tau = T d\alpha/dT$ the Thomson coefficient, and $\tau \mathbf{j} \cdot \nabla T$ the Thomson heat per volume. Note that the material properties α, σ, and κ are in general temperature- and position-dependent quantities and measurable under certain constraints.

The TE material properties α, σ, and κ are in general temperature- and position-dependent quantities and measurable under certain constraints. Often the approach of decoupled dependencies is applied, that is, either a temperature or spatial dependence of the material coefficients; for more information, see [72]. Here we focus on temperature-dependent material. Within the framework of a 1D steady-state model (for details, see Chapter 2 and [72]), the thermal energy balance reads

$$\kappa(T) \frac{\partial^2 T}{\partial x^2} + \frac{d\kappa}{dT} \left(\frac{\partial T}{\partial x} \right)^2 - jT \frac{d\alpha}{dT} \frac{\partial T}{\partial x} = -\frac{j^2}{\sigma(T)}, \tag{6.29}$$

where it is supposed that a constant current density j flows through the element or device in x-direction. Eq. (6.29) is a nonlinear differential equation in T with nonconstant coefficients, which has to be supplemented by boundary conditions. An example calculation based on Eq. (6.29) is presented in Ref. [73] for a Peltier cooler where experimental thermoelectric properties of a p-type bismuth antimony telluride semiconducting crystal have been used. From Ref. [74], sets of material parameters $\alpha(T)$, $\sigma(T)$, and $\kappa(T)$ measured perpendicular to the crystallographic c-axis on single crystalline Bridgman grown materials are available in the temperature range 80 K $\leq T \leq$ 400 K. For a representative sample, temperature-dependent curves for $\alpha(T)$, $\kappa(T)$, $\sigma(T)$ are shown in Figure 6.19 where a quadratic fit to the experimental data (with T in K) is given by

$$\alpha(T) = 224.571 - 0.0024381(T - 349.902)^2 \text{ in } \mu V/K, \tag{6.30a}$$

$$\sigma(T) = 54.8333 + 0.00244286(T - 333.48)^2 \text{ in } 10^3 (\Omega m)^{-1}, \tag{6.30b}$$

$$\kappa(T) = 1.32177 + 0.0000331905(T - 290.136)^2 \text{ in } W/(mK). \tag{6.30c}$$

An important result of [73] shall be pointed out as that, for practical consideration of device performance, volume average values of the material parameters seem to provide quantitatively quite sufficient results for material of moderate performance even in the ∇T_{max} situation where the influence of the temperature dependence is significant. For exact calculation, however, in particular for an exact calculation of maximum cooling, an iterative strategy is required to find ΔT_{max} at the optimal electrical current. For the TE material properties listed in (6.30a), $T_{c,min} = 237.7$ K was found numerically.

A more direct and elegant temperature-based method is available within the compatibility approach.

We present here, first, the derivation of the differential equation for $u(T)$ using nabla calculus. From the definition of the relative current density u as the ratio of electric current density ($\mathbf{j}$) to Fourier heat flux ($-\kappa \nabla T$), it follows:

$$\frac{1}{u}\mathbf{j} = -\kappa \nabla T \quad \text{resp.} \quad \mathbf{j} \cdot \mathbf{j} = -\kappa u \mathbf{j} \cdot \nabla T. \tag{6.31}$$

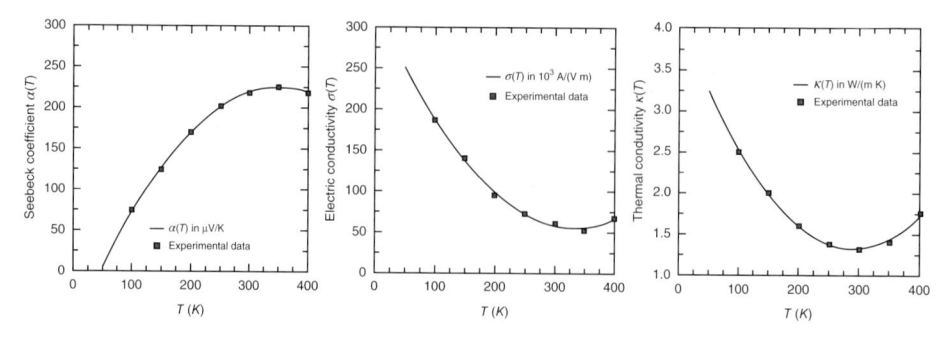

Figure 6.19 Polynomial approximation of experimental data $\alpha(T)$, $\kappa(T)$, $\sigma(T)$ in the temperature range from 80 K to 400 K (figures taken from Ref. [75]).

6.5 Nonlinear Material Parameters

Inserting the relations (6.31) into Eq. (6.28) gives

$$\nabla \cdot \left(\frac{1}{u}\mathbf{j}\right) = -\rho\kappa u\mathbf{j} \cdot \nabla T - \tau\mathbf{j} \cdot \nabla T. \tag{6.32}$$

Nabla calculus yields $\nabla \cdot \left(\frac{1}{u}\mathbf{j}\right) = \nabla\left(\frac{1}{u}\right)\cdot\mathbf{j} + \frac{1}{u}\nabla\cdot\mathbf{j}$, where the last term vanishes due to particle or charge conservation: $\nabla \cdot \mathbf{j} = 0$. Thus, we get

$$\nabla \left(\frac{1}{u}\right) \cdot \mathbf{j} = -\rho\kappa u\mathbf{j} \cdot \nabla T - \tau\mathbf{j} \cdot \nabla T. \tag{6.33}$$

When $u(T)$ is a pure function of temperature, we can write $\nabla\left(\frac{1}{u}\right) = \frac{d}{dT}\left(\frac{1}{u}\right)\nabla T$, which finally leads to

$$\frac{d}{dT}\left(\frac{1}{u}\right)\nabla T \cdot \mathbf{j} = -\rho\kappa u\mathbf{j} \cdot \nabla T - \tau\mathbf{j} \cdot \nabla T. \tag{6.34}$$

Note that $\mathbf{j}$ and ∇T are vectors; their scalar product is commutative. So, we get

$$\frac{d}{dT}\left(\frac{1}{u}\right) = -u\rho\kappa - \tau \quad \text{or, alternatively,} \quad u'(T) = \frac{\alpha^2}{z}u^3 + \tau u^2, \tag{6.35}$$

where $z = \alpha^2/(\rho\kappa)$. An analytical solution exists for constant material properties where $z =$ const. and $\tau = 0$. In this case, Eq. (6.35) reduces to [76]

$$u'(T) = \rho\kappa u^3, \tag{6.36}$$

with the analytic solution

a) for the "pump down" situation $[u > 0, u_s = u(T_s)]$:

$$u(T) = +\frac{1}{\sqrt{\frac{1}{u_s^2} - 2\rho\kappa(T-T_h)}},$$

b) for the "pump up" situation $[u < 0, u_s = u(T_s)]$:

$$u(T) = -\frac{1}{\sqrt{\frac{1}{u_s^2} - 2\rho\kappa(T-T_h)}}. \tag{6.37}$$

Alternatively, the boundary condition $u_a = u(T_a)$ at the heat absorbing side T_a can be used (for a cooler, we have $T_a = T_c$ and $T_s = T_h$).[3]

By assuming the material to be dependent on temperature, Eq. (6.35) presents the relative current density u as a pure temperature-dependent variable.[4] For more general comments, see the introduction to Chapter 5.

---

3) A CPM version (Mathematica notebook) for TEG and TEC can be found at http://library.wolfram.com/infocenter/MathSource/6641.
4) Ultimately, Snyder's compatibility approach [77] is a conceptual model, which only approximates the actual material behavior.

6 Numerical Simulation

For a given temperature dependence of the material properties, Eq. (6.35) has to be solved numerically. The following section offers a *Mathematica*[5] code, which presents the calculation of the TEC's coefficient of performance (COP $\equiv \varphi$) for boundary temperatures $T_c = 280\text{K}$, $T_h = 300\text{K}$ based on the TE material properties given by the polynomial approximation (6.30a). To enable a direct comparison with the published results (see [73, 75]), the element length[6] was set to $L = 1$ mm.

The algorithm for calculating the coefficient of performance (COP) φ as a function of the electrical current density is straightforward: Numerically, solve the first-order differential equation for $u(T)$, Eq. (6.35), by means of the *Mathematica* function NDSolve, evaluate the scaling integral[7] [77]

$$j = \frac{1}{L}\int_{T_h}^{T_c} u(T)\kappa(T)\,dT \tag{6.38}$$

to achieve the corresponding electric current density j (notation "jel" in the *Mathematica* notebook, see next section), and integrate for φ according to

$$\ln\left(1+\frac{1}{\varphi}\right) = \int_{T_c}^{T_h} \frac{1}{T}\,\frac{u\frac{\alpha}{z}(1-u\frac{\alpha}{z})}{u\frac{\alpha}{z}+\frac{1}{zT}}\,dT, \quad T_c \le T \le T_h. \tag{6.39}$$

Alternatively, the COP can be calculated by means of the thermoelectric potential $\Phi = \alpha T + 1/u$ and the relation (see also Chapter 5 in this book)

$$\varphi = \frac{1}{\Phi[T_h]/\Phi[T_c] - 1}. \tag{6.40}$$

All this is done in a loop over varying boundary values of u, which translates to a range of j via the scaling integral.

The Mathematica code in Section 6.5.2 is self-explanatory; helpful comments are embedded in the active code using (∗ comment ∗). The COP is calculated with Eq. (6.40) and, for the proof of concept, with Eq. (6.39) additionally.[8] Both variants naturally yield the same performance results (COP=COP2); they are listed at the end of the code. Further postprocessing has been omitted.

The COP as a function of j (notation "jel" in the *Mathematica* notebook) for different cooling temperatures T_c from 250 K to 290 K (in 10 K steps) is shown in

---

5) *Mathematica* is a registered trademark of Wolfram Research, Inc.
6) Note that the length acts as a scaling parameter in an ideal single-element device.
7) Snyder has specified the scaling integral for TEG where $T_a = T(x=0) > T_s = T(x=L)$ and $u > 0$. In a cooler element, however, we have $T_a = T(x=0) < T_s = T(x=L)$ and $u < 0$. Note that $T(x)$ peaks in the interior of a Peltier cooler element well above the optimum current. Then, u has a pole, and the scaling integral must be split.
8) Doubtless, the variant with Eq. (6.39) provides more insight into the local performance of the TE material; the price to pay for is another numerical integration.

6.5 Nonlinear Material Parameters

Coefficient of performance φ for T_c = 280 K

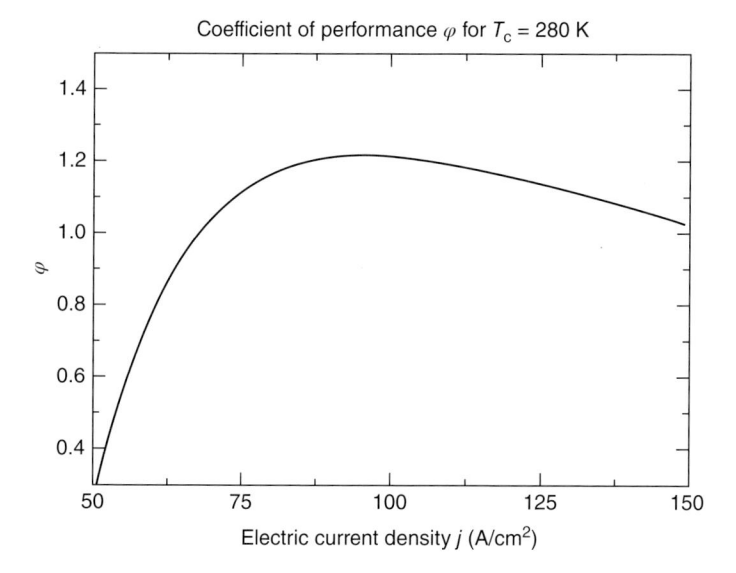

Figure 6.20 Coefficient of performance as a function of j for fixed boundary temperatures $T_c = 280$ K, $T_h = 300$ K and an element length of $L = 1$ mm. Maximum COP = 1.22 at optimal current density $j_{opt} = 95.8$ Acm^2. The data set for this figure has been created by means of the algorithm presented in the next section where u_h has been varied from −20 to −180 in steps of −2.5.

Ref. [75, Figure 4]; we calculate and plot here only the curve for $T_c = 280$ K (in the range 50 A/cm$^2 \le j \le$ 150 A/cm^2, see Figure 6.20), which is the result after a run of the *Mathematica* code listed. The code was tested with Mathematica Version 7.

Note that maximum cooling ($T_c = T_{c,\mathrm{min}}$) is reached when $\varphi_{\max}(j_{opt}) = 0$. For the TE material properties used (see (6.30a)), $T_{c,\mathrm{min}} = 237.7$ K was found numerically [73].

Further note that the (inverse) temperature profile can be determined by a modified scaling integral. When considering a cooler (TEC), the heat absorbing side is the cold side ($T_a \equiv T_c$), and we have the boundary conditions $x(T_h) = L, x(T_c) = 0$, usually with a fixed hot side temperature T_h. Then, the modified scaling integral takes the form:

$$x(T) = -\frac{1}{j}\int_{T_a}^{T} u\kappa\, dT = -\frac{1}{j}\int_{T_c}^{T_h} u\kappa\, dT - \frac{1}{j}\int_{T_h}^{T} u\kappa\, dT \qquad (6.41)$$

$$= L - \frac{1}{j}\int_{T_h}^{T} u\kappa\, dT. \qquad (6.42)$$

Having calculated $T(x) = \mathrm{Inverse}[x(T)]$ from the resulting $u(T)$ and $\kappa(T)$, spatial material profiles can be derived by the transformation $\mathrm{mat}(x) = \mathrm{mat}(T(x))$ with $\mathrm{mat} = \alpha, \sigma, \kappa$.

6.5.2
Temperature-Dependent Material Properties: Mathematica Model Using the Compatibility Approach

```
Clear["Global`*"] ;  Off[General::"spell"];

(* data input *)

    L = 0.001;     (* element length [scales the electri-
cal current !] *)
    T2 = 300;      (* hot side temperature  *)
    T1 = 280;      (* cold side temperature *)

(* polynomial fits of the TE material properties [note the
dimensions!] *)

    alpha[T_] = 224.571 - 0.0024381*(T - 349.902)^2    (* in
10^(-6) V/K *)
    sigma[T_] = 54.8333 + 0.00244286*(T - 333.48)^2
(* in 10^3 S/m   *)
    kappa[T_] = 1.32177 + (0.0000331905*(T - 290.136)^2
(* in W/(K m)   *)

(* data processing  *)

    (* material's figure of merit *)
    FoM[T_] = (alpha[T])^2*sigma[T]/kappa[T]/10^9;

    (* Thomson coeff. in V/K *)
    Thomson[T_] = T*10^(-6)*D[alpha[T],T];

    (* calculate ratio kappa(T)/sigma(T) *)
    kappaoversigma[T_] = kappa[T]/(sigma[T]*10^3);

      zT[T_] = FoM[T]*T;

    (* comp. factor for maximum COP *)
    sTEC[T_] = (-1-Sqrt[1+zT[T]])/(10^(-6)*alpha[T]*T);

(* Solve the nonlinear deq for u[T] using NDSolve in a
loop over u *)
```

```
     Nmax = 65;   res=Array[dat,{Nmax,4}];    (* 65 calcula-
tion steps *)
     uT2 = 17.5;     (* u at the hot side for 1. calcula-
tion step *)
                       (* possibly change uT2 if T2 changes *)
     stepwidth = 2.5;  (* stepwidth controls accuracy   *)

  For[inx=1, inx<(Nmax+1),inx++,    (* loop over u *)

     uini = -(uT2+stepwidth *inx);   (* u<0, note the minus
sign! *)

     Clear[u];

(* define and solve the diff. equation for u (T) *)
     deq = u'[T]-kappaoversigma[T]*(u[T])^3-Thomson[T]*
(u[T])^2 == 0;
     u[T_] = u[T] /. eqsol;

  (* scaling integral yields electrical current density
jel *)
     jel = NIntegrate[kappa[T]*u[T],{T,T2,T1}]/L;

  (* variant using the TE potential *)
     Phi[T_] = 10^(-6)*alpha[T]*T + 1/u[T];
          COP = 1/(Phi[T2]/Phi[T1]-1);

  (* alternative variant using the reduced COP *)
        func[T_] = u[T]*10^(-6)*alpha[T]/FoM[T];
      etared[T_] = 1/func[T]/(1-func[T])*(func[T]+1/zT[T]);
             eta = 1-Exp[- NIntegrate[1/etared[T]/
T,{T,T2,T1}]];
             COP2 = -1/eta;

(* storage of results *)
    res[[inx,1]] = uini;
    res[[inx,2]] = jel/10^4;     (* jel in A/cm^2 *)
    res[[inx,3]] = COP ;
    res[[inx,4]] = COP2
    ];

    Print["u_{i}nitial  jel      COP      COP with eta_{r}  " ];
    res  //TableForm    (* print results in table form *)
```

A
Numerical Data and Illustrative Cases

A.1
Coefficients of the Polynomials of the Material Properties

Table A.1 Coefficients of the polynomial fit of the material properties of the p-type/n-type skutterudite.

Quantity	Unit	A	B	C	D
			p-Type skutterudite		
			Thermal conductivity		
$\kappa_p(T)$	W/(m K)	1.62114	$2.24319 \cdot 10^{-3}$	$-1.00511 \cdot 10^{-5}$	$9.62799 \cdot 10^{-9}$
			Electrical conductivity		
$\sigma_p(T)$	S/cm	375.272	0.889833	$-2.21557 \cdot 10^{-3}$	$1.53044 \cdot 10^{-6}$
			Seebeck coefficient		
$\alpha_p(T)$	µV/K	115.037	-0.208174	$1.14915 \cdot 10^{-3}$	$-9.90329 \cdot 10^{-7}$
			n-Type skutterudite		
			Thermal conductivity		
$\kappa_n(T)$	W/(m K)	6.96792	$-1.70663 \cdot 10^{-2}$	$1.91766 \cdot 10^{-5}$	$-5.08864 \cdot 10^{-9}$
			Electrical conductivity		
$\sigma_n(T)$	S/cm	1204.78	-0.770307	$-1.355275 \cdot 10^{-3}$	$1.7746 \cdot 10^{-6}$
			Seebeck coefficient		
$\alpha_n(T)$	µV/K	-47.99	-0.414487	$-9.62441 \cdot 10^{-5}$	$4.27512 \cdot 10^{-7}$

A.2
Material Properties: p-Type Skutterudite and Averages

(a) Thermal conductivity $\kappa_p(T)$

(b) Electrical conductivity $\sigma_p(T)$

(c) Seebeck coefficient $\alpha_p(T)$

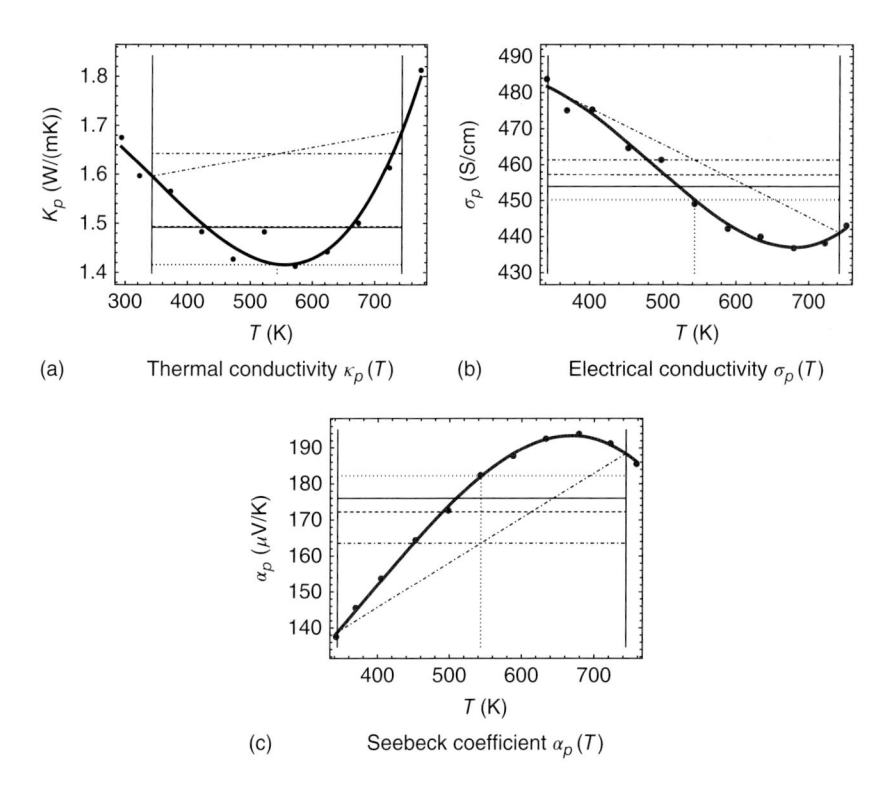

Figure A.1 Material properties: p-type skutterudite and averages.

Table A.2 Values for the different averages of the p-type skutterudite.

Quantity	Linestyle	κ_p	σ_p	α_p
Unit		W/(m K)	S/cm	µV/K
TAv	Solid	1.49226	453.908	176.099
VMT	Dotted	1.41710	450.198	182.293
MBV	Dashed–dotted	1.64259	461.327	163.711
WAv	Dashed	1.49313	457.243	172.300

A.3
Material Properties: n-Type Skutterudite and Averages

(a) Thermal conductivity $\kappa_n(T)$

(b) Electrical conductivity $\sigma_n(T)$

(c) Seebeck coefficient $\alpha_n(T)$

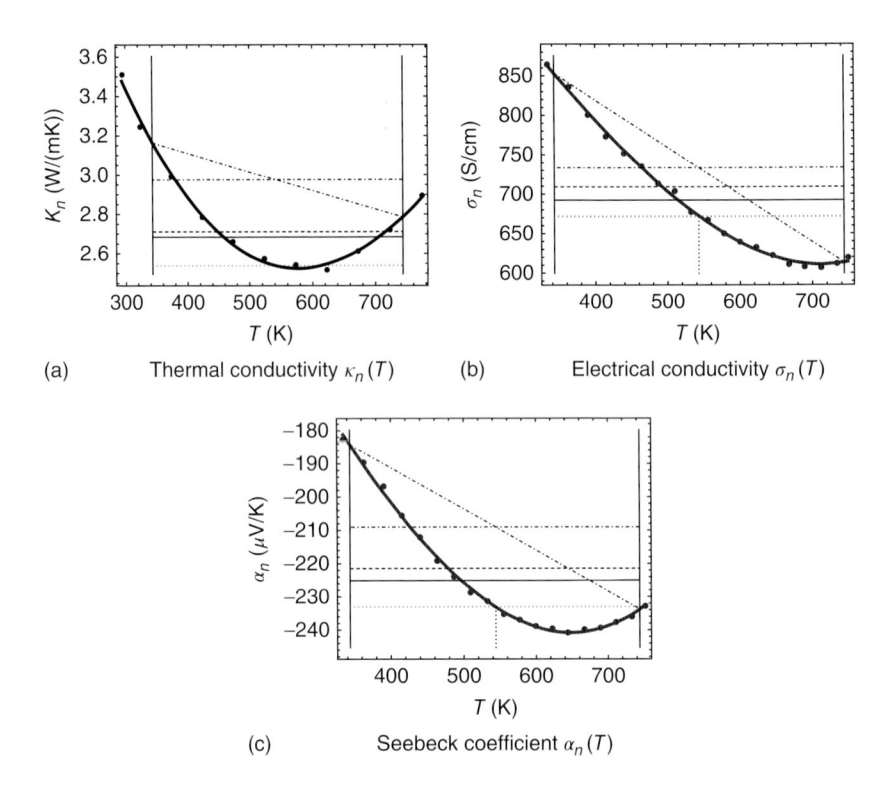

Figure A.2 Material properties: n-type skutterudite and averages.

Table A.3 Values for the different averages of the n-type skutterudite.

Quantity	Linestyle	κ_n	σ_n	α_n
Unit		W/(m K)	S/cm	µV/K
TAv	Solid	2.68543	692.186	−225.001
VMT	Dotted	2.54029	671.667	−233.006
MBV	Dashed–dotted	2.97569	733.222	−208.992
WAv	Dashed	2.71308	708.949	−221.368

A.4
Power Factor and Figure of Merit of the n-Type Material

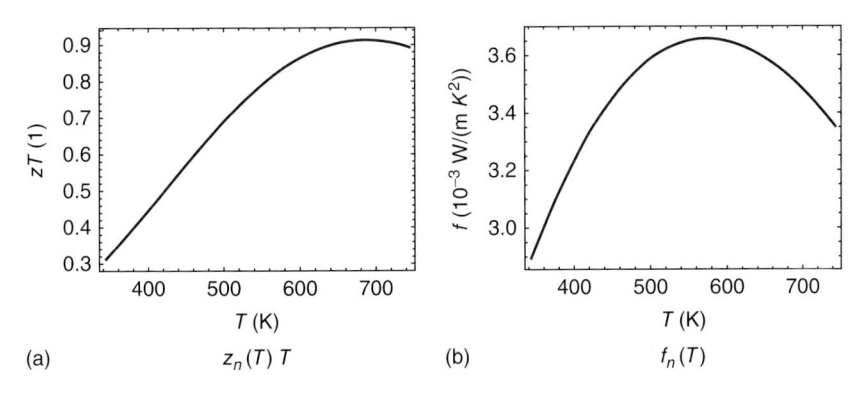

Figure A.3 Figure of merit $z_n(T)\,T$ and power factor $f_n(T)$ of the n-type element.

Table A.4 Optimal performance values for the different averages within the framework of Ioffe-CPM theory.

zT	$T\,(K)$	Value
$zT(T_c)$	343.15 K	0.314099
$zT(T_h)$	743.15 K	0.893509
$zT(T_m)$	543.15 K	0.779697
$(zT)_{max}$	686.34 K	0.913212

$f_n(T)$	$T\,(K)$	$(10^{-3}\,W/(m\,K^2))$
$f(T_c)$	343.15 K	2.89621
$f(T_h)$	743.15 K	3.35123
$f(T_m)$	543.15 K	3.64661
f_{max}	572.93 K	3.6579

	Unit	TAv	VMT	MBV	WAv	Max.
zT_m	1	0.71	0.77	0.58	0.70	0.91
f	$10^{-3}\,W/(m\,K^2)$	3.50	3.65	3.20	3.47	3.66
p_{max}	W/m^2	28033.9	29172.9	25620.3	27792.9	29263.2
$j_{opt,p}$	A/m^2	622971	626011	612949	627754	618276
η_{max}	%	9.35	10.01	8.10	9.22	11.18
$j_{opt,\eta}$	A/m^2	540024	536416	542723	545371	622571

For the n-type element, a length of 5mm and a fixed temperature difference of 400K is assumed.

A.5
Performance Parameters in dependence on the Current Density

A.5.1
Program code

```
For[i = 0, i <= 130, i++,
    (* Numerical solution for different j *)
    s = NDSolve[{kapfit[Ts[x]]*D[Ts[x], {x, 2}]+
        D[kapfit[Ts[x]],Ts[x]]*D[Ts[x], x]^{2} -
        jc*Ts[x]*D[alpfit[Ts[x]],Ts[x]]*10^(-6)
*D[Ts[x], x]==
            -(jc^{2}/(sipfit[Ts[x]]*100)),Ts[0] == 743.15,
Ts[0.005] == 343.15},
        Ts[x], {x, 0, 0.005}, AccuracyGoal -> 10,Precision
Goal -> 10,
        MaxStepSize -> 0.00001] /. jc -> 5000*i;
    (* Def. of the temperature with respect to the precal-
culated result *)
    Tp[x_] = Ts[x] /. s;
    (* First derivative of the temperature with respect to
x *)
    dTp[x_] = D[Tp[x], x];
    (* Calculation of the terminal voltage *)
    U[i] = -NIntegrate[jc/(siT[Tp[x]]*100) +
            alT[Tp[x]]*10^(-6)*dTp[x], {x,0, 0.005}]
/.jc -> 5000*i // Flatten;
    (* Calculation of the power output density *)
    p[i] = -NIntegrate[jc^{2}/(siT[Tp[x]]*100) +
        jc*alT[Tp[x]]*10^(-6)*dTp[x], {x, 0, 0.005}]
/.jc -> 5000*i // Flatten;
    (* Calculation of the hot side heat flux at x=0 *)
    qh[i] = -kaT[Tp[x]]*dTp[x] + i*5000*alT[Tp[x]]*10^
(-6)*Tp[x] /. x -> 0;
    (* Calculation of the cold side heat flux at x=L *)
    qc[i] = -kaT[Tp[x]]*dTp[x] + i*5000*alT[Tp[x]]*10^
(-6)*Tp[x] /. x -> 0.005;
    (* Calculation of the efficiency *)
    eta[i] = p[i]/qh[i]*100;
    (* Output of the results *)
    Print[i*5000, U[i], p[i], qh[i], qc[i], eta[i]];
    Clear[Tp, s, dTp];
    ]
```

A.5.2
Results

See Table A.5

Table A.5 Performance parameters in dependence on the current density.

j A/cm²	V_a mV	p_{el} W/cm²	$\dot{q}_h$ W/cm²	$\dot{q}_c$ W/cm²	p_{el} W/cm²	η %
0.	70.4397	0.	11.9381	11.9381	0	0.
0.5	69.8884	0.0349442	12.0052	11.9703	0.0349442	0.291075
1.	69.3371	0.0693371	12.0721	12.0027	0.0693371	0.57436
1.5	68.7857	0.103179	12.1386	12.0355	0.103179	0.850002
2.	68.2344	0.136469	12.2049	12.0685	0.136469	1.11815
2.5	67.6831	0.169208	12.2709	12.1017	0.169208	1.37893
3.	67.1317	0.201395	12.3367	12.1353	0.201395	1.63249
3.5	66.5803	0.233031	12.4021	12.1691	0.233031	1.87896
4.	66.0289	0.264116	12.4673	12.2032	0.264116	2.11846
4.5	65.4775	0.294649	12.5322	12.2376	0.294649	2.35112
5.	64.926	0.32463	12.5969	12.2722	0.32463	2.57707
5.5	64.3746	0.35406	12.6612	12.3072	0.35406	2.79641
6.	63.8231	0.382939	12.7253	12.3424	0.382939	3.00926
6.5	63.2716	0.411265	12.7891	12.3779	0.411265	3.21574
7.	62.7201	0.43904	12.8527	12.4136	0.43904	3.41595
7.5	62.1685	0.466264	12.9159	12.4496	0.466264	3.61
8.	61.6169	0.492936	12.9789	12.486	0.492936	3.79798
8.5	61.0654	0.519055	13.0416	12.5225	0.519055	3.98
9.	60.5137	0.544624	13.104	12.5594	0.544624	4.15616
9.5	59.9621	0.56964	13.1662	12.5965	0.56964	4.32654
10.	59.4104	0.594104	13.228	12.6339	0.594104	4.49125
10.5	58.8587	0.618017	13.2897	12.6716	0.618017	4.65036
11.	58.307	0.641377	13.351	12.7096	0.641377	4.80397
11.5	57.7553	0.664185	13.412	12.7478	0.664185	4.95216
12.	57.2035	0.686442	13.4728	12.7864	0.686442	5.09501
12.5	56.6517	0.708146	13.5333	12.8252	0.708146	5.23261
13.	56.0998	0.729298	13.5935	12.8642	0.729298	5.36503
13.5	55.548	0.749897	13.6535	12.9036	0.749897	5.49235
14.	54.9961	0.769945	13.7132	12.9432	0.769945	5.61463
14.5	54.4441	0.78944	13.7726	12.9831	0.78944	5.73196
15.	53.8922	0.808382	13.8317	13.0233	0.808382	5.84441
15.5	53.3402	0.826772	13.8906	13.0638	0.826772	5.95203
16.	52.7881	0.84461	13.9492	13.1046	0.84461	6.05491
16.5	52.2361	0.861895	14.0075	13.1456	0.861895	6.1531
17.	51.684	0.878627	14.0655	13.1869	0.878627	6.24667
17.5	51.1318	0.894807	14.1233	13.2285	0.894807	6.33568
18.	50.5796	0.910433	14.1808	13.2704	0.910433	6.42019
18.5	50.0274	0.925507	14.238	13.3125	0.925507	6.50026
19.	49.4752	0.940028	14.295	13.3549	0.940028	6.57594
19.5	48.9229	0.953996	14.3516	13.3976	0.953996	6.6473

Table A.5 (Continued)

j A/cm²	V_α mV	p_{el} W/cm²	$\dot{q}_h$ W/cm²	$\dot{q}_c$ W/cm²	p_{el} W/cm²	η %
20.	48.3706	0.967411	14.408	13.4406	0.967411	6.71439
20.5	47.8182	0.980273	14.4642	13.4839	0.980273	6.77725
21.	47.2658	0.992581	14.52	13.5274	0.992581	6.83595
21.5	46.7133	1.00434	14.5756	13.5713	1.00434	6.89053
22.	46.1608	1.01554	14.6309	13.6154	1.01554	6.94104
22.5	45.6083	1.02619	14.686	13.6598	1.02619	6.98753
23.	45.0557	1.03628	14.7407	13.7045	1.03628	7.03005
23.5	44.5031	1.04582	14.7952	13.7494	1.04582	7.06865
24.	43.9504	1.05481	14.8495	13.7947	1.05481	7.10336
24.5	43.3977	1.06324	14.9034	13.8402	1.06324	7.13423
25.	42.845	1.07112	14.9571	13.886	1.07112	7.16131
25.5	42.2922	1.07845	15.0105	13.9321	1.07845	7.18463
26.	41.7393	1.08522	15.0637	13.9784	1.08522	7.20424
26.5	41.1864	1.09144	15.1165	14.0251	1.09144	7.22018
27.	40.6335	1.0971	15.1691	14.072	1.0971	7.23248
27.5	40.0805	1.10221	15.2215	14.1192	1.10221	7.24118
28.	39.5274	1.10677	15.2735	14.1667	1.10677	7.24633
28.5	38.9743	1.11077	15.3253	14.2145	1.11077	7.24794
29.	38.4212	1.11421	15.3768	14.2626	1.11421	7.24607
29.5	37.868	1.11711	15.4281	14.311	1.11711	7.24074
30.	37.3148	1.11944	15.479	14.3596	1.11944	7.23199
30.5	36.7615	1.12122	15.5297	14.4085	1.12122	7.21985
31.	36.2081	1.12245	15.5802	14.4577	1.12245	7.20436
31.5	35.6547	1.12312	15.6303	14.5072	1.12312	7.18553
32.	35.1012	1.12324	15.6802	14.557	1.12324	7.16341
32.5	34.5477	1.1228	15.7299	14.6071	1.1228	7.13803
33.	33.9941	1.12181	15.7792	14.6574	1.12181	7.1094
33.5	33.4405	1.12026	15.8283	14.708	1.12026	7.07756
34.	32.8868	1.11815	15.8771	14.759	1.11815	7.04255
34.5	32.3331	1.11549	15.9256	14.8102	1.11549	7.00437
35.	31.7793	1.11228	15.9739	14.8616	1.11228	6.96307
35.5	31.2254	1.1085	16.0219	14.9134	1.1085	6.91866
36.	30.6715	1.10417	16.0697	14.9655	1.10417	6.87117
36.5	30.1175	1.09929	16.1171	15.0178	1.09929	6.82063
37.	29.5635	1.09385	16.1643	15.0705	1.09385	6.76706
37.5	29.0094	1.08785	16.2113	15.1234	1.08785	6.71048
38.	28.4552	1.0813	16.2579	15.1766	1.0813	6.65091
38.5	27.901	1.07419	16.3043	15.2301	1.07419	6.58838
39.	27.3467	1.06652	16.3504	15.2839	1.06652	6.52291
39.5	26.7924	1.0583	16.3963	15.338	1.0583	6.45451
40.	26.238	1.04952	16.4419	15.3923	1.04952	6.38322
40.5	25.6835	1.04018	16.4872	15.447	1.04018	6.30904
41.	25.129	1.03029	16.5322	15.5019	1.03029	6.232
41.5	24.5744	1.01984	16.577	15.5572	1.01984	6.15212
42.	24.0197	1.00883	16.6215	15.6127	1.00883	6.06941

(continued overleaf)

Table A.5 (Continued)

j A/cm²	V_α mV	p_{el} W/cm²	$\dot{q}_h$ W/cm²	$\dot{q}_c$ W/cm²	p_{el} W/cm²	η %
42.5	23.465	0.997261	16.6657	15.6685	0.997261	5.9839
43.	22.9102	0.985138	16.7097	15.7246	0.985138	5.8956
43.5	22.3553	0.972456	16.7534	15.781	0.972456	5.80452
44.	21.8004	0.959216	16.7969	15.8376	0.959216	5.71069
44.5	21.2454	0.945419	16.84	15.8946	0.945419	5.61412
45.	20.6903	0.931063	16.8829	15.9519	0.931063	5.51482
45.5	20.1352	0.916149	16.9256	16.0094	0.916149	5.41282
46.	19.5799	0.900677	16.9679	16.0672	0.900677	5.30812
46.5	19.0247	0.884647	17.01	16.1254	0.884647	5.20074
47.	18.4693	0.868058	17.0518	16.1838	0.868058	5.0907
47.5	17.9139	0.85091	17.0934	16.2425	0.85091	4.978
48.	17.3584	0.833203	17.1347	16.3015	0.833203	4.86267
48.5	16.8028	0.814937	17.1757	16.3608	0.814937	4.74471
49.	16.2472	0.796112	17.2165	16.4203	0.796112	4.62413
49.5	15.6915	0.776728	17.2569	16.4802	0.776728	4.50096
50.	15.1357	0.756784	17.2972	16.5404	0.756784	4.37519
50.5	14.5798	0.736281	17.3371	16.6008	0.736281	4.24685
51.	14.0239	0.715218	17.3768	16.6616	0.715218	4.11594
51.5	13.4679	0.693596	17.4162	16.7226	0.693596	3.98247
52.	12.9118	0.671413	17.4554	16.784	0.671413	3.84646
52.5	12.3556	0.648671	17.4942	16.8456	0.648671	3.70791
53.	11.7994	0.625368	17.5329	16.9075	0.625368	3.56683
53.5	11.2431	0.601505	17.5712	16.9697	0.601505	3.42324
54.	10.6867	0.577082	17.6093	17.0322	0.577082	3.27714
54.5	10.1302	0.552097	17.6471	17.095	0.552097	3.12855
55.	9.57368	0.526553	17.6846	17.1581	0.526553	2.97746
55.5	9.01706	0.500447	17.7219	17.2215	0.500447	2.82389
56.	8.46036	0.47378	17.7589	17.2851	0.47378	2.66784
56.5	7.90358	0.446552	17.7957	17.3491	0.446552	2.50933
57.	7.34672	0.418763	17.8321	17.4134	0.418763	2.34836
57.5	6.78978	0.390412	17.8683	17.4779	0.390412	2.18494
58.	6.23276	0.3615	17.9043	17.5428	0.3615	2.01907
58.5	5.67566	0.332026	17.94	17.6079	0.332026	1.85076
59.	5.11847	0.30199	17.9754	17.6734	0.30199	1.68002
59.5	4.56121	0.271392	18.0105	17.7391	0.271392	1.50685
60.	4.00387	0.240232	18.0454	17.8051	0.240232	1.33127
60.5	3.44644	0.20851	18.08	17.8715	0.20851	1.15326
61.	2.88893	0.176225	18.1143	17.9381	0.176225	0.972848
61.5	2.33134	0.143377	18.1484	18.005	0.143377	0.790029
62.	1.77367	0.109967	18.1822	18.0722	0.109967	0.604808
62.5	1.21591	0.0759944	18.2157	18.1397	0.0759944	0.417191
63.	0.65807	0.0414584	18.249	18.2075	0.0414584	0.227182
63.5	0.100145	0.00635924	18.282	18.2756	0.00635924	0.0347841
64.	−0.457863	−0.0293032	18.3147	18.344	−0.0293032	−0.159998
64.5	−1.01596	−0.0655291	18.3472	18.4127	−0.0655291	−0.357162
65.	−1.57413	−0.102319	18.3794	18.4817	−0.102319	−0.556703

References

1. Domenicali, C.A. (1953) Irreversible thermodynamics of thermoelectric effects in inhomogeneous, anisotropic media. *Phys. Rev.*, **92** (4), 877–881.
2. Domenicali, C.A. (1954) Irreversible thermodynamics of thermoelectricity. *Rev. Mod. Phys.*, **26** (2), 237–275.
3. Domenicali, C.A. (1954) Stationary temperature distribution in an electrically heated conductor. *J. Appl. Phys.*, **25** (10), 1310–1311.
4. Strutynsky, M.M. (2009) Computer technologies in thermoelectricity. *J. Thermoelectric.*, (4), 29–44.
5. Hangos, K. and Cameron, I. (2001) *Process Modelling and Model Analysis*, Academic Press.
6. Hogan, T.P. and Shih, T. (2006) Modeling and characterization of power generation modules based on bulk materials, in *CRC Handbook of Thermoelectrics: Macro to Nano*, Chapter 12 (ed. D.M. Rowe), Taylor and Francis, Boca Raton, FL.
7. Özişik, M.N. (1993) *Heat Conduction - Finite-Difference Methods*, Chapter 12, John Wiley & Sons, Inc., pp. 436–501.
8. Özişik, M.N. (1994) *Finite Difference Methods in Heat Transfer*, CRC Press.
9. Tzou, D.Y. (2015) *Macro- to Microscale Heat Transfer - The Lagging Behaviour*, John Wiley & Sons.
10. Edsberg, L. (2008) *Introduction to Computation and Modeling for Differential Equations*, John Wiley & Sons, Inc.
11. Edsberg, L. (2008) *Introduction to Computation and Modeling for Differential Equations - Numerical Methods for Parabolic Partial Differential Equations*, Chapter 6, John Wiley & Sons, Inc., pp. 107–126.
12. Edsberg, L. (2008) *Introduction to Computation and Modeling for Differential Equations - Numerical Methods for Elliptic Partial Differential Equations*, Chapter 7, John Wiley & Sons, Inc., pp. 127–146.
13. Edsberg, L. (2008) *Introduction to Computation and Modeling for Differential Equations - Numerical Methods for Hyperbolic Partial Differential Equations*, Chapter 8, John Wiley & Sons, Inc., pp. 147–162.
14. Mitchell, R.A. (1969) *Computational in Partial Differential Equations*, Introductory Mathematics for Scientists & Engineers, John Wiley & Sons, Inc.
15. Patankar, S.V. (1980) *Numerical Heat Transfer and Fluid Flow*, Series in Computational Methods in Mechanics and Thermal Sciences, Hemisphere Publishing Corporation.
16. Farlow, S.J. (1993) *Partial Differential Equations for Scientists and Engineers*, Dover Publications.
17. Stephenson, G. (1996) *Partial Differential Equations for Scientists and Engineers*, World Scientific Publishing.
18. Langtangen, H.P. (1999) *Computational Partial Differential Equations: Numerical Methods and Diffpack Programming*, Lecture Notes in Computational Science and Engineering, vol. **2**, Springer-Verlag, Berlin Heidelberg.
19. Polyanin, A.D. (2002) *Handbook of Linear Partial Differential Equations for Engineers and Scientists*, Chapman & Hall/CRC Press, Boca Raton, FL.
20. Cai, X., Acklam, E., and Langtangen, H.P. (2003) *Advanced Topics in Computational Partial Differential Equations: Numerical Methods and Diffpack Programming*, Lecture Notes in Computational Science and Engineering, vol. **33**, Springer-Verlag, Berlin Heidelberg.
21. Lynch, D.R. (2005) *Numerical Partial Differential Equations for Environmental Scientists and Engineers: A First Practical Course*, Springer-Verlag.
22. Myint-U, T. and Debnath, L. (2007) *Linear Partial Differential Equations for Scientists and Engineers*, 4th edn, Birkhäuser.
23. Tang, K.-T. (2007) *Mathematical Methods for Engineers and Scientists 3 - Fourier Analysis, Partial Differential Equations and Variational Methods*, Springer-Verlag.
24. Mathew, T. (2008) *Domain Decomposition Methods for the Numerical Solution of Partial Differential Equations*, Lecture Notes in Computational Science and

Engineering, vol. **61**, 1st edn, Springer-Verlag.

25. Glowinski, R. and Neittaanmäki, P. (2008) *Partial Differential Equations: Modelling and Numerical Simulation*, Computational Methods in Applied Sciences, vol. **16**, 1st edn, Springer-Verlag.

26. Brio, M., Webb, G.M., and Zakharian, A.R. (2010) *Numerical Time-Dependent Partial Differential Equations for Scientists and Engineers*, Mathematics in Science and Engineering, vol. **213**, 1st edn, Academic Press.

27. Fitzgibbon, W., Kuznetsov, Y., Neittaanmaki, P., Periaux, J., and Pironneau, O. (eds) (2010) *Applied and Numerical Partial Differential Equations: Scientific Computing in Simulation, Optimization and Control in a Multidisciplinary Context*, Computational Methods in Applied Sciences, 1st edn, Springer-Verlag.

28. Debnath, L. (2011) *Nonlinear Partial Differential Equations for Scientists and Engineers*, Birkhäuser.

29. Hesthaven, J.S. and Ronquist, E.M. (eds) (2011) *Spectral and High Order Methods for Partial Differential Equations: Selected Papers from the ICOSAHOM '09 Conference, June 22-26, Trondheim, Norway*, Lecture Notes in Computational Science and Engineering, vol. **76**, Springer-Verlag, Berlin Heidelberg.

30. Iliev, O.P., Margenov, S.D., Minev, P.D., Vassilevski, P.S., and Zikatanov, L.T. (eds) (2013) *Numerical Solution of Partial Differential Equations: Theory, Algorithms, and Their Applications: In Honor of Professor Raytcho Lazarov's 40 Years of Research in Computational Methods and Applied Mathematics*, Springer Proceedings in Mathematics & Statistics, vol. **45**, Springer-Verlag, New York.

31. Azaïez, M., Fekih, H.E., and Hesthaven, J.S. (2014) *Spectral and High Order Methods for Partial Differential Equations - ICOSAHOM 2012: Selected Papers from the ICOSAHOM Conference, June 25-29, 2012, Gammarth, Tunisia*, Lecture Notes in Computational Science and Engineering, vol. **95**, Springer International Publishing.

32. Bartels, S. (2015) *Numerical Methods for Nonlinear Partial Differential Equations*, Springer Series in Computational Mathematics 47, vol. **47**, Springer International Publishing.

33. O'Brien, G.G., Hyman, M.A., and Kaplan, S. (1951) A study of the numerical solution of partial differential equations. *J. Math. Phys.*, **29**, 223–251.

34. Peaceman, D.W. and Rachford, H.H. Jr. (1955) The numerical solution of parabolic and elliptic differential equations. *J. Soc. Ind. Appl. Math.*, **3** (1), 28–41.

35. Crank, J. and Nicolson, E. (1996) A practical method for numerical evaluation of solutions of partial differential equations of the heat-conduction type. *Adv. Comput. Math.*, **6**, 207–226. Reprinted from Proceedings of the Cambridge Philosophical Society, 43 (1947), 50–67.

36. Trefethen, L.N. (1996) Finite difference and spectral methods for ordinary and partial differential equations. unpublished text, Available at http://people.maths.ox.ac.uk/trefethen/pdetext.html (accessed 29 July 2015).

37. Lee, H.J. and Schiesser, W.E. (2004) *Ordinary and Partial Differential Equation Routines in C, C++, Fortran, Java, Maple, and MATLAB*, Chapman & Hall/CRC.

38. LeVeque, R.J. (2007) *Finite Difference Methods for Ordinary and Partial Differential Equations Steady-State and Time-Dependent Problems*, Society for Industrial and Applied Mathematics.

39. Croft, D.R. and Lilley, D.G. (1977) *Heat Transfer Calculations Using Finite Difference Equations*, Elsevier Science & Technology.

40. Kutz, M. (ed.) (2006) *Heat-Transfer Calculations*, McGraw-Hill.

41. Minkowycz, W.J., Sparrow, E.M., and Murthy, J.Y. (eds) (2006) *Handbook of Numerical Heat Transfer*, 2nd edn, John Wiley & Sons, Inc.

42. Oliveira, K.S.M., Cardoso, R.P., and Hermes, C.J.L. (2014) Two-dimensional modeling of thermoelectric cells. International Refrigeration and Air Conditioning Conference, Purdue. Paper 1357, p. 10–,

http://docs.lib.purdue.edu/iracc/1357 (accessed 29 July 2015).

43. Yan, D. (2011) *Modeling and Application of a Thermoelectric Generator*. Master's thesis. University of Toronto.

44. Yan, D., Dawson, F.P., Pugh, M.C., and El-Deib, A. (2012) Time dependent finite volume model of thermoelectric devices. Energy Conversion Congress and Exposition (ECCE), 2012 IEEE, pp. 105–112.

45. Yan, D., Dawson, F.P., Pugh, M., and El-Deib, A.A. (2014) Time-dependent finite-volume model of thermoelectric devices. *IEEE Trans. Ind. Appl.*, **50** (1), 600–608.

46. Courant, R. (1943) Variational methods for the solution of problems of equilibrium and vibrations. *Bull. Am. Math. Soc.*, **49**, 1–23.

47. Zlámal, M. (1968) On the finite element method. *Numer. Math.*, **12**, 394–409.

48. Zienkiewicz, O.C. and Morgan, K. (1983) *Finite Elements & Approximation*, Dover Publications.

49. Huang, H.-C. and Usmani, A.S. (1994) *Finite Element Analysis for Heat Transfer - Theory and Software*, Springer-Verlag.

50. Lewis, R.W., Nithiarasu, P., and Seetharamu, K.N. (2004) *Fundamentals of the Finite Element Method for Heat and Fluid Flow*, John Wiley & Sons, Ltd.

51. Bergheau, J.-M. and Fortunier, R. (2008) *Finite Element Simulation of Heat Transfer*, John Wiley & Sons, Inc.

52. Zienkiewicz, O.C., Taylor, R.L., and Zhu, J.Z. (2013) *The Finite Element Method: Its Basis and Fundamentals*, Butterworth Heinemann.

53. Zienkiewicz, O.C., Taylor, R.L., and Fox, D.D. (2014) *The Finite Element Method for Solid and Structural Mechanics*, 7th edn, Butterworth Heinemann.

54. Zienkiewicz, O.C., Taylor, R.L., and Nithiarasu, P. (2014) *The Finite Element Method for Fluid Dynamics*, Butterworth Heinemann.

55. Baskharone, E.A. (2014) *The Finite Element Method with Heat Transfer and Fluid Mechanics Applications*, Cambridge University Press.

56. Antonova, E.E. and Looman, D.C. (2005) Finite elements for thermoelectric device analysis in ANSYS. 24th International Conference on Thermoelectrics, ICT 2005.

57. Buist, R.J. (1995) The extrinsic Thomson effect, in *Proceedings of the 14th International Conference on Thermoelectrics*, A.F. Ioffe Physical-Technical Institute, St. Petersburg, pp. 301–304.

58. Lau, P.G. and Buist, R.J. (1996) Temperature and time dependent finite-element model of a thermoelectric couple. 15th International Conference on Thermoelectrics, ICT 1996.

59. Lau, P.G. and Buist, R.J. (1997) Calculation of thermoelectric power generation performance using finite element analysis. 16th International Conference on Thermoelectrics, ICT 1997.

60. Jägle, M. (2007) Simulating thermoelectric effects with finite element analysis using COMSOL. Proceedings of the 5th European Conference on Thermoelectrics, Odessa.

61. Freunek, M., Müller, M., Ungan, T., Walker, W., and Reindl, L.M. (2009) New physical model for thermoelectric generators. *J. Electron. Mater.*, **38** (7), 1214–1220.

62. Ziolkowski, P., Pionas, P., Leszczynski, J., Karpinski, G., and Müller, E. (2010) Estimation of thermoelectric generator performance by finite element modeling. *J. Electron. Mater.*, **39** (9), 1934–1943.

63. Sandoz-Rosado, E. and Stevens, R. (2010) Robust finite element model for the design of thermoelectric modules. *J. Electron. Mater.*, **39** (9), 1848–1855.

64. Strutinsky, M.N. (2010) Computer technologies of solution of the inverse problems of thermoelectricity. *J. Thermoelectric.*, **2**, 5–16.

65. Zabrocki, K., Müller, E., and Seifert, W. (2010) One-dimensional modeling of thermogenerator elements with linear material profiles. *J. Electron. Mater.*, **39**, 1724–1729, doi: 10.1007/s11664-010-1179-3.

66. Zabrocki, K., Ziolkowski, P., Dasgupta, T., de Boor, J., and Müller, E. (2013) Simulations for the development of thermoelectric measurements. *J. Electron. Mater.*, **42** (7), 2402–2408.

67. Ioffe, A.F. (1957) *Semiconductor Thermoelements and Thermoelectric Cooling*, Infosearch, Ltd., London.
68. Draper, N.R. and Smith, H. (1998) *Applied Regression Analysis*, John Wiley & Sons, Inc.
69. Everitt, B.S. and Skrondal, A. (2010) *The Cambridge Dictionary of Statistics*, Cambridge University Press.
70. Freedman, S.I. (1966) Thermoelectric power generation, in *Direct Energy Conversion*, Inter-University Electronic Series, vol. **3**, Chapter 3 (ed. G.W. Sutton), McGraw-Hill Book Company, pp. 105–180.
71. Ure, R.W. Jr. and Heikes, R.R. (1961) *Theoretical Calculation of Device Performance*, Chapter 15, Interscience Publishers, Inc., New York, pp. 458–517.
72. Müller, E., Zabrocki, K., Goupil, C., Snyder, G.J., and Seifert, W. (2012) Functionally graded thermoelectric generator and cooler elements, in *CRC Handbook of Thermoelectrics: Thermoelectrics and its Energy Harvesting*, Chapter 4 (ed. D.M. Rowe), CRC Press, Boca Raton, FL.
73. Seifert, W., Ueltzen, M., and Müller, E. (2002) One-dimensional modelling of thermoelectric cooling. *Phys. Status Solidi A*, **1** (194), 277–290.
74. Süßmann, H. and Müller, E. (1995) Verification of a transport model for p-type $(Bi_{0.5} Sb_{0.5})_2 Te_3$ and $(Bi_{0.25} Sb_{0.75})_2 Te_3$ mixed crystals by means of temperature dependent thermoelectric properties below room temperature, in *14th International Conference on Thermoelectrics* (eds M.V. Vedernikov, M.I. Federov, and A.E. Kaliazin), International Thermoelectric Society, St. Petersburg, pp. 1–7.
75. Seifert, W., Snyder, G.J., Toberer, E.S., Goupil, C., Zabrocki, K., and Mueller, E. (2013) The self-compatibility effect in graded thermoelectric cooler elements. *Phys. Status Solidi A*, **210** (7), 1407–1417.
76. Seifert, W., Müller, E., and Walczak, S. (2006) Generalized analytic description of one-dimensional non-homogeneous TE cooler and generator elements based on the compatibility approach, in *25th International Conference on Thermoelectrics, Vienna, Austria, 06-10 August, 2006* (ed. P. Rogl), IEEE, Piscataway, NJ, pp. 714–719.
77. Snyder, G.J. and Ursell, T.S. (2003) Thermoelectric efficiency and compatibility. *Phys. Rev. Lett.*, **91** (14), 148301.

Index

a
Anisotropic thermoelectric elements 147, 148
Archimedes' principle. 191

b
Benedicks effect 12
Bridgman effect 13

c
Cascaded thermoelectric generators 230
CHI *See* Constant heat input (CHI) model
Classical Ioffe method, thermal conductivity measurement
- CTEM 193
- error estimation 194
- heat equation 194
- inverse Laplace transformation 202
- inversion theorem, LT 203
- Laplace transformation 197
- original equation LT 198
- Seebeck coefficient 193
- temperature profile inversion 205
- temperatures, s-domain 199
- theoretical basis 194

Compatibility
- and segmented thermogenerators
- – Carnot efficiency 230
- – efficiency 230
- – interface temperature 230
- – negative reduced efficiency 231
- – p-type materials 232
- – TAGS or skutterudite 231
- cascaded generator 230
- criterion $u = s$ and calculus of variations 243
- optimal material grading 241
- power-related
- – CPM 241
- – electrical power output 239
- – maximum power output 240, 241
- – power output 238, 239
- – self-compatible elements 239
- reduced efficiencies and self-compatibility
- – efficiency 233
- – local efficiency dependence, current 235
- – performance integrals, efficiency and COP 233
- relative current density
- – advantage 229
- – definition 227
- – FGM 229
- – temperature dependence, chemical potential 228
- self-compatibility and optimum material grading
- – constant and variable Lagrange multiplier 248
- – constraint variational problems 248
- – Gerstenmaier/Wachutka approach 249
- – inhomogeneous generators and coolers 248
- – optimal TE potential 247
- – optimal temperature and optimal Seebeck profile relation 247, 248
- – optimization strategies 246–248
- – reduced "efficiencies" 247
- – Snyder's criterion 246
- – TEG efficiency improvement 247
- self-compatible TEG and TEC elements
- – optimal figure of merit 253
- – performance 251
- – Seebeck coefficients 255

Continuum Theory and Modelling of Thermoelectric Elements, First Edition. Edited by Christophe Goupil.
© 2016 Wiley-VCH Verlag GmbH & Co. KGaA. Published 2016 by Wiley-VCH Verlag GmbH & Co. KGaA.

Compatibility (contd.)
– – temperature profile, $u = s$ material 256
– thermodynamics 249
– Thomson cooler see Thomson cooler
– vs. device optimization 276
Constant heat input (CHI) model 100
Continuum theory
– anisotropic thermoelectric elements 147
– contacts and contact resistances
– – electrical contact resistance 130
– – thermoelectric element, contacting bridge 126
– convection losses and benefits 144
– CPM devices
– – inverse performance equations 91
– – single-element device 82
– – TEC see Thermoelectric cooler (TEC)
– – TEG see Thermoelectric generator (TEG)
– – TEH see Thermoelectric heater (TEH)
– – thermoelectric element, performance parameters 84
– Domenicali's heat balance equation
– – heat balance and source terms 76
– – spatial and temperature averaging 79
– – tensorial character, material properties 75
– radiation losses and benefits 146
– temperature dependence see Temperature dependent materials
– thermogenerator element
– – CHI model 100
– – load resistance 96
– – power and efficiency 93
– transferred heat balance 80

d
Dirichlet boundary conditions 160, 163
Dissipative coupling
– finite-time thermodynamics optimization 133
– power and efficiency maximization 137
– temperature difference calculation 136
– thermal flux and electrical current 135
– thermoelectric generator model 133
Domenicali's heat balance equation
– heat balance and source terms 76
– spatial and temperature averaging 79
– tensorial character, material properties 75

e
eddy currents 157
Ettingshausen coefficient 21
Euler–Lagrange differential equation 243, 251
Explicit Euler method 286–288

f
FDM See Finite difference methods (FDM)
FEM See Finite element method (FEM)
Finite difference methods (FDM)
– description 284
– difference quotient approximation 284
– 1D temperature distribution 286
– explicit Euler method 286, 287
– PDEs 286
– Taylor series expansion, function $f = f(x)$ 284
Finite element method (FEM)
– algorithm scheme 291
– description 290
– steps 291
Finite volume method (FVM)
– control volume, 2D 288
– description 288
– Fourier's heat equation 289
– PDEs 290
– procedure 289
Forces and fluxes, thermoelectric systems
– and kinetic coefficients 38
– coefficients
– – coupled processes 41
– – decoupled processes 40
– dimensionless figure of merit 43
– energy flux and heat flux 39
– entropy per carrier 41
– kinetic coefficients and transport parameters 42
– thermoelectric effects 37
Fourier heat conduction 195
Fourier heat divergence 259, 271
Fourier's law 50, 171
Fourier–Mellin integral 204
Functionally graded materials (FGM) 229, 261
FVM See Finite volume method (FVM)

g
Galvanomagnetic and thermomagnetic effects
– devices 22
– Ettingshausen coefficient 21
– Hall coefficient 21
– Nernst coefficient 21

– Righi–Leduc coefficient 22
Gerstenmaier/Wachutka approach 249
Green's function (GF) approach
– continuity equations 208
– linear Onsager theory 207
– NEGF 207
– one-dimensional, steady state 209
– perturbative approach (1D) 210
– spin Seebeck effect 207
– steady state 208
– time dependent *see* Time dependent Green's function approach

h

Hall coefficient 21
Heat and entropy
– entropy production density 45
– heat flux and Peltier–Thomson coefficient 46
– local energy balance 47
– Onsager expressions 44
– Peltier–Thomson term 46
– volumetric heat production 45

i

Ideal Fermi gas
– characteristics 28
– electron gas, thermoelectric cell 29
– entropy, carrier 30
– equation of state, ideal electron gas 32
– mass renormalization 28
– temperature dependence, chemical potential 34

l

Laplace transformation (LT) 197
Laser flash analysis (LFA)
– calculation routine 191
– definition 190
– Fourier series expansion 191
– mass density, sample 191
– measurement principle 191
– radiation heat losses 192
– thermal diffusivity determination 192
Linear non-equilibrium thermodynamics
– forces and fluxes 35
– linear response and reciprocal relations 36
Linear Onsager theory 207
Linear transient approach
– relaxation time 212
– response approximation 214
– transient field equations 213

m

Magnus' law 10, 13
Mathematica model, compatibility approach 320
Millman theorem 158, 169, 170, 172

n

Nernst coefficient 21
Networking
– discretization 171
– implementation 173
– Millman theorem 172
– numerical illustration 174
– presentation 169
– useful expressions 171
Non-equilibrium thermodynamics 56
Nonlinear material parameters
– Mathematica model, compatibility approach 320
– temperature dependent properties
– – coefficient of performance (COP) 318
– – decoupled dependencies 315
– – 1D steady-state model 315
– – Mathematica code 318
– – maximum cooling calculation 316
– – Nabla calculus 317
– – nabla calculus 316
– – relative current 317
– – spatial material profiles 319
– – temperature dependent curves 316
– – temperature profile 319
– – Thomson coefficient 315
Numerical simulation
– FDM *see* Finite difference methods (FDM)
– FEM *see* Finite element method (FEM)
– FVM *see* Finite volume method (FVM)
– nonlinear materials *see* Nonlinear material parameters
– TEG performance calculation
– – averages and performance values 297
– – averages, material properties 294
– – cross-sectional areas adjustment 310
– – FEM 311
– – finite difference scheme, TE element 300
– – optimal performance, averaged material properties 299
– – p-n thermocouple 309
– – performance parameters, current density 326
– – power factor and figure of merit 325
– – processing, material properties 295

o

Ohm's law 171
Onsager approach 158, 169
Onsager force-flux derivation 38
Onsager theory 13
Onsager–de Groot–Callen theory 26, 141

p

Pauli exclusion principle 28, 29
Peltier cooler 318
Peltier coolers
– applications 259
– characteristics 259
– CPM 273
– heat production 265
– multi-stage 260
– self-compatibility effect 261
Peltier effect 8
Peltier–Thomson coefficient and term 46
Performance calculation, TEG
– average, material properties 294
– averages and performance values 297
– cross-sectional areas adjustment 310
– FEM simulation 311
– finite difference scheme, TE element 300
– material properties 293
– optimal performance, averaged material properties 299
– p-n thermocouple calculation 309
– parameters, current density 326
– power factor and figure of merit 325
– processing, material properties 295
– thermocouple 292
Phonon-glass electron-crystal (PGEC) 16
Prigogine's principle 59

q

Quasi-stationary processes
– bismuth telluride 182
– lunation 179
– response time 182
– temperature difference and absolute value, lunation 180
– temperature, moon's surface 180
– timescales, transient electrical systems 181
– transient phenomena, electrical systems 181

r

Righi–Leduc coefficient 22

s

Seebeck coefficient 157, 159, 161, 171, 174, 246
Seebeck effect
– thermomagnetism 4
– electromagnetism 6
– electrometer (Schweigger) 4
– electromotive forces 3
– galvanothermal effect (GTE) 4
– holography invention 5
– mechanical stress, amorphous transparent material 5
– power generation 11
– thermocouple 3
– thermomagnetism 3, 5, 6
– unequal heating, metals 6
Segmented devices
– double-segmented element 160
– electrical conductivity 161
– graded elements, numerical parameter studies 166
– gradient configurations, usefulness 158
– material inhomogeneity, effects 157
– material properties, spatial variation 157
– multi-segmented elements, algorithm 164
– Onsager approach 158
– optimal grading strategies 159
– Seebeck coefficient 159
– self-compatibility approach 158
– thermal conductivity 163
– Thomson and distributed Peltier heat 159
Self-compatible elements
– optimal figure of merit 253
– Seebeck coefficients 255
– TEG and TEC elements performance 251
– temperature profile, $u = s$ material 256
Shaped thermoelectric elements
– element shape factor 142
– conical and cylindrical elements 141
– disk-shaped thermocouples 142
– entropy production 141
– hyperbolic heat conduction model 143
– non-cylindrical thermoelements 141
– non-equilibrium thermodynamics 142
– TEC 142
Snyder's criterion 245, 246, 271
Solar thermoelectric generator (STEG) 147
6-Stage cooler of Marlow 261
Supercooling
– metrics, transient response 182, 183

– parameters 187
– steady state operation, thermoelectric cooler
– – absorbed thermal power, mixed boundary conditions 185
– – bismuth telluride sample material properties 185
– – boundary conditions (BC) 183
– – CPM 184
– – Dirichlet BC 184
– – maximum absorbing heat flux 186
– – optimum current, maximum cooling case 185
– – TEC 183
– – thermal power, sink side 186

t
Temperature dependent materials
– algebraic and general
– – algebraic and general 119
– Constant Thomson coefficient
– – averaged coefficients and geometric optimization 114
– – linear temperature dependence of resistivity 121
– – TEG performance 116
– inverse temperature dependence
– – maximum power output 108
– – performance equations 106
– – thermal conductivity 103
– linear temperature dependence of resistivity 123
Thermodynamic engine
– efficiency goal 27
– endoreversible engine 27
– energy budget 27
Thermoelectric coefficients
– coupled processes 41
– decoupled processes 40
Thermoelectric cooler (TEC) 89–91
Thermoelectric engine
– efficiency 49
– electrochemical potential difference 50
– entropy production, non-adiabatic branches 50
– Fourier's law 50
– heating and cooling mode 48
– traditional steam engines 48
Thermoelectric generator
– construction 15
– for space 16, 17
Thermoelectric generator (TEG) 86, 88, 89
– transient behavior 189

Thermoelectric heater (TEH) 91
Thermoelectric potential
– non-equilibrium thermodynamics 56
– relative current and dissipation ratio 51
– TEG, TEC and TEH 53
Thermoelectricity
– applications
– – 1920–1970 15
– – 1970–2000 16
– – 2000- 17
– Benedicks effect 12
– Bridgman effect 13
– Magnus' law 10
– Peltier effect 8
– Seebeck effect 2
– semiconductors as thermoelectric materials 14
– thermoelectric devices 11
– thermoelectric generator 16
– Thomson effect 9
Thermopiles 12
Thomson cooler
– advantages 266
– COP 262
– limits 266
– phase-space 265
– self-compatible material, cooling 268
– – $u = s$ material approximation 272
– – compatible Seebeck coefficient 269
– – cooler models comparison 273
– – CPM 273
– – CPM Peltier cooler 274
– – dependence 273
– – heat flux 271
– – maximum heat pumping 270
– – maximum temperature difference 270
– – optimal current density 269
– – Peltier cooler 275
– – scaling integral, TEG 269
– – temperature difference 274
– – temperature profile 271
– – varying gap energies estimation 275
– – Wiedemann–Franz-law 269
Thomson effect 9
Time dependent Green's function approach
– boundary conditions, Robinson type 216, 217
– 1D transient problem 217
– heat conduction equation 215

Time dependent Green's function approach (*contd.*)
– inhomogeneous heat equation 215
– Laplace transformation 217
– three dimensions 216
– time-dependent heat equation 216
Transient behavior, TEG 189

w

Wiedemann–Franz law 232, 269

z

Zur Farbenlehre 5